Advances in

MICROBIAL PHYSIOLOGY

VOLUME 46

Contents

Microarray Analysis of Bacterial Pathogenicity
Gary K. Schoolnik

The Extracytoplasmic Function (ECF) Sigma Factors
John D. Helmann

How Oxygen Damages Microbes: Oxygen Tolerance and Obligate Anaerobiosis
James A. Imlay

Drug Resistance in Yeasts – an Emerging Scenario
Rajendra Prasad, Sneh Lata Panwar and Smriti

The Physiology and Collective Recalcitrance of Microbial Biofilm Communities
Peter Gilbert, Tomas Maira-Litran, Andrew J. McBain,
Alexander H. Rickard and Fraser Whyte

Biochemistry, Regulation and Genomics of Haem Biosynthesis in Prokaryotes
Mark R. O'Brian and Linda Thöny-Meyer

Global Adjustment of Microbial Physiology During Free Radical Stress
Pablo J. Pomposiello and Bruce Demple

Contributors to Volume 46

Bruce Demple, Department of Cancer Cell Biology, Harvard School of Public Health, Boston, MA 02115, USA (Bdemple@hsph.harvard.edu)

Peter Gilbert, School of Pharmacy and Pharmaceutical Sciences, Coupland III Building, Coupland Street, University of Manchester, Manchester, M13 9PL, UK (Peter.gilbert@man.ac.uk)

John D. Helmann, Department of Microbiology, Wing Hall, Cornell University, Ithaca, NY 14853-8101, USA (jdh9@cornell.edu)

James Imlay, Department of Microbiology, University of Illinois, Urbana, IL 61801, USA (jimlay@uiuc.edu)

Tomas Maira-Litran, Channing Laboratory, Harvard Medical School, Longwood Avenue, Boston, MA 02115, USA

Andrew J. McBain, School of Pharmacy and Pharmaceutical Sciences, Coupland III Building, Coupland Street, University of Manchester, Manchester, M13 9PL, UK

Mark R. O'Brian, Department of Biochemistry, 140 Farber Hall, State University of New York at Buffalo, Buffalo, NY 14214, USA (mrobrian@buffalo.edu)

Snehlata Panwar, Department of Genetics, Cell Biology and Development, 250 Biological Sciences Center, 1445 Gortner Avenue, St. Paul, MN 55108, USA

Pablo J. Pomposiello, Department of Microbiology, Morrill Science

Center IV North, University of Massachusetts, Amherst, MA 01003, USA; and Department of Cancer Cell Biology, Harvard School of Public Health, Boston, MA 02115, USA

RAJENDRA PRASAD, Membrane Biology Laboratory, School of Life Sciences, Jawaharlal Nehru University, New Delhi, 110067 India (rp47@hotmail.com)

ALEXANDER H. RICKARD, School of Pharmacy and Pharmaceutical Sciences, Coupland III Building, Coupland Street, University of Manchester, Manchester, M13 9PL, UK

GARY SCHOOLNIK, Beckman Centre, Room B241A, 300 Pasteur Drive, Stanford, CA 94305-5428, USA (schnoolni@cmgm.stanford.edu)

SMRITI, Membrane Biology Laboratory, School of Life Sciences, Jawaharlal Nehru University, New Delhi, 110067 India

LINDA THÖNY-MEYER, Institute of Microbiology, ETH Zürich, Schmelzbergstrasse 7, CH-8092 Zürich, Switzerland

FRASER WHYTE, School of Pharmacy and Pharmaceutical Sciences, Coupland III Building, Coupland Street, University of Manchester, Manchester, M13 9PL, UK

Microarray Analysis of Bacterial Pathogenicity

Gary K. Schoolnik

*Departments of Medicine, Microbiology and Immunology
Stanford University Medical School,
Stanford, CA 94305, USA*

ABSTRACT

The DNA microarray, a surface that contains an ordered arrangement of each identified open reading frame of a sequenced genome, is the engine of functional genomics. Its output, the expression profile, provides a genome wide snap-shot of the transcriptome. Refined by array-specific statistical instruments and data-mined by clustering algorithms and metabolic pathway databases, the expression profile discloses, at the transcriptional level, how the microbe adapts to new conditions of growth – the regulatory networks that govern the adaptive response and the metabolic and biosynthetic pathways that effect the new phenotype. Adaptation to host microenvironments underlies the capacity of infectious agents to persist in and damage host tissues. While monitoring the whole genome transcriptional response of bacterial pathogens within infected tissues has not been achieved, it is likely that the complex, tissue-specific response is but the sum of individual responses of the bacteria to specific physicochemical features that characterize the host milieu. These are amenable to experimentation *in vitro* and whole-genome expression studies of this kind have defined the transcriptional response to iron starvation, low oxygen, acid pH, quorum-sensing pheromones and reactive oxygen intermediates. These have disclosed new information about even well-studied processes and provide a portrait of the adapting bacterium as a 'system', rather than the product of a few genes or even a few regulons. Amongst the regulated genes that compose this adaptive system are transcription factors. Expression profiling experiments of transcription factor mutants delineate the corresponding regulatory cascade. The genetic basis for pathogenicity

ADVANCES IN MICROBIAL PHYSIOLOGY VOL 46
ISBN 0-12-027746-8

can also be studied by using microarray-based comparative genomics to characterize and quantify the extent of genetic variability within natural populations at the gene level of resolution. Also identified are differences between pathogen and commensal that point to possible virulence determinants or disclose evolutionary history. The host vigorously engages the pathogen; expression studies using host genome microarrays and bacterially infected cell cultures show that the initial host reaction is dominated by the innate immune response. However, within the complex expression profile of the host cell are components mediated by pathogen-specific determinants. In the future, the combined use of bacterial and host microarrays to study the same infected tissue will reveal the dialogue between pathogen and host in a gene-by-gene and site- and time-specific manner. Translating this conversation will not be easy and will probably require a combination of powerful bioinformatic tools and traditional experimental approaches – and considerable effort and time.

ABBREVIATIONS

EST   Expressed sequence tag
INH   Isonicotinic acid hydrazide
IPTG Isopropyl β-D-thiogalactopyranoside
LPS   Lipopolysaccharide
MOI   Multiplicity of infection
PT   Pertussis toxin
ROI   Reactive oxygen intermediates
SDS   Sodium dodecyl sulfate

1. INTRODUCTION

1.1. Microarray Experimental Paradigms

The rapidly expanding number of genome sequence databases of pathogenic bacteria and their hosts provides an unprecedented opportunity for the study of pathogenesis from a whole-genome perspective through the use of DNA microarrays. For the purposes of this discussion, a microarray will be considered to be any device that provides a surface containing representations of all (or most) of the identified open reading frames (ORFs) of a sequenced and annotated genome. The surface used and the method by which ORF-specific DNA is bound, oriented and arranged on the surface vary with the system employed. However, all array formats share in common the capacity for parallel recognition, by complementary base pairing, of each of the DNA or RNA species in a complex sample which have corresponding ORF-specific sequences on the array surface. This simple experimental system provides the basis for two, quite distinct applications: comparative genomics and functional genomics by expression profiling. The distinction between functional and comparative genomics has led to the use of the terms 'genome' and 'transcriptome' to differentiate and classify information coming, respectively, from the study of an organism's genome content and transcriptional repertoire. Below, these two experimental paradigms are discussed in general and in greater detail in subsequent sections. Not discussed in this chapter are two related areas: 'proteome' and 'phenotome', which refer to the protein coding capacity of the genome and the phenotype diversity these proteins confer.

1.2. Comparative Genomics

Two genomes can be rapidly compared at the ORF-content level of resolution by hybridizing labelled DNA from a non-sequenced strain to a microarray containing representations of each ORF from a sequenced, reference strain. Despite its limitations (discussed in section 5), this application can be used to deduce the evolutionary relatedness of strains and species by providing a quantitative measure of diversity and information about how that diversity arose. It can also illuminate possibly crucial differences between pathogenic and non-pathogenic strains by identifying pathogen-specific genes that might code for virulence determinants.

1.3. Functional Genomics

The value of microarray-based functional genomics comes from the recognition that the transcriptional state of each gene of a genome under multiple, different conditions of growth is far more informative than the genome sequence itself or even the direct base-by-base comparison of two sequenced genomes (Ferea and Brown, 1999). Microarray-based functional genomic studies use RNA prepared from an intact micro-organism in a particular growth state. Hybridization of the labelled RNA (or corresponding labelled cDNA) to the array surface affords access to an organism's transcriptome by providing a quantitative measurement of mRNA. The hybridization intensity directly reflects gene-specific RNA abundance and serves as a reasonable measure of the gene's transcriptional activity for the condition of growth at the time just preceding lysis of the organism (providing methods and reagents are employed that instantaneously stop RNA transcription and prevent RNA degradation). In contrast to gene-by-gene assays of transcript abundance that have long been used by molecular microbiologists (Eisen and Brown, 1999), whole genome DNA microarrays measure the transcript abundance of each ORF simultaneously (i.e., in parallel), thus providing a condition-specific and time-specific genome-scale snapshot of transcriptional activity. The resulting information is commonly termed the 'expression profile' and differences between expression profiles illuminate the regulatory networks and metabolic and biosynthetic pathways and are active during the assayed growth state. Here it must be emphasized that the power of this method does not reside in the absolute quantitative measurement of transcript number per cell, but rather in the comparison of expression profiles obtained from carefully selected, different experimental conditions. Commonly studied comparisons include expression profiles from contrasting conditions of growth, at different time-points in a time-course or at different stages in a multi-stage developmental program.

The relatively small number of microarray expression studies of bacteria pathogens published thus far and reviewed below are surprisingly diverse with respect to experimental question and design. Not evident from the examination of individual studies are two trends that emerge from a survey of the field in aggregate.

First, the expanding microarray-based expression database for a particular strain or species has generated an archive that in the future will serve as a whole-genome transcriptional reference of an organism's physiological repertoire. Comparison of new expression profiles obtained from bacteria living is complex and undefined microenvironments with these archived, condition-specific profiles from *in vitro* experiments should help define the physicochemical features of host tissues and cells. The expression profile thus serves as a bioprobe of an organism's microenvironment from the perspective of the microbe and in circumstances not amenable to direct measurement with physical sensors.

Second, the comparison of expression profiles obtained from multiple conditions of growth and the analysis of these profiles with various clustering tools has led to the identification of sets of genes with common expression patterns. Providing the number of compared conditions is large, the resulting clusters are likely to contain genes that are co-regulated, including of course, operons, but also non-operonic sets of genes controlled by the same transcription factor. Such gene sets can be proposed as regulons. By contrast, a 'stimulon' is the non-clustered expression profile as a whole that was evoked by a particular condition of growth and probably encompasses several regulons (Tao *et al.*, 1999). Membership in a cluster not only implies co-regulation, but also suggests that the products of the clustered genes carry out some common metabolic, biosynthetic or pathogenic function (Eisen *et al.*, 1998; Ferea and Brown, 1999). Moreover, because approximately one-half of the ORFs composing most sequenced genomes code for proteins of unknown function (usually annotated as hypothetical or conserved hypothetical ORFs), it can be anticipated that the same proportion of genes in any cluster will also be of unknown function. However, because genes in a cluster are likely to be functionally related, those with known functions suggest functions for the others (Ferea and Brown, 1999). In this way, the cluster analysis of expression profiles can contribute to the annotation process.

A powerful additional application of expression profiling is the dissection of regulatory networks. While such networks are implied by the cluster analysis described above, expression profiling can be used to experimentally identify genes whose regulation depends on a particular transcription factor. One approach comes from the recognition that transcription factors are often co-regulated with the genes they in turn regulate. Therefore, cluster analysis can lead to the identification of a gene encoding a putative transcription factor within a cluster of other genes. This hypothesis can be tested by determining if disruption of the transcription factor abolishes expression of the gene cluster of which it is a member. In this case, both the hypothesis and its test come from expression experiments. Alternatively, genome sequence projects typically classify genes according to functional categories and amongst these are various classes of regulators. Expression experiments that identify conditions that change the expression level of a putative transcription factor can be followed by the use of these same conditions to study transcription factor mutants. The resulting expression profiles, compared with those obtained with the wild-type parent, should disclose gene sets regulated by the transcription factor. If used systematically, this approach can in principle identify genes governed by, for example, each annotated alternative sigma factor or two-component regulator in the genome. Because some of the regulated genes will be other transcription factors, it should be possible to reconstruct hierarchies of regulatory networks from these data.

While the combined use of mutational and microarray studies will identify transcription factors and the genes they regulate, in most cases the results will not distinguish between direct and indirect mechanisms of regulation. Indirect regulation can simply be a consequence of a multi-tiered regulatory cascade and the position of the transcription factor high in the hierarchy. Here, the distinction between a direct and indirect effect can only be made with certainty by the use of assays that determine if the transcription factor physically binds the promoter of the gene it is presumed to directly regulate. Alternatively, indirect regulation can come from the use of conditions and strains that cause pleotrophic downstream effects. This source of ambiguity can be mitigated by the use of short time-courses and conditions that do not alter the overall fitness or growth rate of a mutant compared with the wild-type parent. Accordingly, to avoid identifying differentially expressed growth phase or growth rate-specific genes in the expression profile, most workers now recommend the use of experimental conditions that do not cause differences in growth between two tested conditions. This often poses subtle and difficult experimental design problems and is a frequent cause of expression profiles that contain a combination of specific and non-specific effects.

Microarray expression profiling *per se* will not necessarily identify a transcription factor whose effect on expression is due to a post-translational event. Examples of these include phospho relays through two component systems and the activation of preformed transcription factors by changes in their conformation. Additionally, because most microarray formats do not represent regions of the genome that specify small, non-translated RNAs, their increasingly recognized regulatory functions may fail to be identified by this method (Wassarman *et al.*, 1999).

2. MICROARRAY FORMATS

The studies described in the following sections have employed one of three array formats. For the purposes of this discussion, the three formats will be referred to as 'high density oligonucleotide arrays', 'microarrays' and 'membrane macroarrays'.

Bacterial expression studies, independent of the array format, entail the preparation of labelled cDNA probes from total RNA because, unlike eukaryotes, bacterial mRNA is not polyadenylated and thus cannot be purified or selectively primed in the same manner as eukaryotic mRNA (Tao *et al.*, 1999; Arfin *et al.*, 2000). Since ~85% of total RNA is composed of rRNA and tRNA, the use of random hexamers by most protocols during cDNA synthesis will mainly label non-mRNA. While it is not clear if this decreases sensitivity by increasing the background, the signal coming from rRNA spots on membrane arrays may be so intense that adjacent spots are obscured. Thus, new methods to preferentially label mRNA in bacterial systems have been examined and two studies have used primers specific for the 3′ end of each *E. coli* ORF to prime first-strand cDNA synthesis from total RNA (Tao *et al.*, 1999; Arfin *et al.*, 2000). One of these studies compared the two priming methods and found that the hybridization success rate with probes produced with ORF-specific primers was only one-third that of probes produced with random hexamers (Arfin *et al.*, 2000).

2.1. High-Density Oligonucleotide Arrays

High-density oligonucleotide arrays, manufactured by Affymetrix (Santa Clara, CA, USA) consist of 15 to 20 different 25-base oligonucleotides for each ORF of a sequenced genome. Also represented in the same manner are intergenic regions greater than 200 bps (Lockhart *et al.*, 1996; Lipshutz *et al.*, 1999; Harrington *et al.*, 2000). The selection of gene-specific oligonucleotides is based in part on sequence uniqueness in order to reduce cross-hybridization artifacts between paralogs, i.e. other genes in the genome that contain related sequences. Each oligonucleotide is paired with a so-called 'mismatch' control oligonucleotide that differs from its 'perfect match' partner by only one, centrally located base. Comparison of the hybridization intensity of the perfect match and mismatch oligonucleotide provides a method for determining and subtracting background fluorescence.

Affymetrix arrays are manufactured by combinatorial solid-phase, light-directed oligonucleotide synthesis that is performed on a 1.28×1.28 cm glass surface. Synthetic linkers containing a protecting group that can be photochemically removed are attached to glass. Light directed through predetermined windows of an otherwise opaque 'mask' causes localized deprotection leaving behind a reactive group through which a reactive deoxynucleoside can be coupled. The stepwise application of a new mask, deprotection of the growing oligonucleotide and coupling with an added deoxynucleoside yields a microarray containing thousands of different, but predetermined oligonucleotides each complementary to a sequence of interest. The complete repertoire of an array consisting of 25 base oligonucleotides and representing each ORF and qualifying intergenic region of a bacterial genome sequence, can be synthesized by a maximum of 100 cycles ($4N$, where N is the oligonucleotide length).

To obtain expression profiles from bacteria using a high-density oligonucleotide array, a recent study used biotin-labelled cDNA from total RNA that was not heat-denatured before reverse transcription in order to reduce priming of rRNA (De Saizieu *et al.*, 2000). Hybridization of biotin-labelled cDNA to the array was followed by fluorescent staining with streptavidin–phycoerythrin. If necessary, the signal can be further amplified with a biotinylated antistreptavidin antibody followed by the streptavidin–phycoerythrin conjugate (DeSaizieu *et al.*, 2000). The pattern and intensity of the resulting signal is detected by laser scanning using the Affymetrix gene chip scanner. In contrast to the two-colour hybridization microarray system described in the next section, RNA from each sample is hybridized to a separate array (Harrington *et al.*, 2000), and thus an experiment comparing the expression profile from two contrasting conditions requires two high-density oligonucleotide arrays.

2.2. Glass-Spotted DNA Microarrays

The DNA format first described by P. Brown and colleagues at Stanford University (Schena *et al.*, 1996; DeRisi *et al.*, 1997; Eisen and Brown, 1999) consists of a microscope slide whose surface contains an x by y matrix of printed spots, each spot containing a polymerase chain reaction (PCR)-derived double-strand amplicon that corresponds to all or part of an ORF of the sequenced genome. Alternatively, single-strand synthetic 50 to 70 base oligonucleotides containing unique ORF-specific sequences have also been printed instead of PCR products. In either case, each ORF of the genome is represented on the array as a separate spot, its location designated by its xy address. Additional spots are added and correspond to internal controls which monitor the printing and hybridization steps. Thus far, most such arrays do not contain representations of intergenic regions.

The principal innovation in gene expression profiling introduced by Brown and colleagues is 'two-colour' hybridization (Eisen and Brown, 1999). This method employs two populations of cDNAs that have been differentially labelled with two different fluorochromes (ordinarily Cy3- and Cy5-dUTP) during a first-strand reverse transcription reaction using random hexamers as primers. The resulting cDNAs are usually derived from RNA prepared from the same organism cultivated under, or exposed to, two contrasting conditions. Equal masses of the two differentially labelled populations of cDNAs are combined, applied to the array surface and allowed to hybridize to their corresponding ORF-specific targets. The array is then scanned and the intensity of each label for each ORF-specific spot is quantified. These values are compared, yielding ratios that serve as a measure of the relative degree of expression or repression of each ORF for the two tested conditions.

2.3. Membrane Macroarrays

Membrane macroarrays also contain robotically printed PCR products corresponding to each of the annotated ORFs of a genome. However, unlike the DNA glass-spotted microarrays described above, membrane macroarrays are produced by printing the double-strand amplicons on to positively charged nylon membranes (Tao *et al.*, 1999). In one commercially available membrane macroarray (Sigma-GenoSys Biotechnologies, Woodland, TX, USA), PCR products corresponding to 4290 of the ORFs identifed during the *E. coli* K-12 genome sequencing project were printed in duplicate on 12- by 24-cm membranes. Rather than the fluorescently labelled cDNA probes that are used with glass-spotted DNA microarrays, the macroarray format employs ^{33}P-labelled probes and the hybridized array is scanned by phosphorimaging. Once scanned, the membrane can be stripped and rehybridized with probes from a second experimental condition, thus controlling for membrane-to-membrane variation and reducing the cost per experiment.

2.4. Comparison of Array Formats

In several respects, differences between the high-density oligonucleotide array format and the DNA-spotted microarray and macroarray formats are significant and affect experimental design and data interpretation (Harrington *et al.*,

2000). Bacterial genomes ordinarily contain many paralogues (genes containing shared nucleotide sequences) and gene duplications and highly repetitive sequences are common. High-density oligonucleotide arrays employ predetermined, short sequences for each ORF and these are selected to be ORF-specific. By contrast, PCR-derived amplicons, often containing most or all of an ORF, necessarily contain shared paralogous sequences. Even if primer design programs are used that preferentially exclude common sequences in amplified fragments, the inclusion of cross-hybridizing sequences is nearly inevitable. However, the increasing use of 50–70 base ORF-specific oligonucleotides instead of PCR products will certainly reduce cross-hybridization artefacts.

High-density oligonucleotide arrays, in contrast to DNA spotted micro- or macroarrays, contain representations of intergenic regions, including both the clockwise (Watson) and counterclockwise (Crick) strands (Wassarman *et al.*, 2001). This is a considerable advantage for comparative genomic studies aimed at deducing phylogenetic relationships between strains, because such regions may be under different selective constraints than protein-coding regions. Furthermore, intergenic regions may code for small ORFs not identified during the annotation of the genome and for non-translated RNAs, some of which have important roles in functions as diverse as RNA processing, mRNA stability, translation, protein stability and secretion (Wassarman *et al.*, 1999). DNA-spotted microarrays and macroarrays could be expanded by the inclusion of intergenic regions and since double-stranded PCR products are printed, both strands would be represented.

DNA glass-spotted microarrays employ the two-colour hybridization system described above and thus each experiment serves as its own control and the hybridization intensities are normalized for background effects in a pairwise manner (Eisen and Brown, 1999). While DNA membrane macroarrays can be stripped and reprobed, one study found that this practice led to more variability than the two-colour hybridization, glass microarray method (Richmond *et al.*, 1999). By contrast, a separate high-density oligonucleotide array is used for each sample, potentially increasing the possibility of array-to-array variability. This potential disadvantage is partly offset by more consistent quality control standards that come from the photolithographic manufacturing process.

The two-colour hybridizaton system provides relative levels of RNA abundance between the two tested samples. In contrast, by determining the hybridization intensity of an individual RNA sample, the high-density oligonucleotide microarray system provides a measure of the actual level of RNA abundance (Harrington *et al.*, 2000). However, this potential advantage is reduced by the recognition that hybridization intensity is the product of several effects, in addition to RNA abundance, including base composition and transcript size (Richmond *et al.*, 1999).

Of the array formats considered here, the DNA glass-spotted microarray system is the most flexible because amplicons can easily be added or deleted by the investigator. This feature, their capacity for fabrication by small laboratories, relatively low cost and the simultaneous comparison of two samples on the same array surface are strong reasons for their use by academic laboratories. However, the evolution of array formats is rapid and many innovations can be expected in the near future. Considerable additional information is available about each of these array formats via the internet and a useful list of website addresses has been published (Ferea and Brown, 1999). Detailed protocols for the fabrication and use of DNA glass-spotted microarrays for bacterial expression studies have been published (Schoolnik *et al.,* 2001; Wilson *et al.,* 2001) and are available at http://schoolniklab.stanford.edu.

3. ANALYSIS OF MICROARRAY DATA

Like all assays of biological systems, results obtained from microarray expression experiments are subject to multiple sources of variability – inherent biological variability and variability due to experimental and technical factors including array-to-array variation, slight differences in media, growth and time-course and different RNA preparation and labelling efficiencies (Long *et al.,* 2001). In turn, these are likely to be compounded by the multi-step nature of array protocols.

3.1. Statistical Analysis of Microarray Data

Each of the array formats discussed above generates ratios of signal intensities between control and experimental samples (Richmond *et al.,* 1999), either directly from the same array (DNA glass-spotted microarray format) or by comparing intensity values between two separate arrays (high-density oligonucleotide array format). It has been the practice of many investigators to arbitrarily select a particular ratio, often termed a 'fold-difference', that is believed to imply a biologically significant difference in a gene's transcriptional activity between two conditions. Most frequently selected is a two-fold difference. This practice generates lists of genes above this value (or below 0.5, if repressed) and implies that they are significantly regulated by the experimental condition compared with the control condition. However, a recent study persuasively points out the error of this logic, by noting that each microarray experiment generates thousands of measurements and that errors due to technical and experimental variations are likely to assume a Gaussian distribution (Arfin *et al.,* 2000). Consequently, for an array that provides mRNA abundance data for 5000 genes for two conditions of growth, a standard t test distribution will detect 250 genes (5%) whose expression levels between the two conditions will differ simply by chance alone, assuming a P value <0.05 (Arfin *et al.*, 2000). Therefore, if the expression experiment actually detects 500 differentially regulated genes between the two conditions, then fully one-half of these differences (250 genes) will be false positives at a 95% confidence interval (Arfin *et al.*, 2000). From this argument, it is self-evident that a fold-difference criterion alone is inadequate and that the statistical analysis of multiple biological replicates of the same experiment is required. Recognition of this issue has led to the development of new statistical instruments and their use has clearly demonstrated that no arbitrarily selected fold-difference reliably identifies all significantly regulated genes. Nor does it exclude genes whose apparent regulation is due to the sources of experimental variability described above (Arfin *et al.*, 2000; Tusher *et al.*, 2001). A comprehensive study by Hatfield and colleagues used a statistical instrument based on a linear analysis of variance model (available at http://www.genomics.uci.edu/) that showed that some genes with fold-induction levels less than two are in fact significantly regulated, whereas other genes with large fold-differences are not (Arfin *et al.*, 2000). These findings were confirmed by Tusher and coworkers who developed a different statistical instrument, designated the significance analysis of microarrays (available at http://www-stat-class.stanford/SAM/SAMSerlet/) and used it to analyse microarray studies of the effects of ionizing radiation (Tusher *et al.*, 2001).

Both statistical methods require the analysis of biological experimental replicates and the current consensus in the field suggests that at least three and optimally five are sufficient, providing microarray-specific statistical tools are used (Arfin *et al.*, 2000; Lee *et al.*, 2000; Long *et al.*, 2001). If a t test alone is used, >10 replicates may be required, a number that is unrealistic for microarray work (Long *et al.*, 2001). Having demonstrated that, the combined use of experimental replicates and appropriate statistical instruments will identify genes whose regulation is statistically significant, wo recent studies then conducted Northern analysis of mRNA abundance and these data corroborated results from the statistically analysed microarray results (Arfin *et al.*, 2000; Tusher *et al.*, 2001). Thus, the use of these statistical methods may make it unnecessary to routinely employ Northern, slot blot or RT-PCR assays of mRNA abundance to validate microarray expression results. Clearly however, the question of their biological significance must be addressed through the use of other methods.

3.2. Analysis of Microarray Data: Clustering Algorithms

Cluster algorithms are now commonly used to data mine microarray expression profiles to identify co-regulated genes that perform common metabolic or biosynthetic functions. Based on the foregoing discussion, these algorithms are best applied to genes identified as significantly regulated by the statistical methods referred to above.

A brief discussion follows of the two most commonly used unsupervised clustering algorithms.

Hierarchical clustering has been employed in the analysis of microarray expression data in order to place genes into clusters based on sharing similar patterns of expression (Eisen *et al.*, 1998). This method yields a graphical display that resembles a kind of phylogenetic tree, where the relatedness of the expression behaviour of each gene to every other gene is depicted by branch lengths. The programs Cluster and TreeView, both written by Michael Eisen at Stanford University, are available at http://rana.stanford.edu/software/.

Self-organizing maps (SOMs), a non-hierarchical method, have also been used to analyse microarray expression data (Tamayo *et al.*, 1999). This method involves selecting a geometry of nodes, where the number of nodes defines the number of clusters. Then, the number of genes analysed and the number of experimental conditions that were used to provide the expression values of these genes are subjected to an iterative process (20,000–50,000 iterations), which maps the nodes and data-points into multidimensional gene expression space. After the identification of significantly regulated genes, the expression level of each gene is normalized across experiments. As a result, the expression profile of the genome is highlighted in a manner that is relatively independent of each gene's expression magnitude. Software for the GENECLUSTER SOM program for microarray expression analysis can be obtained from the Whitehead/MIT Center for Genome Research at http://www-genome.wi.mit.edu/MPR/software.html.

Supervised clustering algorithms have also been proposed (Cummings and Relman, 2000).

3.3. Analysis of Microarray Data: Metabolic Pathway Databases

Further data mining can be carried out by analysing genes within a particular cluster through the use of metabolic pathway databases (Karp *et al.*, 1999). This is a logical second step because, as noted above, clusters often contain genes that mediate common cellular functions and the unit of analysis for a pathway database is a metabolic or biosynthetic pathway. From a physiological and systems perspective, such pathways are the functional unit of the microbe and thus changes in the expression of pathway genes underlie many of the adaptive responses of the organism to changes in growth conditions, to different microenvironments of the host and to the presence of antimicrobial compounds. Accordingly, the availability of pathway databases is likely to facilitate microarray data analysis. In a reciprocal manner, microarray experiments that probe the transcriptional response of an organism to different metabolic and physiological conditions of growth will probably yield data that can be used to refine the pathway database.

A comprehensive, empirically grounded *E. coli* metabolic database, suitable for the interpretation of microarray expression data, is available as EcoCyc (Karp *et al.*, 2000) at http://ecocyc.org/ecocyc/metacyc.html. Each pathway in the EcoCyc pathway database is characterized by its enzymes and the reactions they catalyse, its products, metabolites and precursors and the co-factors these reactions require. The processes by which EcoCyc determined if a particular metabolic pathway could be encoded by the *E. coli* genome and by which it assigned an ORF to the pathway, depended in part on the possible function of the ORF, predicted by reference to genes of proven function in other species. However, these initial functional and pathway assignments were then reviewed by a pathway curator who reconciled the initial pathway assignments (based on functional annotations from the *E. coli* sequencing project) with the pre-existing *E. coli* empirical database of biochemically proven metabolic pathways.

The development of pathway databases for bacterial species other than *E. coli* has relied on the information about *E. coli* metabolic pathways that is embedded in the EcoCyc database. Functionally annotated genes in a particular species' genome database are matched to homologues of the *E. coli* K12 genome database. The EcoCyc program then identified those genes of the species which have *E. coli* homologues that have been identified by EcoCyc to be components of an *E. coli* metabolic pathway. This method has been used to develop pathway databases for over 30 bacterial species; collectively these comprise the MetaCyc database (Karp *et al.*, 2000). The process by which these non-*E. coli* pathway databases were derived, as exemplified for *M. tuberculosis*, is as follows. The *E. coli* K-12 genome is predicted to have 4668 ORFs which specify 138 metabolic pathways. *M. tuberculosis*, which has 3924 ORFs, is predicted to have 103 pathways in common with *E. coli*, but several putative *M. tuberculosis* pathways were found to be incomplete in comparison with the *E. coli* pathways. The resulting *M. tuberculosis* pathway database, though a valuable resource for the analysis of expression data, therefore did not identify pathways present in *M. tuberculosis*, but absent in *E. coli*. Moreover, the *M. tuberculosis* pathway designations were based entirely on the identification of *M. tuberculosis* genes which were found to share sequence homology with *E. coli* genes and thus were predicted to be functionally homologous as well. The addition of empirically derived data to this database from biochemical studies of *M. tuberculosis* metabolic pathways will address these deficiencies.

In summary, starting with log-transformed, normalized intensity values, the analysis of microarray data begins with the statistical analysis of experimental replicates and the evaluation of significantly regulated genes by cluster algorithms. Further analysis of the resulting clusters by reference to the organism's pathway database leads to a physiological interpretation of the expression profile.

4. MICROARRAY EXPRESSION PROFILING OF PATHOGENIC BACTERIA

The focus of this chapter is pathogenicity. Thus, microarray studies of the non-pathogenic, model micro-organisms *E. coli* K-12, *Bacillus subtilis*, and *Caulobacter crescentus* will not be discussed. However, these studies are summarized in Table 1 and recommended to the reader because they demonstrate how microarray expression profiling can reveal new information about previously well-studied physiological responses; the value of comparing the expression profiles of carefully selected mutant and wild-type strains; and the use of expression profiling to analyse complex, multistage developmental programs.

Table 1 Microarray expression studies of non-pathogenic *E. coli* strains and non-pathogenic bacterial species.

Strain/Species	Pathogenic status	Condition studies	Reference
E. coli K-12, MG1655	Non-pathogenic	Heat shock, IPTG	Richmond *et al.*, 1999
E. coli K-12, MG1655	Non-pathogenic	H_2O_2	Zheng *et al.*, 2001
E. coli K-12, MG1655	Non-pathogenic	Rich medium, minimal medium Exponential phase, transition phase, ITPG	Wei *et al.*, 2001a
E. coli K-12, MG1655	Non-pathogenic	Rich medium, log-phase, stationary phase	Selinger *et al.*, 2000
E. coli K-12, MG1655	Non-pathogenic	Late log-phase, minimal medium, rich medium	Tao *et al.*, 1999
E. coli K-12, TG1	Non-pathogenic	Stationary phase, metal resistance (ZnII, CdII, CoII or NiII)	Brocklehurst and Morby 2000
E. coli K-12, MG1655	Non-pathogenic	Exponential phase, nitrogen limitation	Zimmer *et al.*, 2000
E. coli W3110	Non-pathogenic	Tryptophan excess or starvation	Khodursky *et al.*, 2000
B. subtilis 168	Non-pathogenic species	Anaerobic growth	Ye *et al.*, 2000
C. crescentus	Non-pathogenic species	Cell cycle	Laub *et al.*, 2000

The ultimate goal of whole genome expression studies of pathogenic bacteria is the identification of bacterial genes which are differentially regulated in the host. Within this class of genes are those that mediate adaptation of the microbe to host-specific microenvironments or encode virulence determinants. Ideally, studies of this kind would compare expression profiles of bacteria within infected tissues with profiles from bacteria cultured under standardized *in vitro* conditions of growth. Unfortunately, this technically formidable goal has not been achieved using non-amplification methods, because the number of organisms within infected tissues is normally small, RNA from host cells is vastly more abundant than bacterial RNA and no efficient method to differentially extract stabilized bacterial RNA from tissues has been described.

Due to these considerations, most microarray studies of pathogenic bacteria have employed conditions of growth achievable *in vitro* using media that do not contain eukaryotic cells. *In vitro* growth conditions are selected that are believed to simulate host microenvironments with respect to one or more of the following parameters: growth phase, temperature, oxygen tension, carbon source, pH, osmolarity, divalent cation and iron concentrations and the presence of products of the activated macrophage, including reactive oxygen and nitrogen intermediates. Inherent in this experimental design is the assumption that the gene expression profile of a microbe in tissue is the sum of its responses to individual physicochemical parameters that characterize the tissue microenvironment. If this assumption is valid, then the complex transcriptional response of the organism in tissue can be deduced from a series of microarray expression studies that capture the transcriptional response to a series of individual *in vitro* conditions where each of the tested conditions simulates a particular *in vivo* value. While this assumption has never been thoroughly tested, many examples now exist showing that the expression of genes encoding virulence determinants can be induced *in vitro* by simple modifications of laboratory media (Mekalanos, 1992). Studies of this kind are summarized in Table 2 and reviewed below.

4.1. *Pasteurella multocida*: Iron Limitation

This experimental approach was undertaken by Paustian and colleagues in their study of the transcriptional response of *Pasteurella multocida* growing under iron-limited conditions *in vitro* (Paustian *et al.*, 2001). *P. multocida* is a Gram-negative bacillus that causes economically important infections of fowl (cholera), cattle (septicaemia), swine (atrophic rhinitis) and rabbits (snuffles). Available free iron is limited in host tissues by the presence of the iron-binding proteins lactoferrin and transferrin and accordingly iron limitation induces a complex iron-scavenging response involving the production of iron-binding compounds and transport systems. To characterize the *P. multocida* low iron

Table 2   Microarray expression studies of pathogenic bacteria.

Species and strain	Infection syndrome	Growth condition/treatment	Reference
E. coli O157:H7	Haemorrhagic colitis/ Haemolytic uraemic syndrome	Acetate-induced acid tolerance	Arnold et al., 2000
H. pylori NTU-D1	Gastritis, peptic ulcer disease, gastric cancer, gastric lymphoma	Acid-induced gene expression	Ang et al., 2001
Pasteurella multocida	Fowl cholera, haemorrhagic sepsis (cattle), rhinitis (swine), snuffles (rabbits)	Iron starvation	Paustian et al., 2001
S. pneumoniae	Pneumonia, meningitis, sepsis, otitis media, septic arthritis	Induction of competence by CSP	Rimini et al., 2000
S. pneumoniae	See above	Response to auto-inducer peptide BlpC*	De Saizieu et al., 2000
M. tuberculosis	Tuberculosis	Response to antibiotic isoniazid	Wilson et al., 1999
M. tuberculosis	Tuberculosis	Growth at reduced oxygen tension	Sherman et al., 2001
M. tuberculosis	Tuberculosis	Stress response: SDS, ROI	Manganelli et al., 2001

response, log phase cultures of the organism were grown either in the iron-replete medium, brain–heart infusion (BHI) broth, or in BHI containing the iron chelator 2,2′-dipyridyl; bacteria were harvested and total RNA prepared 15, 30, 60 and 120 min later, labelled and the transcriptional response determined using a DNA microarray containing 96% of the organism's 2014 ORFs. The expression levels of 135 genes (~7% of the genome) were found to be altered by two-fold or more during iron-limited growth; of these, 85 were up-regulated and 50 down-regulated (Paustian *et al.*, 2001). This set of regulated genes includes many known in other organisms to be iron-regulated, including the Yfe, Fbp and Fec operons that participate in the transportation of iron from the periplasmic space across the cytoplasmic membrane into the cytoplasm. However, the response was considerably more diverse including the down-regulation of genes involved with energy metabolism and the up-regulation of genes involved with DNA and central intermediary metabolism and with amino acid biosynthesis. Further, although *fnr*, a regulator of anaerobic metabolism was up-regulated, some Fnr-regulated genes, including fumarate reductase and formate dehydrogenase, were down-regulated. The complexity of the transcriptional response to iron limitation raises the following question, typical of many whole genome expression results: which genes in the overall expression profile compose the primary or direct response to iron limitation and which are secondary, downstream consequences of an intracellular deficiency in iron which might affect multiple pathways? For example, iron limitation might disrupt iron-sulphur centre-dependent enzyme systems, including electron transport proteins and those that monitor the redox state of the cell. The pleotrophic nature of the transcriptional response to an apparently simple change in the concentration of only one metal, also points to the interdependency of multiple metabolic and biosynthetic pathways. This in turn encourages a systems view of cellular homeostasis which contrasts dramatically with the reductionist focus on a single component of the whole.

4.2. Enterohaemorrhagic *E. coli* O157:H7: Adaptation to Acidic Microenvironments

The pH of some host microenvironments, like the concentration of available iron, challenges the capacity of bacteria to adapt, especially in strongly acidic organs. The most extreme of these, the stomach, normally reaches pH values as low as 2.0 and virtually all bacteria that ultimately compose the rich and varied microflora of the colon and distal small intestine have successfully transited the 'gastric acidity barrier' before coming to reside in the more alkaline environment of the intestine. Some pathogenic species, particularly *Shigella* sp. and the enterohaemorrhagic and enteroinvasive *E. coli* biotypes are particularly acid-resistant, explaining why infections with these organisms can be initiated by fewer than 30 bacteria (Gorden and Small, 1993). Most remarkable of the acid-tolerant group is *Helicobacter pylori*, which chronically colonizes gastric mucous membranes and thus is normally exposed, often for decades, to an acidic environment. By contrast, many other enteric pathogens are less acid tolerant and require infectious doses exceeding 100,000 organisms.

Growth conditions that precede acid exposure can have a striking effect on acid tolerance. Amongst the most potent of these is exposure to short chain fatty acids, particularly acetate, the protonated form of which crosses the cytoplasmic membrane, dissociates in the cytoplasm and thus lowers cellular pH, even when the pH of the growth medium is neutral (Arnold *et al.*, 2001). Acetate-treated *E. coli* transferred from neutral to acidic media survive significantly better than non-pretreated bacteria. To learn more about the mechanism of acetate-induced acid tolerance, log phase cultures of enterohaemorrhagic *E. coli* serotype O157:H7 were incubated in M63 glucose medium, pH 7.0 with or without 100 mM acetate for 30 min, the bacteria harvested and the total RNA prepared, labelled and analysed using an *E. coli* K-12 membrane macroarray containing representations of all of the ORFs of the prototype sequenced non-pathogenic strain MG1655. Therefore, the 1,387 ORFs that are present in the *E. coli* O157:H7 genome, but absent in the genome of *E. coli* K-12 (Perna *et al.*, 2001) would not have been monitored by this K-12-specific membrane macroarray. Despite this limitation, of the *E. coli* O157:H7 ORFs represented on the K-12 array, 26 were found to be up-regulated two-fold or greater in acetate-treated bacteria (Arnold *et al.*, 2001). Amongst these were six genes previously known to be required for acid resistance including *gadA* and *gadB*, whose products catalyse the conversion of glutamate to γ-aminobutyrate and *gadC*, which encodes a γ-aminobutyrate antiporter. Taken together, these genes specify functions that defend the intracellular pH during exposure of the organism to acidic conditions of growth. Moreover, this transcriptional response is consistent with the findings of prior studies showing that disruption of *gadA, gadB* or *gadC* results in an acid-sensitive phenotype. Also induced by acetate was *cfa*, whose product cyclopropanates unsaturated fatty acids in the inner membrane, possibly decreasing its permeability to protons. The gene coding for HdeA was also induced. This periplasmic chaperon probably prevents acid-induced denaturation and aggregation of periplasmic proteins. Three oxidative stress genes, *dps*, *katE* and *grxB* were induced, as well indicating that exposure to acetate generates reactive oxygen species. Some of the genes induced by acetate are regulated by RpoS, an alternative sigma factor previously known to be required for acid survival. Although disruption of *rpoS* prevented acetate-induced acid tolerance, its induction did not confer acid tolerance. This information, in combination with the acetate-induced expression results, shows that the induction of acid tolerance by acetate requires genes that are not part of the RpoS regulon (Arnold *et al.*, 2001). Acetate reduced by two-fold the mRNA abundance of 60 genes, of which 48 encode components of the transcription–translation machinery, whose down-regulation reflects an acetate-induced decline in growth rate.

4.3. *Helicobacter pylori*: Adaptation to an Acidic Microenvironment of the Host

To identify the transcriptional response of *H. pylori* to acidic conditions of growth, Ang and colleagues grew a recent gastric isolate from a patient with peptic ulcer disease on two batches of Columbia agar containing 5% sheep blood and an antibiotic supplement. One batch had been titrated to pH 7.2 and the other to pH 5.5 (Ang *et al.*, 2001). Forty-eight hours later, total RNA was prepared and the transcriptional profile obtained using a membrane macroarray containing representations of 96% of the 1534 predicted ORFs of *H. pylori* strain 26695. Compared with the expression profile of bacteria grown at pH 7.2, the mRNA steady-state levels of 80 ORFs was increased five-fold or greater by growth for 48 h at pH 5.5. Sixteen of these were previously known to be acid-induced, either in *H. pylori* or in other species. These include Omp11, a proton-translocating ATPase believed to contribute to acid tolerance by extruding cytoplasmic protons and arginase whose homologue in *B. subtilis* is involved with urea cycle regulation and is essential for acid tolerance. The expression of *ceuE*, which codes for a component of an iron (III) ABC transporter was also strongly induced, indicating a possible convergence of the acid stress and iron-scavenging responses. An additional 43 up-regulated ORFs were identified that code for functionally annotated proteins that had not been previously known to be acid-regulated. Finally, 21 up-regulated ORFs specified proteins of unknown function. Only four ORFs were down-regulated by acid and two of these compose part of an operon encoding two subunits of NADH-ubiquinone oxidoreductase, a component of the respiratory chain that transfers periplasmic protons across the inner membrane. Their reduced expression suggests that proton pumping is suppressed during acidic conditions of growth. Taken together, these results show that only a minority of the acid-regulated genes can be directly associated with acid tolerance by a plausible physiological mechanism or by reference to prior work (Ang *et al.*, 2001). In part, this probably reflects the complex and relatively unexplored nature of the acid-resistant phenotype in this species and the large number of its ORFs that code for proteins of unknown function. However, in part this may also come from two aspects of the study's experimental design. First, the use of organisms grown on a solid medium leads to self-generated gradients, due to the consumption of nutrients and the production of metabolic by-products, that potentially create a heterogeneous population with respect to metabolic state. Moreover, it is possible that the pH of the agar adjacent to areas of heavy bacterial growth was titrated by physiological processes to a different and unknown pH. Second, sampling the transcriptome after a protracted period of growth (48 h) at pH 5.5 is likely to characterize a new steady state rather than the adaptive process itself. That is, acid-induced stress probably generates an immediate transcriptional response leading to the accumulation of relatively stable protein products that mediate the adaptive process. Once translation of these proteins has occurred, the transcriptional activity of their cognate genes may decline to near-basal levels, too low to be identified amongst the regulated set of genes.

4.4. *Streptococcus pneumoniae*: Induction of Competence by ComC

S. pneumoniae, a common cause of pneumonia, sepsis and meningitis, normally colonizes the pharyngeal mucous membranes of asymptomatic persons. Genetic diversity in this species is thought to arise by transformation through the uptake of DNA released from the other strains and species within this ecosystem that have undergone autolysis. Competence, the capacity to bind and take up extraneous DNA, is a transient physiological state that entails the co-ordinate regulation of gene groups whose expression exhibits a predetermined and predictable temporal profile. Rimini and co-workers used a membrane macroarray containing PCR products from 4301 clones covering most of the genome of *S. pneumoniae*, capsular serotype 19F, to profile the kinetics of gene expression during competence (Rimini *et al.*, 2000). Competence was initiated by the addition of synthetic competence stimulating peptide (CSP), a quorum-sensing pheromone encoded by *comC*. Total RNA was obtained 0, 5, 10, 15 and 30 min later. This time-course was chosen because prior work had showed that one group of competence genes is induced at 5 min (designated the 'early' set), a second group is induced at 10 min (designated the 'late' set) and that transcripts corresponding to both sets disappear by 20 min. After addition of CSP, the genome-wide expression profile corroborated the kinetics of the early and late transcriptional patterns and within these groups were 17 genes previously described to be up-regulated during competence. However, more important was the identification of 23 other up-regulated genes not previously recognized to be associated with competence (Rimini *et al.*, 2000). Amongst these was a pair of adjacent genes coding for two putative pheromones and preceded by a *cin*-box consensus sequence, a motif that is also present in the promoter regions of several late competence genes. By contrast, the promoter region of an up-regulated, operonic-like cluster of two genes, orthologous to *yhaQ* and *yhaP* in *B. subtilis*, was found to contain a ComE-binding site. ComE is the cognate response regulator of ComD, the histidine kinase CSP receptor, and thus controls the expression of early competence genes. While most of the CSP up-regulated genes exhibited either early or late expression kinetics, *ciaH*, which encodes a sensor kinase, was up-regulated much later in the time-course (30 min after the addition of CSP) suggesting a possible role in the termination of competence. In contrast to prior studies of competence, Rimini and co-workers also discovered seven CSP down-regulated genes, most of which encode ribosomal proteins, thus indicating that competence is associated with lower levels of protein synthesis. This study demonstrates how microarray expression profiling can illuminate a programmed physiological process, by relating the transcriptional state of genes to a time-course that marks its initiation, manifestation and decline. It also shows that a genome-wide study will nearly always reveal new information, even about a well-studied phenomenon. Finally, the use of promoter motifs, such as the *cin*-box consensus sequence and ComE-binding site described above, powerfully inform expression results by grouping regulated genes into distinct functional sets and regulons, based on the presence of these motifs, and by helping to distinguish direct effects (genes with conserved upstream motifs) from indirect effects (genes with diverse upstream sequences).

4.5. *S. pneumoniae*: Cell Density-Dependent Gene Regulation

Microarray expression analysis of cell density-dependent gene regulation in *S. pneumoniae* was also used by De Saizieu and co-workers to study a bacteriocin-like peptide, two-component system that is remarkably similar with respect to its regulation, processing, export and signal transduction to the competence system described above (De Saizieu *et al.*, 2000). As in the preceding study, a synthetic bacteriocin-like peptide (designated BlpC*), corresponding to the sequence deduced from the cognate gene in strain KNR.7/87, was added to exponentially growing pneumococcal strains R6 and KNR.7/87. RNA was prepared from these cultures and the transcriptional profile obtained using a high-density oligonucleotide Affymetrix array containing 1973 ORFs and 323 intergenic regions. Sixteen genes were significantly up-regulated in strain KNR.7/87 10 min after addition of the peptide and their expression persisted until stationary phase. The expressed genes were clustered on the chromosome near *blpHR*, which encodes the cognate two-component system, which regulates expression of *blpC*. These data also showed that *blpC* is autoinduced. However, peptide BlpC* did not induce this regulon when added to a culture of pneumococcal strain R6, showing that the peptide exhibits strain specificity (De Saizieu *et al.*, 2000). The demonstration by Throup *et al.* that a *blpHR* mutant exhibits attenuated virulence in a murine model of pneumococcal pneumonia shows that this quorum-sensing system is expressed *in vivo* and contributes to pathogenicity (Throup *et al.*, 2000).

4.6. *Mycobacterium tuberculosis*: Low-Oxygen Gene Regulation and Induction of Dormancy

Like *S. pneumoniae*, *M. tuberculosis* mainly infects the lung. However, unlike patients with pneumococcal pneumonia, an acute infectious process, most people infected with *M. tuberculosis* have a latent form of the disease and are non-infectious and asymptomatic. However, these persons harbour viable, but non-replicating bacilli in granulomatous lesions. Immunosuppression owing to age, illness, cancer therapy or organ transplantation substantially increases the probability that latent foci of infection will reactivate leading to clinical disease and possible spread of *M. tuberculosis* to other people. Although ~90% of persons with latent tuberculosis never reactivate, the minority who do infect others and today one-third of the world's population (~2 billion people) harbour this organism.

The nature of latency is not only a serious clinical and epidemiological issue, but the most compelling and poorly understood aspect of this species' biology. *In vitro*, a state of non-replicating persistence (termed 'dormancy' to distinguish it from 'latency', which is a clinical phenomenon) can be produced by allowing a non-stirred culture to generate a low oxygen gradient as the respiring bacteria settle to the bottom of a culture tube (Wayne, 1976). Non-replicating cultures of this kind, viable for months and perhaps years, can be resuscitated to the replicating state by re-introduction of oxygen. This simple *in vitro* dormancy model is believed to simulate aspects of *in vivo* latency because mature granulomas are often surrounded by calcified fibrous tissue and thus may be hypoxic. Moreover, reactivation most commonly occurs in the apices of the lung where oxygen levels are high. To identify *M. tuberculosis* genes differentially regulated by hypoxia, an early exponential phase culture of *M. tuberculosis* was shifted from growth in air (~20% O_2) to growth in a hypoxic atmosphere (0.2% O_2, 99.8% N_2); 2 h later, RNA was prepared and the genes induced by hypoxia were identified using a DNA microarray containing >97% of the 3924 ORFs predicted from the genome sequence of strain H37Rv (Sherman *et al.*, 2001). The expression of 47 genes was found to be significantly up-regulated. They were clustered in eight chromosomal locations and within seven of these clusters were three or more contiguous ORFs and/or operonic-like groups with opposing transcriptional orientations indicating the presence of at least two separate 5′ upstream promoter regions per cluster. This chromosomal arrangement, immediately evident from the inspection of whole genome expression data, indicates a kind of super-operonic organization for functionally linked sets of genes.

Although approximately two-thirds of the induced genes are of unknown function, several with annotated functions are plausibly involved with adaptation to hypoxia and possibly to dormancy. Amongst these is *acr*, which encodes α-crystallin, a 14-kDa protein with chaperone activity that was previously shown to accumulate during dormancy. Also induced were genes that indicate a shift from aerobic to microaerophilic metabolism, including *narX* and *narK2*, which encode a fused nitrate reductase and nitrite extrusion protein, respectively. Several genes coding for putative transcription factors were also identified within the induced gene set including two within a three-gene operon. One of these two is predicted to encode a membrane-bound sensor, histidine kinase, and the other a two-component response regulator. Mutational analysis of this operon showed that disruption of the response regulator, but not of the adjacent sensor histidine kinase, prevented hypoxic induction of *acr* whose expression was thought to reflect the transcriptional state of other genes within the low oxygen stimulon (Sherman *et al.*, 2001). Compared with the wild-type parent strain, this mutant survived less well in late stationary phase. The expression of 60 genes was found to be reduced during low oxygen growth; their annotated functions point to a decline in aerobic metabolism, amino acid, protein and lipid biosynthesis and functions involved with cell division. The expression profile described above was evident after only 2 h of low oxygen growth. This time-course, considered in the context of clinical latency (which in most people persists for a lifetime) suggests that this set of low O_2-regulated genes is but the first transcriptional stage in a multistage developmental program that leads to a stable non-replicating state *in vivo*.

The identification of the cognate two-component response regulator of this low oxygen regulon exemplifies how microarray expression analysis can be used to identify transcription factors and dissect regulatory networks. Other such studies are summarized in Table 3.

4.7. *M. tuberculosis*: σ Factor Regulatory Cascades

Developmental programs of the kind exemplified by *M. tuberculosis* dormancy frequently entail temporally ordered transcriptional events governed by a hierarchy of transcription factors. σ factors, located near the top of this hierarchy, confer different promoter specificities by associating with the RNA polymerase core enzyme. *M. tuberculosis* is particularly well endowed with alternative σ factors, having 13 genes which are predicted to be of the σ^{70} class, of which 10 belong to the ECF (extracytoplasmic) group. The ECF subset of alternative σ factors are of particular interest because in other species some have been found to confer adaptive responses to environmental factors and stress or to be required for virulence.

M. tuberculosis is a facultative intracellular parasite and in host tissues it resides mainly within the macrophage. The resting macrophage offers a milieu that is vastly different from *in vitro* growth with respect to iron availability, pH and carbon source. Activation of the macrophage with IFNγ, an event that

Table 3 Regulon identification by microarray expression profiling.

Species	Transcription factor	Growth condition	Reference
M. tuberculosis	σ^E	Stress response: SDS	Manganelli *et al.*, 2001
M. tuberculosis	*nrpR1* (regulates low O_2-induced expression of *acr*, encoding α-crystallin)	Early exponential phase, low/normal O_2 tension, wild-type/$\Delta nrpR1$	Sherman *et al.*, 2001
E. coli	*trpR* (tryptophan repressor)	Minimal medium + Trp/indole acrylate	Khodursky *et al.*, 2000
E. coli	*marA*-positive regulator of the Mar (multiple-antibiotic resistance) regulon	Exponential phase, *marA*-deleted/ *marA*-expressing strains	Barbosa and Levy, 2000
E. coli	NtrC-nitrogen regulatory protein C	Exponential phase, NH_4Cl or Gln nitrogen source; *ntrC* null allele or NtrC-activated expressing strains	Zimmer *et al.*, 2000
E. coli	*sdiA* (positive regulator of *ftzQAZ*) and thus of septation	Exponential phase, LB broth, *sdiA* overexpressing strain compared with wild-type	Wei *et al.*, 2001b
E. coli	*oxyR* (H_2O_2-regulated genes)	Exponential phase, LB broth, H_2O_2-treated wild-type and $\Delta oxyR$ mutant	Zheng *et al.*, 2001

coincides *in vivo* with the onset of cell-mediated immunity, leads to the prompt generation of reactive oxygen intermediates (ROIs) owing to the respiratory burst, reactive nitrogen intermediates (RNIs) due to expression of macrophage inducible nitric oxide synthase (iNOS) and the production of antimicrobial peptides. Adaptation to these stresses presumably occurs *in vivo* because, as discussed above, *M. tuberculosis* can persist for years within the tissues of immunocompetent hosts.

Microarray expression profiling would seem an ideal means by which to identify the gene set governed by ECF σ factors required for successful survival in the macrophage phagosome. Manganelli and colleagues used a combination of mutational and microarray methods to define the function and characterize the regulon of the *M. tuberculosis* ECF σ factor, σ^E (Manganelli *et al.*, 2001). Disruption of *sigE* yielded a strain that was more sensitive than the wild-type parent to heat shock, the ionic detergent sodium dodecyl sulphate (SDS) and to the oxidants hydrogen peroxide, cumene hydroperoxide and plumbagine (a superoxide generator). To determine if the sensitivity of this mutant to these stresses affected its survival within macrophage cell lines, the growth rates of mutant and wild-type in the human THP-1 monocytic cell line and in the murine J774A.1 cell line, were compared. The mutant grew less well than the wild-type strain in each of these cell lines and was more rapidly killed in J774A.1 cells that had been activated by the addition of IFNγ and LPS, than the parent. These findings provided a sound basis for the selection of growth conditions to identify genes that, directly or indirectly, are regulated by σ^E. To identify σ^E-regulated genes under non-stress, physiological conditions of growth, total RNA was obtained for expression profiling from the mutant and wild-type from mid exponential phase cultures (Manganelli *et al.*, 2001). This comparison was valid because the growth rates of both strains in this medium were equivalent. Thus, the differential regulation of growth-rate dependent genes would not be confused with σ^E-dependency. The expression levels of 38 genes were increased in the wild-type compared with the *sigE* mutant; amongst these was *sigB*, a principal σ factor of the σ^{70} class. Exposure of a mid-exponential culture of mutant and wild-type to 0.05% SDS for 90 min revealed 23 genes whose SDS-induced expression required σ^E. When this gene set and the non-stressed σ^E-dependent mid-exponential set were compared, σ^B was the only gene in common indicating that the genes showing σ^E dependence during physiological, non-stressed growth (except σ^B) are indirectly regulated by σ^E, possibly through an indirect effect of σ^E on the expression of housekeeping genes. By contrast, many of the σ^E-dependent, SDS-induced genes seem likely to directly require σ^E. Besides σ^B, these include *aceA* (encoding isocitrate lyase of the glyoxalate shunt) and *fadB2* (encoding 3-hydroxyacyl CoA dehydrogenase, a component of the fatty acid β oxidation pathway), both of which are known to be induced by mycobacteria within macrophages. Their induction suggests that host fatty acids are the main source of carbon and energy *in vivo*.

The expression of *aceA* and *fadB2* by exposure to SDS may point to damage of cell envelope lipids by this detergent and their recycling and catabolism via β oxidation. Similarly, the σ^E-dependent induction by SDS of *fadE23* and *fadE24*, two genes that are also induced by isoniazid-dependent inhibition of mycolic acid biosynthesis (Wilson *et al.*, 1999), suggests that SDS also damages biosynthetic pathways required for the production and/or maintenance of cell envelope complex lipids.

4.8. *M. tuberculosis*: Inhibition of Biosynthetic Pathways and the Identification of New Drug Targets

M. tuberculosis is not only a formidable human pathogen, but is also increasingly difficult to treat because of emerging resistance to one or more antitubercular drugs. To identify genes differentially regulated upon exposure of the organism to isoniazid (isonicotinic acid hydrazide, INH), a compound discovered in 1952 that is still commonly used in combination with other antibiotics to treat tuberculosis, Wilson and co-workers prepared a microarray containing representations of ~97% of the ORFs of the recently sequenced *M. tuberculosis* genome (strain H37Rv) (Wilson *et al.*, 1999). It has long been known that INH blocks the biosynthesis of mycolic acids, an essential component of the mycobacterial cell envelope, and recent biochemical studies indicate that it does so by inhibiting the type II fatty acid synthase (FAS-II) complex that is required for the production of the full-length meromycolate chain. Two mechanisms of action have been proposed: binding of NADH within the active site of enoyl-acyl carrier protein reductase (InhA) (Rozwarski *et al.*, 1998) or the formation of a ternary complex with β-ketoacyl-ACP-synthase (KasA) and the acyl carrier protein, AcpM (Mdluli *et al.*, 1998).

The microarray study of INH was undertaken to learn if a drug-induced expression profile could be used to predict a compound's mode of action. Mid-log phase cultures of *M. tuberculosis* were treated with concentrations of INH at or below the minimal inhibiting concentration of the drug for this strain or with a drug-free buffer. Bacteria were isolated and total RNA prepared 40 min, 1, 4 and 8 h thereafter (Wilson *et al.*, 1999). Since this slow-growing species divides in culture approximately once every 18 h, this time-course preceded any discernable effect on bacterial growth. Remarkably, of the 3834 ORFs whose RNA abundance was monitored by DNA expression analysis, only 14 were found to be differentially regulated by INH at these time-points and drug concentrations. Amongst these genes was the induction by INH of an operon-like cluster that encodes components of the FAS-II complex including AcpM and KasA. This result, evident at the earliest monitored time-point (40 min), indicates that an expression result can provide useful information about a compound's mode of action and demonstrates that inhibition of a biosynthetic pathway is sensed and responded to at the transcriptional level within minutes. Also induced was *fbpC*, which encodes trehalose-dimycolyl transferase, an activity at the end of the mycolate biosynthetic pathway that esterifies mycolic acids with cell wall carbohydrates. Induction of this gene probably follows INH-induced depletion of mature mycolates and the need of the pathway to anchor its product into the cell wall. This result shows that an expression result can illuminate components of a multicomponent pathway which are remote from its direct site of action.

Inhibition of a biosynthetic pathway logically results in the accumulation of precursor molecules and biochemical studies demonstrate that INH-treated mycobacteria accumulate fatty acids as the levels of mature mycolates decline. Congruent with these biochemical data, microarray expression analysis identified the increased abundance of transcripts coding for two acyl-CoA dehydrogenases, FadE23 and FadE24, likely to be involved in the β-oxidation of the accumulated fatty acids. Induction of this shunt pathway is thus an early, but indirect consequence of INH-mediated inhibition of mycolate biosynthesis. Also induced was *efpA*, predicted to encode a proton-energized, membrane-bound transporter which might export mycolate precursors or expel drug or toxic by-products. Finally, induction of *ahpC*, the gene encoding alkyl-CoA dehydrogenase, probably signals increased levels of reactive oxygen intermediates, and in particular, lipid peroxides. Taken together, *efpA* and *ahpC* compose a set of genes in the INH-induced expression profile induced by the toxic effects of mycolate pathway inhibition. Thus, the entire expression profile can be deconvoluted into three components: pathway-specific effects on gene expression, including a likely site of INH action; induction of a shunt pathway, an indirect effect secondary to pathway inhibition; and the induction of genes responding to the toxic by-products that accumulate owing to pathway inhibition (Wilson *et al.*, 1999).

This study illustrates how microarray expression profiling can contribute to the antibiotic discovery process. The rationale for a microarray-based drug discovery strategy comes from the recognition that the mode of action of most antibiotic classes is the inhibition of a vital metabolic or biosynthetic pathway. Inhibitors of this kind predictably cause a decrease in pathway products downstream of the point of inhibition and an accumulation of pathway precursors upstream of the site of inhibition. This model predicts that the resulting fluctuations in the abundance of products and precursors are sensed by the genome and result in increased expression of genes coding for pathway enzymes distal to the point of inhibition and decreased expression of genes proximal to the point of inhibition. As shown by the INH experiment described above, increased expression of genes in associated shunt pathways may be expected to occur as well, including those that degrade or expel toxic by-products that have accumulated as a result of pathway inhibition.

This paradigm predicts that exposure of a micro-organism to a drug or compound of unknown mode of action should elicit an expression profile that incriminates the affected pathway and perhaps even the target in the pathway. Thus, expression profiling can help define the mode of action of leads identified during high-throughput screens for bacteriocidal compounds. Moreover, through the use of inhibitors whose sites of action are known, it should be possible to identify signature profiles that are pathway-specific. Matching the expression profile evoked by a new compound with an archive of pathway-specific profiles should greatly accelerate mode of action studies.

The study described above also showed that the use of a site-specific inhibitor (e.g. INH) can illuminate other components of the affected pathway (e.g. FbpC) beyond the target itself. If pathway inhibition is lethal, then inhibition of any of the critical enzymes that compose the pathway should be lethal as well. They, therefore, constitute new drug target candidates. This strategy is particularly attractive when mutations of the original target in a critical pathway have led to antibiotic resistance. In this case, expression profiling may lead to the identification of other targets in the same pathway.

5. MICROARRAY-BASED COMPARATIVE GENOMIC STUDIES

Genetic variability and natural selection yield strains and species adapted to particular microenvironments of the host and result in phenotypic differences between non-pathogenic commensals and virulent biotypes. As a result, genomic comparisons between pathogenic and non-pathogenic strains of the same species can be particularly informative because genes exclusively present in the former may be required for infectivity and virulence. Thus far however, with respect to pathogenic bacteria, only five complete genome sequences for two or more strains of the same species have been published. These include *H. pylori* (Tomb *et al.*, 1997; Aim *et al.*, 1999), *Chlamydia trachomatis* (Read *et al.*, 2000), *C. pneumoniae* (Shirai *et al.*, 2000), meticillin-resistant *S. aureus* (Kuroda *et al.*, 2001) and *E. coli* (two K-12 strains and enterohaemorrhagic *E. coli* O157:H7) (Perna *et al.*, 2001). Until additional complete intra-species sequences become available, microarray-based comparisons between a fully sequenced genome and an unsequenced, but related genome can provide valuable information about the diversity and evolution of pathogens and symbionts (Ochman and Jones, 2000). Comparisons of this kind employ a microarray containing representations of all the ORFs of the sequenced, reference strain and labelled DNA from the unsequenced, experimental strain. The resulting hybridized array will disclose genes common to both strains and genes that are present in the reference strain, but absent in the experimental strain. This method, however, cannot

Table 4   Microarray-based comparative genomics.

Sequenced species and strain	Interrogated species and strain	Reference
M. tuberculosis; strain H37Rv	*M. bovis* and 13 variants of the bacille Calmette-Guerin (BCG) attenuated *M. bovis* vaccine strain	Behr *et al.*, 1999
M. tuberculosis; strain H37Rv	19 clinical *M. tuberculosis* strains representing 15 clones	Kato-Maeda *et al.*, 2001
H. pylori, two sequenced strains J99 and 26695	15 strains of *H. pylori*, including J99 and 26695; five of these were from patients with documented gastritis or peptic ulcer disease	Salama *et al.*, 2000
H. pylori, two sequenced strains J99 and 26695	*H. pylori* gastric ulcer strain B128, and duodenal ulcer strain G1.1	Israel *et al.*, 2001
E. coli K-12, strain MG1655	Four *E. coli* non-pathogenic strains W3110, ECOR 21, 37 and 40	Ochman and Jones, 2000
E. coli K-12, strain MG1655	*Wigglesworthia brevipalpis*, an endosymbiont of the tsetse fly	Akman and Aksoy, 2001
E. coli K-12, strain MG1655	*Klebsiella pneumoniae* 342, an endophyte of maize	Dong *et al.*, 2001

detect genes present in the experimental strain, but absent in the reference strain; point mutations, including frame-shift mutations; small deletions and deletions in homologous repetitive elements; rearrangements of the genome that have not resulted in deletion of a gene; or differences in the number of multicopy genes (Salama *et al.*, 2000; Ochman and Moran, 2001). Additionally, in contrast to high-density oligonucleotide arrays, DNA-spotted microarrays and membrane macroarrays ordinarily do not include representations of intergenic regions of the genome and thus cannot detect deletions within these non-coding segments, even though these specify promoter elements and small, non-translated RNAs and thus could be functionally important. Finally, results falsely indicating deletion of a gene in the experimental strain can arise if that gene's sequence has diverged sufficiently from its orthologue in the reference strain to prevent hybridization. Conversely, cross-hybridization between ancestrally unrelated genes can yield a false-positive result. Despite these limitations, the few published studies of this kind (Table 4) have been quite informative, in part because events leading to gene acquisition and gene loss are a major source of diversity in bacterial pathogens and symbionts (Ochman and Moran, 2001) and many changes of this kind are readily detected by microarray methods.

5.1. Comparative Genomics of *M. tuberculosis, M. bovis* and BCG Vaccine Strains

Behr and co-workers, in perhaps the first example of a study of this kind, used a DNA microarray to compare the genome composition of the recently sequenced *M. tuberculosis* laboratory strain H37Rv with the closely related pathogenic species *M. bovis* and with several strains of the bacille Calmette-Guerin (BCG) vaccine variant that was produced by serial *in vitro* passage of *M. bovis* between 1908 and 1921 (Behr *et al.*, 1999). Immediately following its development 80 years ago, the BCG vaccine strain was provided to the public health ministries of many countries and these descendants of the original strain are now widely used for the prevention of tuberculosis. However, the original BCG strain was lost and the vaccine efficacy of its contemporary derivatives varies widely. This investigation was conducted using currently available BCG strains in order to reconstruct their genealogy at the genomic level of resolution. Additionally, the study was designed to identify the genetic differences between the attenuated BCG strains and the pathogenic *M. bovis* parent strain and thus to discover genes that might contribute to the virulence of the latter.

Compared with the sequenced *M. tuberculosis* strain, 11 regions containing 91 ORFs have been deleted from one or more of the tested pathogenic *M. bovis* strains. Compared with pathogenic *M. bovis* strains, five additional regions containing 38 ORFs have been deleted from one or more of the tested BCG strains. Each of the identified deletions was confirmed by PCR amplification and sequencing across the deleted region, thus validating the microarray hybridization results. Analysis of the regions deleted from BCG, but present in the sequenced *M. tuberculosis* strain, showed that genes classified as transcriptional regulators were lost disproportionately and some of these might govern the expression of genes required for virulence (Behr *et al.*, 1999).

5.2. Comparative Genomics of Multiple *M. tuberculosis* Clinical Isolates

M. tuberculosis, unlike many other pathogens that inhabit mucosal surfaces (Ochman and Moran, 2001), has not been demonstrated to undergo horizontal gene transfer, and nucleotide sequence analysis of representative housekeeping genes in a variety of strains has revealed comparatively few single nucleotide polymorphisms. These findings have led to the notion that *M. tuberculosis* has a highly conserved clonal population structure and that differences in the apparent virulence of clinical isolates probably reflect differences in the immune status of the host rather than strain-specific properties. To test this hypothesis and to learn more about the natural variability within this species, Kato-Maeda and co-workers used a high-density oligonucleotide array representing the H37Rv genome to compare 19 recent clinical isolates of *M. tuberculosis* (Kato-Maeda *et al.*, 2001). Compared with the sequenced, reference strain, each unique clinical isolate was found to have lost, on average, ~17 ORFs corresponding to ~0.3% of the H37Rv genome. In all, 25 deleted sequences were detected, including 22 intergenic segments and all or part of 93 ORFs. Fifty-two of these ORFs were functionally annotated and of these 25 were phage-related and eight were insertion sequences. Several of the other deleted genes could conceivably affect virulence including three encoding phospholipase-C, one encoding a polyketide synthase and three encoding putative transcriptional regulators. Remarkably, strains that had sustained the most deletions were less likely to have been isolated from patients with pulmonary cavitation (Kato-Maeda *et al.*, 2001). Cavity formation is a hallmark of tuberculosis and essential for the efficient transmission of the organism to susceptible hosts. Thus, degradation of the genome may be associated in this species with a trend to decreased infectivity.

5.3. *H. pylori* Strain Diversity

H. pylori infection of the upper gastrointestinal tract causes a spectrum of conditions ranging from asymptomatic infection to gastritis, gastric and duodenal peptic ulcer disease and gastric cancer. Comparison of two complete *H. pylori* sequences revealed that ~6% of each genome was not present in the other genome and that recombinations, insertions and deletions, changes in repetitive elements and single nucleotide substitutions had created considerable diversity (Aim *et al.*, 1999). To further explore the genomic diversity of this species, Salama and colleagues used a microarray containing representations of 98.6% of all ORFs of both sequenced species to examine the genomic content of 15 *H. pylori* strains (Salama *et al.*, 2000). They identified 1281 ORFs that were common to all the tested strains; these represent the 'functional core' of this species' genome. Amongst these are genes coding for metabolic and biosynthetic pathways and for cellular and regulatory functions. By contrast, 362 ORFs, comprising 22% of the genome, were absent from one or more of the tested strains; these comprise strain-specific genes and were hypothesized to encode functions that adapt the organisms to a particular host microenvironment. Strain-specific genes tended to be located in two regions containing horizontally transferred blocks of genes and included those coding for an adhesin that might promote gastric colonization of people expressing the Lewis B blood group antigen, surface-exposed components of the outer membrane, including outer membrane proteins and lipopolysaccharide, components of a restriction modification system and ATPases that might function in the assembly of a type IV secretion apparatus. An intriguing aspect of this study was the use of a cluster program for the analysis of strain-specific genes and the identification of several genes that may have been co-inherited with genes in the pathogenicity island and therefore might also encode virulence determinants (Salama *et al.*, 2000). In a separate study, Israel and co-workers used the same *H. pylori* microarray to compare the genomic content of two strains from gastric and duodenal ulcer patients that produce significantly different levels of gastritis, cellular proliferation and apoptosis in the gerbil gastritis model. The microarray results showed that the less proinflammatory strain had sustained a large deletion of the *cag* pathogenicity island, providing a genetic explanation for its relative attenuation (Israel *et al.*, 2001).

5.4. Microarray Methods to Compare Distantly Related Species

In addition to studies using microarray methodology to compare closely related strains and species, two studies have used an *E. coli* K-12 microarray to characterize the genomes of distantly related species. Akman and Aksoy examined *Wigglesworthia*, an endosymbiont of the tsetse fly that has a genome smaller than 770 kb (compared with the 4.6 Mb genome of the *E. coli* K-12 reference strain), lives in specialized epithelial cells termed bacteriocytes and whose presence is required by the fly for normal growth and reproduction (Akman and Aksoy, 2001). *Wigglesworthia* is believed to comprise a clade within the tribe *Enterobacteriaceae*, thus justifying the use of an *E. coli* microarray to study its genomic composition. Akman and Askoy were able to identify 650 genes in *Wigglesworthia* orthologous with *E. coli* genes and based on the estimated size of its genome and assuming an average ORF size of 1 kb, these orthologues comprised fully 85% of its genome. Analysis of these orthologues, based on the functional annotations provided by the *E. coli* genome project, showed that this symbiont has retained genes from its larger ancestral free living, pathogenic or symbiotic ancestors that mediate the replication, transcription and translation machinery of the microbe, the production of some co-factors and amino acids, and the proper folding and transport of proteins. They also carried out expression analysis of *Wigglesworthia* using the *E. coli* K-12 array and identified 61 expressed genes whose putative functions suggest it to be a facultative anaerobe that uses ammonium as a nitrogen source.

Dong and coworkers used a microarray containing 4098 of the 4290 ORFs of *E. coli* K-12 to characterize the genome composition of *Klebsiella pneumoniae* 342, a diazotrophic endophyte of maize with an estimated genome size of 4.8 Mb (Dong *et al.*, 2001). Microarray hybridization experiments using labelled DNA from *K. pneumoniae* strain 342 identified ~3000 *E. coli* orthologues, indicating that ~1500 genes of strain 342 are either derived from a different lineage or have diverged significantly. Of these, the authors estimate that ~200 might mediate the interaction of *K. pneumoniae* strain 342 with maize. Most of the genes common to *E. coli* K-12 and *K. pneumoniae* strain 342, were predicted to code for conserved energy, amino acid and fatty acid metabolic pathways and for proteins required for cell division, replication, transcription, translation and transport. By contrast, *E. coli* genes that were not identified in the strain *K. pneumoniae* 342 genome included some genes predicted to specify regulatory proteins, chaperones, outer membrane proteins and surface-exposed organelles including those mediating motility and, as expected, genes likely to have been acquired recently in evolutionary time from phage, plasmids and transposons.

6. MICROARRAY EXPRESSION PROFILING OF HOST CELLS

Pathogenesis entails not only the differential expression of bacterial genes, but also responses by the host. In principle then, microarray expression analysis of bacterially infected cells and tissues can identify, simultaneously and in the same sample, host and pathogen genes that are regulated during the infectious process. With respect to studies using cultured cells, the results of the analysis will be influenced by the cell type employed (primary or transformed) and their activation state, the ratio of the number of infecting bacteria to the number of host cells, i.e. the multiplicity of infection (MOI), the time-course, and, of

Table 5 Microarray transcriptional profiling of host cells *ex vivo* in response to pathogenic bacteria and their toxins.

Bacterial species mutants and toxins	Cell type	MOI	Time-course	Array	Reference
Salmonella typhimurium (wild-type) purified LPS	RAW 264.7 murine Macrophage cell line +/– IFNγ	20	4 h	Mouse Clontech 588 cDNAs	Rosenberger *et al.*, 2000
S. typhimurium (wild type) *phoP::Tn10* mutant	U-937 human monocytic cell line	4–12	0.5, 1, 3, 4 h	Human 22,571 cDNAs	Detweiler *et al.*, 2001
Listeria monocytogenes	THP1 human promyelocytic cell line	10	2 h	Human 6800 high-density Affymetrix array; 18,376 cDNAs colony filter Genome Systems array; 588 cDNAs Clontech array	Cohen *et al.*, 2000
Pseudomonas aeruginosa PAK non-piliated mutant PAK-NP	A549 human lung carcinoma-derived type II pneumocyte cell line	50	0, 3 h	1506 human cDNAs	Ichikawa *et al.*, 2000
Bordetella pertussis (wild-type) pertussis toxin (PT) mutant expressing catalytically inactive PT; purified PT	BEAS-2B human bronchial cell line	50	3 h	Human 6800 high-density Affymetrix array	Belcher *et al.*, 2000

course, the bacterial strain studied and how it was grown for the experiment. Further information about the microbial molecules responsible for induction of the host cell transcriptional response can be obtained through the use of purified bacterial products, including pro-inflammatory cell wall constituents and toxins, and mutants carrying inactivated genes coding for virulence determinants. As the studies described below illustrate, experiments using transformed cell lines are informative, but results from the use of primary cells freshly isolated from tissues, e.g. primary bone marrow-derived macrophages, may better simulate *in vivo* conditions. However, studies using single cell types, whether primary or transformed, lack the context of a multicellular milieu where signalling between cells of different lineages modulates the response of individual cells to an infectious agent. This kind of complexity is difficult to simulate *in vitro* and will require technical innovations that permit expression profiling of individual cell types within infected tissues. In view of these experimental design issues, it is not surprising that the small number of studies thus far published (Table 5) have used transformed cell lines, large MOIs and short time-courses. As a result, all such studies reflect early events and those employing macrophage-like cells are thus exploring aspects of innate rather than acquired immunity. Finally, it will be noted that none of the studies reviewed here used a microarray format containing most mouse or human genes and in one study as few as 588 genes or expressed sequence tags (ESTs) were surveyed.

6.1. Macrophage Genes Induced During Infection with *Salmonella typhimurium*

Rosenberger and coworkers used a Clontech array containing 588 mouse genes/ESTs to study the response of a murine macrophage cell line 4 h after infection with *Salmonella enterica*, subspecies *typhimurium*, an agent that causes a systemic infection in mice resembling typhoid fever in humans (Rosenberger *et al.*, 2000). In this murine model, *S. typhimurium* has been shown to induce the release of IFNγ, which in turn activates macrophages leading to the production of reactive oxygen and nitrogen intermediates and the synthesis of antimicrobial peptides. The microbicidal capacity of the macrophage is thus enhanced. Furthermore, previous studies have demonstrated that components of the bacterial cell surface or cell wall bind macrophage receptors, an event that initiates an innate immune response involving the transcription of genes coding for cytokines, chemokines, cationic peptides and various transcription factors. Lipopolysaccharide, embedded in the *Salmonella* outer membrane, is the most potent of these effector molecules.

Four hours after infection, microarray analysis of the macrophage cell line identified increased expression of 77 macrophage genes/ESTs by two-fold or more, including those coding for proteins previously described to have pro-inflammatory properties. These include inducible nitric oxide synthase (iNOS), the chemokines MIP-1α, MIP-1β and MIP-2α and the cytokine IL-1. Induction of genes coding for TNF-α and CD40 were also observed indicating an enhanced capacity of the macrophage to signal other cells of the immune system. Curiously, the simultaneous expression of genes coding for proteins believed to attenuate the pro-inflammatory response, including induction of κB(I-κB)α and β (the inhibitory subunits of NF-κB) and TGF-β, was also seen indicating a capacity to modulate the intensity of the innate immune response and thus reducing collateral damage to host tissues (Rosenberger et al., 2000).

Salmonella infection is recognized to initiate programmed cell death and the expression profile included genes coding for proteins previously shown to mediate this process including caspase 1 and FAS. Finally, genes involved with myeloid differentiation and maintenance of the macrophage phenotype were identified. These responses were shown to be largely reproduced when purified LPS was used alone confirming that the transcriptional state that characterizes innate immunity lacks specificity and can be provoked by a few, widely shared, pro-inflammatory compounds.

The effect of the activation state of the cell on the transcriptional response to *Salmonella* was examined. IFNγ activation of the macrophage cell line in the absence of bacteria induced the expression of several genes, including MIP-1α and the MIG cytokine. The effect on the transcriptome was dramatically increased when IFNγ-treated cells were infected with bacteria and included the enhanced expression of iNOS, I-κBβ, NF-κB, JunB, JunD and TNF-α, when compared with their expression levels in bacterially infected cells that had not been IFNγ-activated. The expression results of this study were confirmed by using Northern blotting to evaluate the expression state of some genes and immunoassays to confirm that an increase in the abundance of mRNAs coding for three pro-inflammatory cytokines (TNF-α, MIP-1α and IL-1β) was corroborated by increased levels of the corresponding protein in culture supernatants (Rosenberger et al., 2000).

Detweiler and colleagues used a 22,571 gene/EST human cDNA array to study the interaction of *S. typhimurium* with the human monocytic tissue culture cell line J-937 that had been differentiated to a macrophage-like phenotype by treatment with phorbol 12-myristate 13-acetate (Detweiler et al., 2001). An innovative feature of this study's experimental design was the comparison of the macrophage response to the wild-type *S. typhimurium* with the response to an isogenic transposon mutant in *phoP*, the response regulator of a two-component signal transduction system that is activated inside macrophages and regulates a variety of genes that adapt the organism for intracellular survival. Both mutant and wild-type invaded and grew equally well in the macrophage-like tissue culture cells at 2 h and 24 h indicating that strain-specific differences in the expression of macrophage genes cannot be attributed to differences in growth rate or bacterial biomass. Compared with uninfected tissue culture cells, 68 mRNAs were found to be more abundant in the cells infected by the wild-type strain. Amongst the 55 mRNAs coded for by genes with known or putative functions were those regulated by NF-κB, including the proinflammatory cytokines IL-1 and IL-8. Additionally Jun and prostaglandin synthase were induced. Taken together, this profile corroborated prior studies showing induction of the innate immune response by Gram-negative bacteria and LPS. Most of the same genes were induced to the same levels by the wild-type and *phoP* mutant. However, compared with the wild-type strain, the expression of 32 macrophage genes by the *phoP* mutant was significantly reduced. Twenty-one of these 32 genes had been functionally annotated and of these many are involved with programmed cell death or with regulation of the cell cycle. This result was corroborated by the observation that the *phoP* mutant killed approximately 50% fewer macrophages than the wild-type strain. Taken together, these results suggest that *phoP* governs the expression of genes that initiate programmed cell death and demonstrate the utility of contrasting the expression profiles of host cells infected with isogenic mutants of the same strain (Detweiler *et al.*, 2001).

6.2. Macrophage Genes Induced by *Listeria monocytogenes*

L. monocytogenes, like *Salmonella* species, is an important human infectious agent and a facultative intracellular parasite. However, in contrast with *Salmonella*, it is a Gram-positive bacillus and thus lacks LPS. It also differs with respect to the molecular mechanisms by which it invades and persists in host cells. Cohen and colleagues used three different human DNA formats to obtain expression patterns from the human promyelocytic THP1 cell line, differentiated by treatment with 1,25-dihydroxyvitamin D_3 and retinoic acid to a monocyte phenotype, 2 h after having been infected with *L. monocytogenes* (Cohen *et al.*, 2000). Comparative data between the array types demonstrated greater sensitivity for the high-density oligonucleotide Affymetrix format and the value of experimental replicates to identify a core set of reproducibly regulated genes. Seventy-four up-regulated and 23 down-regulated genes were identified amongst the 6800 human genes/ESTs represented on the high-density array. Dramatic up-regulation of cytokines (IL-1 and TNFα), CC chemokines (MIPα, MIP1β, MIP3α, MCP1) and CXC chemokines (IL-8, GROβ and GROα) was observed and probably enhance host defences in response to invading bacteria by increasing the microbicidal capacity of the cell and by promoting the migration of phagocytic cells to sites of infection. Many of these up-regulated genes are regulated by NF-κB and code for the proinflammatory components of innate immunity. Although the foregoing events precede the development of acquired immunity, the initiation of cell-mediated immunity early in the time-course might be indicated by the up-regulation of two T cell surface receptors (4.1BB and CD27) that participate in the activation and stimulation of CD4$^+$ and CD8$^+$ T cells and in the expansion of cytotoxic T cells. To counter induction of innate immunity during invasion, *Listeria* causes down-regulation of genes coding for chemokine receptors (CCR2α and CCR1) and down-regulation of the receptor for TNF. Moreover, to counter induction of acquired immunity, *Listeria* down-regulates genes coding for the major histocompatibility class I and class II proteins and thus disrupts antigen presentation.

In view of the expression results obtained with *Salmonella* and discussed above, most of the genes induced by *Listeria* are probably associated with a generalized innate immune response to bacteria and thus lack species specificity. However, the induction of genes coding for actin-bundling protein and tubulin may reflect the recruitment and reorganization of cytoskeletal proteins by *Listeria* during its entry into, and movement through the cytoplasm of the invaded host cell. If so, then *Listeria* not only exploits pre-existing cytoskeletal components, but in addition modulates their production at the transcriptional level (Cohen *et al.*, 2000). Furthermore, unlike *Salmonella*, *Listeria* induces anti-apoptotic genes, in effect preserving the cells that provide its intracellular niche.

6.3. Regulation of Pneumocyte Genes by *Pseudomonas aeruginosa*

P. aeruginosa is an opportunistic Gram-negative bacillus that causes devastating acute infections in burn patients and chronic pulmonary infections in children with cystic fibrosis. Its pathogenicity has been extensively studied and identified virulence determinants include several secreted toxins and proteases, pili by which it attaches to epithelial cell surfaces and, in mucoid cystic fibrosis strains, an exopolysaccharide composed of alginate. Ichikawa and coworkers used a 1506 human cDNA array to study the transcriptional response of a type II pneumocyte cell line infected for 3 h with the PAK strain of *P. aeruginosa* (Ichikawa *et al.*, 2000). This time-course was selected because bacterial adherence had progressed to completion and >90% of the cultured pneumocytes remained viable. Of the 1506 monitored genes/ESTs, the expression of 22 was increased by >two-fold, whereas the expression of only two was reduced. As in the other studies described above, genes encoding for proinflammatory components of the innate immune response were induced, including MCP-1 and the TNFα-induced protein, A20. While the induction of these genes lacked species specificity, the increased expression of the DPH2L gene product, required for the conversion of histidine to diphthamide in elongation factor 2, was probably due to the action of *Pseudomonas* exotoxin A-mediated ADP ribosylation, and thus represents a species-specific component of the transcriptional response. To determine how much of this transcriptional response required pili-mediated adherence to the cell surface, the expression profiles evoked by the wild-type, piliated strain and a non-piliated isogenic mutant, were compared. Sixteen of the 22 up-regulated genes were classified as adhesion-dependent, that is, they were not expressed during infection with the pili-negative mutant. One of these, the gene encoding the interferon regulatory factor 1 (IRF-1) was studied in greater detail (Ichikawa *et al.*, 2000). IFR-1 is known to be induced by interferon (IFN) and cytokines in response to viral infection of a cell, but was not previously known to be responsive to bacteria. However, the induction of IRF-1 by *P. aeruginosa* and the capacity of IRF-1 to regulate the gene encoding iNOS strongly argues for a role in host defence, since nitric oxide (NO) is a potent microbicidal molecule and cystic fibrosis sputa contain significant levels of NO. IRF-1 is also known to decrease the expression of secretory leukocyte protease inhibitor (SLPI) possibly explaining why proteases released from dying polymorphonuclear leukocytes in sputa might damage lung tissue.

Because adhesion brings the LPS-rich bacterial outer membrane into close apposition with the cytoplasmic membrane of the host cell, the effect of physiological concentrations of purified pseudomonal LPS on the expression of IRF-1 was examined by RT-PCR. LPS did not induce IRF-1 expression and thus its regulation is adhesin-dependent, but LPS-independent.

6.4. Bronchial Epithelial Cell Gene Regulation by *Bordetella pertussis*

B. pertussis is a Gram-negative coccobacillus and the aetiological agent of whooping cough, a sometimes fatal infection of the respiratory tract characterized by the accumulation of mucous and inflammatory cells. Belcher and coworkers used the 6800 high-density Affymetrix oligonucleotide array to identify host cell genes regulated 3 h after *B. pertussis* infection of the SV40-transformed human bronchial epithelial cell line, BEAS-2B (Belcher *et al.*, 2000). Remarkably, only 33 genes were found to exhibit >three-fold increase in mRNA abundance at this time-point. Included in this group are genes encoding proinflammatory cytokines and chemokines whose activities are possibly responsible for the concentration of polymorphonuclear leukocytes and monocytes within infected tissues. These expression results were supported by the detection of IL-6, IL-8 and MCP-1 in the *B. pertussis*-infected tissue culture cells by antibody-based cytokine assays. In addition, the growth-related oncogenes GRO2 and GRO3, previously demonstrated to be associated with the cellular damage inflicted by *B. pertussis* tracheal cytotoxin, were up-regulated.

Finally, two antiapoptotic factors were up-regulated. Their possible significance was supported by immunoassays of the 3-h infected cell culture showing no evidence of apoptosis (Belcher *et al.*, 2000). By contrast to the other studies reviewed here, the expression of more genes was significantly reduced (65) than increased (33) by *B. pertussis* infection. Amongst the repressed genes were 10 transcription factors, several genes coding for signal transduction factors (including STAT2) and the gene coding for iNOS.

Many of the proinflammatory subset of genes induced by *B. pertussis* were hypothesized to require the NFκB transcription factor. To further investigate this possibility, the cell culture was pretreated with either a glucocorticoid or with sodium salicylate before addition of *B. pertussis*. Both compounds, which inhibit NFκB-mediated gene induction, blocked the expression of multiple genes that collectively compose the innate immune, proinflammatory response (Belcher *et al.*, 2000). This interesting result demonstrates how the use of known inhibitors of cellular responses can complement microarray studies of host cells by revealing activation pathways.

To determine how much of the regulation induced by infection with the wild-type *B. pertussis* strain was due to the ADP-ribosyltransferase activity of pertussus toxin, comparative studies were performed with an isogenic mutant that secreted a normally folded, but catalytically inactive variant of pertussis toxin. Additional studies were undertaken using uninfected cells treated with purified, catalytically active pertussis toxin and with the purified, catalytically inactive variant of it. Taken together, these experiments led to the identification of a set of bronchial epithelial cell genes whose regulation by pertussis toxin was both necessary and sufficient. Amongst these are genes known to be G-protein or cAMP-regulated including matrix metalloproteinase-1, βB1-crystallin, basic fibroblast growth factor and a Na^+/Cl^--dependent serotonin transporter (Belcher *et al.*, 2000).

Modest (1.5-fold) elevation of the transcript coding for MUC2, a respiratory tract mucin glycoprotein, led to the use of a luciferase reporter system to study the transcription of the gene encoding MUC2 and of three other genes that encode MUC5AC, also a respiratory tract mucin, MUC3, an intestinal mucin, or MUC6, a gastric mucin. This experiment showed that *B. pertussis* infection induces the expression of genes encoding the two respiratory tract mucins, thus confirming the microarray expression result, but not genes encoding the intestinal or gastric mucins, sites where *B. pertussis* infection does not occur. These results prompted a study of the affinity of *B. pertussis* for bovine salivary mucin and demonstrated a four-fold increased adherence of the bacteria to mucin-coated microtitre wells compared with non-coated wells. Since mucin production is ordinarily thought to protect ciliated epithelia from invading bacteria, its induced production by, and affinity for *B. pertussis* suggests that the bacterium has subverted this component of host defence (Belcher *et al.*, 2000). As a result, bacteria enmeshed in mucous might be expected to accumulate in airways, a prediction supported by pathological examination of lungs obtained from fatal cases of pertussis.

6.5. Summary

Taken together, the studies reviewed in this section have contributed significant new information about the nature of innate immunity while, in general, confirming prior results that were obtained by other methods – this despite the obvious limitations that come from the use of cell lines, short time-courses (and in some studies a single time-point) and arrays that represent only a portion of the human or murine genome. As predicted by current models of innate immunity, diverse bacteria, acting through shared pathways, evoke overlapping sets of proinflammatory genes. Nonetheless, embedded within these expression profiles are features that point to the production of molecules that characterize and separate the pathogenic mechanisms of one species from another. These studies also demonstrate the value of probing the same cell line with isogenic mutants of bacteria. This experimental design can identify a subset of the genes that compose the entire expression profile whose expression depends, directly or indirectly, on the activity encoded by a particular microbial gene. These studies also show that corroborating data can be obtained through the use of purified microbial molecules, including LPS and toxins. It will be informative to follow this line of investigation with studies of primary cells from wild-type and knockout mice lacking the capacity to express components of innate and acquired immunity and with combinations of cells that create some of the complexity that characterizes the multicellular environment of tissue.

ACKNOWLEDGEMENTS

The author gratefully acknowledges the members of his laboratory, past and present, for their dedication, resourcefulness and energy in the development of our microarray capacity and research program. This work is supported by three grants from the National Institutes of Health (AI39521, AI43422 and AI44826) and by a grant from the Glaxo Research Limited – Action TB Program.

REFERENCES

Aim, R.A., Ling, L.-S., Moir, D.T. *et al.* (1999) Genomic-sequence comparison of two unrelated isolates of the human gastric pathogen *Helicobacter pylori. Nature* **397**, 176–180.

Akman, L. and Aksoy, S. (2001) A novel application of gene arrays: *Escherichia coli* array provides insight into the biology of the obligate endosymbiont of tsetse flies. *Proc. Natl Acad. Sci. USA* **98**, 7546–7551.

Ang, S., Lee, C.-Z., Peck, K. *et al.* (2001) Acid-induced gene expression in *Helicobacter pylori*: study in genomic scale by microarray. *Infect. Immun.* **69**, 1679–1686.

Arfin, S.M., Long, A.D., Ito, E.T., *et al.* (2000) Global gene expression profiling in *Escherichia coli* K-12. *J. Biol. Chem.* **275**, 29672–29684.

Arnold, C.A., McElhanon, J., Lee, A. *et al.* (2001) Global analysis of *Escherichia coli* gene expression during the acetate-induced acid tolerance response. *J. Bacteriol.* **183**, 2178–2186.

Barbosa, T.M. and Levy, S.B. (2000) Differential expression of over 60 chromosomal genes in *Escherichia coli* by constitutive expression of MarA. *J. Bacteriol.* **182**, 3467–3474.

Behr, M.A., Wilson, M.A., Gill, W.P. *et al.* (1999) Comparative genomics of BCG vaccines by whole-genome DNA microarray. *Science* **284**, 1520–1523.

Belcher, C.E., Drenkow, J., Kehoe, B. *et al.* (2000) The transcriptional response of respiratory epithelial cells to *Bordetella pertussis* reveal host defensive and pathogen counter-defensive strategies. *Proc. Natl Acad. Sci. USA* **97**, 13847–13852.

Brocklehurst, K.R. and Morby, A.P. (2000) Metal-ion tolerance in *Escherichia coli*: analysis of transcriptional profiles by gene-array technology. *Microbiology* **146**, 2277–2282.

Cohen, P., Bouaboula, M., Bellis M. *et al.* (2000) Monitoring cellular responses to *Listeria monocytogenes* with oligonucleotide arrays. *J. Biol. Chem.* **275**, 11181–11190.

Cummings, C.A. and Relman, D.A. (2000) Using DNA microarrays to study host–microbe interactions. *Emerg. Infect. Dis.* **6**, 513–525.

DeRisi, J.L., Iyer, V.R. and Brown, P.O. (1997) Exploring the metabolic and genetic control of gene expression on a genomic scale. *Science* **278**, 680–686.

De Saizieu, A., Gardes, C., Flint, N. *et al.* (2000) Microarray-based identification of a novel *Streptococcus pneumoniae* regulon controlled by an autoinduced peptide. *J. Bacteriol.* **182**, 4696–4703.

Detweiler, C., Cunanan, D.B. and Falkow, S. (2001) Host microarray analysis reveals a role for the *Salmonella* response regulator *phoP* in human macrophage cell death. *Proc. Natl Acad. Sci. USA* **98**, 5850–5855.

Dong, Y., Glasner, J.D., Blattner, F.R. *et al.* (2001) Genomic interspecies microarray hybridization: rapid discovery of three thousand genes in the maize endophyte, *Klebsiella pneumoniae* 342, by microarray hybridization with *Escherichia coli* K-12 open reading frames. *Appl. Environ. Microbiol.* **67**, 1911–1921.

Eisen, M.B. and Brown, P.O. (1999) DNA arrays for analysis of gene expression. *Meth. Enzymol.* **303**, 179–205.

Eisen, M.B., Spellman, P.T., Brown, P.O. *et al.* (1998) Cluster analysis and display of genome-wide expression patterns. *Proc. Natl Aad. Sci. USA* **95**, 14863–14868.

Ferea, T.L. and Brown, P.O. (1999) Observing the living genome. *Curr. Opin. Gen. Dev.* **9**, 715–722.

Gorden, J. and Small, P.L. (1993) Acid resistance in enteric bacteria. *Infect. Immun.* **61**, 364–367.

Harrington, C.A., Rosenow, C. and Retief, J. (2000) Monotoring gene expression using DNA microarrays. *Curr. Opin. Microbiol.* **3**, 285–291.

Ichikawa, J.K., Norris, A., Bangera, M.G. *et al.* (2000) Interaction of *Pseudomonas aeruginosa* with epithelial cells: identification of differentially regulated genes by expression microarray analysis of human cDNAs. *Proc. Natl Acad. Sci. USA* **97**, 9659–9664.

Israel, D.A., Salama, N., Arnold, C.N. *et al.* (2001) *Helicobacter pylori* strain-specific differences in genetic content, identified by microarray, influence host inflammatory responses. *J. Clin. Invest.* **107**, 611–620.

Karp, P.D., Krummenacker, M., Paley, S. *et al.* (1999) Integrated pathway-genome databases and their role in drug discovery. *Trends Biotechnol.* **17**, 275–281.

Karp, P.D., Riley, M., Saier, M. *et al.* (2000) The EcoCyc and MetaCyc databases. *Nucl. Acids Res.* **28**, 56–59.

Kato-Maeda, M., Rhee, J.T., Gingeras, T.R. *et al.* (2001) Comparing genomes within the species *Mycobacterium tuberculosis. Genome Res.* **11**, 547–554.

Khodursky, A.B., Peter, B.J., Cozzarelli, N.R. *et al.* (2000) DNA microarray analysis of gene expression in response to physiological and genetic changes that affect tryptophan metabolism in *Escherichia coli. Proc. Natl Acad. Sci. USA* **97**, 12170–12175.

Kuroda M., Ohta, T., Uchiyama, I. *et al.* (2001) Whole genome sequence of meticillin-resistant *Staphylococcus aureus. Lancet* **357**, 1225–1240.

Laub, M.T., McAdams, H.H., Feldblyum, T. *et al.* (2000) Global analysis of the genetic network controlling a bacterial cell cycle. *Science* **290**, 2144–2148.

Lee, M.-L., Kuo, F.C., Whitmore, G.A. *et al.* (2000) Importance of replication in microarray gene expression studies: statistical methods and evidence from repetitive cDNA hybridizations. *Proc. Natl Acad. Sci. USA* **97**, 9834–9839.

Lipshutz, R.J., Fodor, S.P., Gingeras, T.R. *et al.* (1999) High density synthetic oligonucleotide arrays. *Nat. Genet.* **21** (Suppl. 1), 20–24.

Lockhart, D.J., Byrne, M.C., Follettie, M.T. *et al.* (1996) Expression monitoring by hybridization to high-density oligonucleotide arrays. *Nat. Biotechnol.* **14**, 1675–1680.

Long A.D., Mangalam, H.J., Chan, B.Y. *et al.* (2001) Improved statistical inference from DNA microarray data using analysis of variance and a Bayesian statistical framework. *J. Biol. Chem.* **276**, 19937–19944.

Manganelli, R., Voskuil, M.I., Schoolnik, G.K. *et al.* (2001) The *Mycobacterium tuberculosis* ECF sigma factor σ^E: role in global gene expression and survival in macrophages. *Mol. Microbiol.* **41**, 423–437.

Mdluli, K., Slayden, R.A., Zhu, Y. *et al.* (1998) Inhibition of a *Mycobacterium tuberculosis* beta-ketoacyl ACP synthase by isoniazid. *Science* **280**, 1607–1610.

Mekalanos, J.J. (1992) Environmental signals controlling the expression of virulence determinants in bacteria. *J. Bacteriol.* **174**, 1–7.

Ochman, H. and Jones, I.B. (2000) Evolutionary dynamics of full genome content in *Escherichia coli. EMBO J* **19**, 6637–6643.

Ochman, H. and Moran, N.A. (2001) Genes lost and genes found: evolution of bacterial pathogenesis and symbiosis. *Science* **292**, 1096–1098.

Paustian, M.L., May, B.J. and Kapur, V. (2001) *Pasteurella multocida* gene expression in response to iron limitation. *Infect. Immun.* **69**, 4109–4115.

Perna, N.T., Plunkett, G., Burland, V. *et al.* (2001) Genome sequence of enterohemorrhagic *Escherichia coli* O157:H7. *Nature* **409**, 529–533.

Read, T.D., Brunham, R.C., Shen, C. *et al.* (2000) Genome sequences of *Chlamydia trachomatis* MoPn and *Chlamydia pneumoniae* AR39. *Nucl. Acids Res.* **28**, 1397–1406.

Richmond, C.S., Glasner, J.D., Mau, R. *et al.* (1999) Genome-wide expression profiling in *Escherichia coli* K-12. *Nucl. Acids Res.* **27**, 3821–3835.

Rimini, R., Jansson, B., Feger, G. *et al.* (2000) Global analysis of transcription kinetics during competence development in *Streptococcus pneumoniae* using high density DNA arrays. *Mol. Microbiol.* **36**, 1279–1292.

Rosenberger, C.M., Scott, M.G., Gold, M.R. *et al.* (2000) *Salmonella typhimurium* infection and lipopolysaccharide stimulation induce similar changes in macrophage gene expression. *J. Immun.* **164**, 5894–5904.

Rozwarski, D.A., Grant, G.A., Barton, D.H. *et al.* (1998) Modification of the NADH of the isoniazid target (InhA) from *Mycobacterium tuberculosis. Science* **279**, 98–102.

Salama, N., Guillemin, K., McDaniel, T.K. *et al.* (2000) A whole-genome microarray reveals genetic diversity among *Helicobacter pylori* strains. *Proc. Natl Acad. Sci. USA* **97**, 14668–14673.

Schena, M., Shalon, D., Davis, R.W. *et al.* (1996) Quantitative monitoring of gene expression patterns with a complementary DNA microarray. *Science* **270**, 467–470.

Schoolnik, G.K., Voskuil, M.I., Schnappinger, D. *et al.* (2001) Whole genome DNA microarray expression analysis of biofilm development by *Vibrio cholerae* O1 El Tor. *Meth. Enzymol.* **336**, 3–18.

Selinger, D.W., Cheung, K.J., Mei, R. *et al.* (2000) RNA expression analysis using a 30 base pair resolution *Escherichia coli* genome array. *Nat. Biotechnol.* **18**, 1262–1268.

Sherman, D.R., Voskuil, M., Schnappinger, P. *et al.* (2001) Regulation of the *Mycobacterium tuberculosis* hypoxic response gene encoding alpha-crystallin. *Proc. Natl Acad. Sci. USA* **98**, 7534–7539.

Shirai, M., Hirakawa, H., Kimoto, M. *et al.* (2000) Comparison of whole genome sequences of *Chlamydia pneumoniae* J138 from Japan and CWL029 from USA. *Nucl. Acids Res.* **28**, 2311–2314.

Tamayo, P., Slonim, D., Mesirov, J. *et al.* (1999) Interpreting patterns of gene expression with self-organizing maps: methods and application to hematopoietic differentiation. *Proc. Natl Acad. Sci. USA* **96**, 2907–2912.

Tao, H., Bausch, C., Richmond, C. *et al.* (1999) Functional genomics: expression analysis of *Escherichia coli* growing on minimal and rich media. *J. Bacteriol.* **181**, 6425–6440.

Throup, J.P., Koretke, K.K. and Bryant, A.P. (2000) A genomic analysis of two-component signal-transduction in *Streptococcus pneumoniae*. *Mol. Microbiol.* **35**, 566–576.

Tomb, J.-F., White, O., Kerlavage, A.R. *et al.* (1997) The complete genome sequence of the gastric pathogen *Helicobacter pylori*. *Nature* **388**, 539–547.

Tusher, V.G., Tibshirani, R. and Chu, G. (2001) Significance analysis of microarrays applied to the ionizing radiation response. *Proc. Natl Acad. Sci. USA* **98**, 5116–5121.

Wassarman, K.M., Zhang, A. and Storz, G. (1999) Small RNAs in *Escherichia coli*. *Trends Microbiol.* **7**, 37–45.

Wassarman, K.M., Repoila, F., Rosenow, C. *et al.* (2001) Identification of novel small RNAs using comparative genomic microarrays. *Genes Dev.* **15**, 1637–1651.

Wayne, L.G. (1976) Dynamics of submerged growth of *Mycobacterium tuberculosis* under aerobic and microaerophilic conditions. *Am. Rev. Respir. Dis.* **114**, 807–811.

Wei, Y., Lee, J.-M., Richmond, C. *et al.* (2001a) High-density microarray-mediated gene expression profiling of *Escherichia coli*. *J. Bacteriol.* **183**, 545–556.

Wei, Y., Lee, J.-M., Smulski, D.R. *et al.* (2001b) Global impact of *sdiA* amplification revealed by comprehensive gene expression of *Escherichia coli*. *J. Bacteriol.* **183**, 2265–2272.

Wilson, M., DeRisi, J., Kristensen, H.-H. *et al.* (1999) Exploring drug-induced alterations in gene expression in *Mycobacterium tuberculosis* by microarray hybridization. *Proc. Natl Acad. Sci. USA* **96**, 12833–12838.

Wilson, M., Voskuil, M.I., Schnappinger, D. *et al.* (2001) Functional genomics of *Mycobacterium tuberculosis* using DNA microarrays. *Meth. Mol. Med.* **54**, 335–358.

Ye, R.W., Tao, W., Bedzyk, L. *et al.* (2000) Global gene expression profiles of *Bacillus subtilis* grown under anaerobic conditions. *J. Bacteriol.* **182**, 4458–4465.

Zheng, M., Wang, X., Templeton, L.J. *et al.* (2001) DNA microarray-mediated transcriptional profiling of the *Escherichia coli* response to hydrogen peroxide. *J. Bacteriol.* **183**, 4562–4570.

Zimmer, D.P., Soupene, E., Lee, H.L. *et al.* (2000) Nitrogen regulatory protein C-controlled genes of *Escherichia coli*: scavenging as a defense against nitrogen limitation. *Proc. Natl Acad. Sci. USA* **97**, 14674–14679.

The Extracytoplasmic Function (ECF) Sigma Factors

John D. Helmann

*Department of Microbiology, Wing Hall, Cornell University, Ithaca,
NY 14853-8101, USA*

ABSTRACT

Bacterial sigma (σ) factors are an essential component of RNA
polymerase and determine promoter selectivity. The substitution of one
σ factor for another can redirect some or all of the RNA polymerase in a
cell to activate the transcription of genes that would otherwise be silent.
As a class, alternative σ factors play key roles in coordinating gene
transcription during various stress responses and during morphological
development. The extracytoplasmic function (ECF) σ factors are small
regulatory proteins that are quite divergent in sequence relative to most
other σ factors. Many bacteria, particularly those with more complex
genomes, contain multiple ECF σ factors and these regulators often
outnumber all other types of σ factor combined. Examples include
Bacillus subtilis (7 ECF σ factors), *Mycobacterium tuberculosis* (10),
Caulobacter crescentus (13), *Pseudomonas aeruginosa* (~19), and
Streptomyces coelicolor (~50). The roles and mechanisms of regulation
for these various ECF σ factors are largely unknown, but significant
progress has been made in selected systems. As a general trend, most
ECF σ factors are cotranscribed with one or more negative regulators.
Often, these include a transmembrane protein functioning as an anti-σ
factor that binds, and inhibits, the cognate σ factor. Upon receiving a
stimulus from the environment, the σ factor is released and can bind to
RNA polymerase to stimulate transcription. In many ways, these anti-
σ:σ pairs are analogous to the more familiar two-component regulatory
systems consisting of a transmembrane histidine protein kinase and a
DNA-binding response regulator. Both are mechanisms of coordinating
a cytoplasmic transcriptional response to signals perceived by protein
domains external to the cell membrane. Here, I review current
knowledge of some of the better characterized ECF σ factors, discuss

ADVANCES IN MICROBIAL PHYSIOLOGY VOL 46
ISBN 0-12-027746-8

the variety of experimental approaches that have proven productive in defining the roles of ECF σ factors, and present some unifying themes that are beginning to emerge as more systems are studied.

ABBREVIATIONS

CAMP   Cationic antimicrobial peptide
LTA   Lipoteichoic acid
MIC   Minimal inhibitory concentration
PE   Phosphatidylethanolamine
PG   Peptidoglycan
RNAP   RNA polymerase
ROMA   Run-off transcription – macroarray analysis
SDS   Sodium dodecyl sulphate
TA   Teichoic acid
WTA   Wall teichoic acid

1. INTRODUCTION

The latter half of the twentieth century has witnessed an information explosion. More than ever, we now appreciate that the utility of information rests both with its content and its accessibility. On the cellular level, the vast storehouse of genetic information, embodied in DNA, must be retrieved in a highly selective and timely manner. This requires the coordinated effort of sensory proteins that function to monitor the prevailing environmental conditions and regulatory proteins that control the flow of genetic information from DNA to RNA to protein.

In bacteria, the primary checkpoint for controlling gene expression is the transcription of DNA into RNA by RNA polymerase. RNA polymerase is a complex multisubunit enzyme that contains a dissociable sigma (σ) subunit responsible for promoter recognition. Typically, the vast majority of transcription in rapidly growing bacteria requires a single, primary σ subunit similar to σ^{70} of *Escherichia coli*. Control of genes transcribed by this dominant RNA polymerase often rests with DNA-binding repressor and activator proteins. In other cases, however, transcription is regulated by a σ switching mechanism in which the primary σ subunit is replaced by an alternative σ factor with a distinct promoter selectivity. Alternative (also called secondary) σ factors function in place of the primary σ factor by binding to core RNA polymerase to allow promoter recognition. In general, alternative σ factors control specialized regulons active during growth transitions, in stationary phase, in response to stress conditions, or during morphological differentiation. With the exception of the σ^{54} subfamily of regulators, alternative σ factors are related, in sequence and presumably in structure, to σ^{70} (Lonetto *et al.*, 1992).

Historically, most alternative σ factors were discovered either through biochemical studies of transcription selectivity (e.g. their initial discovery in *Bacillus subtilis* and its phages; Haldenwang, 1995; Kroos *et al.*, 1999) or by the sequencing of regulatory genes (e.g. the characterization of *E. coli htpR*; later designated σ^{32}; Grossman *et al.*, 1984). However, with the exponential growth in DNA sequence information, including the availability of dozens of complete genome sequences, many putative σ factor genes have now been identified for which we do not yet have functional confirmation. In this review, I focus on one rapidly growing group of σ factors: the ECF subfamily (Lonetto *et al.*, 1994). Related reviews focus more specifically on *Escherichia coli* σ^{E} (Missiakas and Raina, 1998; Ravio and Silhavy, 2001), σ^{FecI} (Braun, 1997), the *Streptomyces coelicolor* ECF σ factors (Paget *et al.*, 2002), and the roles of anti-σ factors (Hughes and Mathee, 1998). Herein, I summarize these systems and also review recent progress in understanding the *Bacillus subtilis* and *Mycobacterium tuberculosis* ECF σ factor regulons.

2. FAMILIES OF BACTERIAL SIGMA (σ) FACTORS

Bacterial σ factors belong to two large, and apparently unrelated, protein families: the σ^{70} and the σ^{54} families (Helmann and Chamberlin, 1988; Gross *et al.*, 1992, 1998; Helmann, 1994). Within the σ^{70} family, there are several phylogenetic groups that often, but not always, correlate with function. Lonetto *et al.* (1992) originally distinguished between the primary (group 1) σ factors, a group of closely related but nonessential paralogues (group 2), and the more divergent alternative σ factors (group 3). To this classification, we can now add the ECF σ factors (group 4) and the newly emerging TxeR family (group 5).

The nomenclature of σ factors and their genes has generated considerable confusion. In general, most σ factors in *E. coli* and other Gram-negative bacteria are given the designation of RNA polymerase subunits, *rpo*. Examples include the primary σ, encoded by the *rpoD* gene, and the heat shock σ factor, encoded by the *rpoH* gene. The σ factors themselves are often identified by a superscript to reflect their molecular mass (in kDa): RpoD is σ^{70}, RpoH is σ^{32}. For many of those σ factors identified genetically, the σ is still identified by the original gene name: examples include σ^{FecI} in *E. coli* and σ^{AlgU} in *P. aeruginosa* (also known now as σ^{E}). In *B. subtilis*, and most other Gram-positive organisms, an alternative scheme has been adopted in which each alternative σ factor is given a letter designation and the corresponding genes are given *sig* designations. By convention, the primary σ factor is σ^{A} and is encoded by the *sigA* gene, and the alternative σ factors are identified by other letters. In some cases, other nomenclatures are still in place: for example in some species group 2 σ factors (see below) carry *hrd* (homologue of *rpoD*) designations. It remains to be seen how the nomenclature will evolve now that at least one species (*S. coelicolor*) has more σ factors than there are letters in the alphabet.

2.1. Group 1. The Primary σ Factors

The Group 1 σ factors include *E. coli* σ^{70} and its orthologues (Lonetto *et al.*, 1992). These σ factors are essential proteins responsible for most transcription in rapidly growing bacterial cells and are thus often referred to as the 'primary' σ factors. As a group, the primary σ factors are usually between 40 and 70 kDa in size and have four characteristic conserved sequence regions: regions 1 to 4 (reviewed in Helmann and Chamberlin, 1988; Gross *et al.*, 1998). In addition, in most species where promoter selectivity is well understood, primary σ factors recognize promoters of similar sequence: TTGaca near –35 and TAtaaT near –10 (where uppercase refers to more highly conserved bases).

2.2. Group 2. Nonessential Proteins Highly Similar to Primary σ Factors

In some species there are σ factors that are closely related to the primary σ, but dispensable for growth. These group 2 σ factors include the *E. coli* σ^{S} (RpoS) protein and three of the four Hrd (Homologue of RpoD) proteins in *Streptomyces coelicolor*: only HrdB is essential and is, by this criterion, a group 1 σ (Buttner and Lewis, 1992). Like the group 1 σ factors, the group 2 proteins contain all four of the conserved sequence regions characteristic of primary σ factors (Lonetto *et al.*, 1992). Moreover, the regions of σ factor that determine promoter selectivity are often nearly identical between the group 1 and 2 σ factors. Thus, it is likely that the group 1 and 2 proteins have extensive overlap in promoter recognition.

The most extensively studied group 2 σ is the *E. coli* RpoS (σ^S) stationary phase σ factor (Hengge-Aronis, 1999, 2000). Many promoters transcribed by the σ^{70}-containing holoenzyme are also recognized by σ^S and only a few truly σ^S-specific promoters have been described. Indeed, it has been quite difficult to discern those features of the DNA sequence that allow selective recognition by σ^S. This was dramatically illustrated when SELEX methods were used to determine the optimal binding sequence for the σ^S holoenzyme: the resulting 'consensus' was identical to that already documented for σ^{70} (T. Gaal and R. Gourse, personal communication). This has led to a model in which consensus promoters, which are extremely rare, can be recognized by both σ factors and the key to selectivity is the differential tolerance of nonconsensus bases. For example, σ^S transcribes efficiently from promoters lacking a consensus -35 element (Wise *et al.*, 1996) or having a C adjacent to the upstream T of the -10 element, whereas these changes can greatly reduce recognition by σ^{70} holoenzyme (Becker and Hengge-Aronis, 2001; Lee and Gralla, 2001).

In the cyanobacteria and in *S. coelicolor* the situation is made more complex by the presence of three or more group 2 σ factors. The functions of these σ factors have remained elusive. They are clearly dispensable and even multiply mutant strains do not display obvious phenotypes (Buttner and Lewis, 1992). In several cases, these group 2 σ factors have been found to be preferentially expressed during nutrient stress conditions (Caslake *et al.*, 1997; Muro-Pastor *et al.*, 2001). One interpretation of these data is that activation of one or more group 2 σ factors can alter, perhaps in subtle ways, the precise set of genes that are expressed while maintaining expression of most housekeeping functions normally dependent on the primary σ.

2.3. Group 3. Secondary σ Factors of the σ^{70} Family

In 1992, Lonetto *et al.* assigned the remaining known alternative σ factors to group 3. These proteins could all be clearly recognized as σ factors based on the presence of the conserved amino acid sequences of regions 2 and 4. However, in many cases conserved region 1 and often region 3 was absent. These group 3 proteins are significantly smaller than their group 1 and 2 paralogues (typically 25 to 35 kDa in molecular mass).

While the majority of the RNA polymerase (RNAP) core enzyme in rapidly growing cells is associated with the primary σ factor (e.g. *E. coli* σ^{70} or *B. subtilis* σ^{A}), the fraction associated with group 3 σ factors can be greatly increased under conditions of stress or during developmental processes (Price, 2000; Hecker and Volker, 2001). By reprogramming RNAP, these σ factors function as global regulators allowing the coordinate activation of numerous unlinked operons. As a class, the group 3 σ factors are regulated in diverse ways: some at the level of synthesis, others by proteolysis, and others by the reversible interaction with an anti-σ factor (Haldenwang, 1995; Hughes and Mathee, 1998; Helmann, 1999; Kroos *et al.*, 1999).

The group 3 σ factors can be divided into several clusters of evolutionarily related proteins, often with conserved or related functions. Thus, there is a heat shock cluster, a flagellar biosynthesis cluster, and a sporulation cluster (Lonetto *et al.*, 1992). In some cases, these clusters of σ factors are associated with conserved promoter sequences. For example, the promoter selectivity of the flagellar (σ^{28}) subfamily is conserved between diverse bacteria (Helmann, 1991) and the *B. subtilis* σ can partially complement the corresponding *E. coli* mutant (Chen and Helmann, 1992). Within the sporulation subfamily, different paralogues within *B. subtilis* display overlapping promoter selectivity, such that some (but not all) target promoters can be recognized by more than one σ factor allowing transcription initiation from coincident start points (Helmann and Moran, 2002).

2.4. Group 4: The ECF Subfamily

In 1994, a convergence of several lines of research led to the initial designation of the extracytoplasmic function (ECF) subfamily of σ factors (Lonetto *et al.*, 1994). Mark Buttner identified the gene for the alternative σ factor σ^{E} in *S. coelicolor* and noted that it had only distant similarity to known σ factors. At about the same time, Mike Lonetto in Carol Gross' laboratory noted that *S. coelicolor* σ^{E}, the recently identified *E. coli* σ^{E}, and several other known regulatory proteins, formed a distinct subfamily within the σ^{70} family of regulators.

Prior to the seminal paper of Lonetto *et al.* (1994), many of the ECF σ factors were known to function as positive activators of gene expression, but were assumed to act as classical transcription activators functioning in conjunction with one or more forms of holoenzyme. This assumption was challenged by Lonetto *et al.* (1994) who predicted that these diverse regulators would all function as σ factors. This prediction has been confirmed for all tested examples. Interestingly, the sequence similarity between one family member, *Pseudomonas aeruginosa* AlgU(AlgT), and *B. subtilis* σ^{H} had been noted prior to the Lonetto study (Martin *et al.*, 1993), but there was no experimental confirmation of the role of this protein as a σ factor.

In keeping with the classification scheme introduced by Lonetto *et al.* (1992), I propose that the ECF σ factors be referred to as 'group 4'. Note that previously Wosten (1998) assigned these σ factors as subgroup 3.2 of the group 3 σ factors. However, ECF σ factors are significantly more divergent in sequence, and in many organisms they equal or exceed in numbers, the group 3 σ factors. Therefore, it seems fitting that they define their own group within the σ^{70} family. This view is further supported by the assignment of ECF σ factors as a unique group within the conserved orthologous groups (COG) database (Tatusov *et al.*, 2000).

As a class, the ECF σ factors share several common features (Fig. 1). First, they often recognize promoter elements with an 'AAC' motif in the −35 region. Second, in many cases the ECF σ factor is cotranscribed with a transmembrane anti-σ factor with an extracytoplasmic sensory domain and an intracellular inhibitory domain. Third, they often control functions associated with some aspect of the cell surface or transport.

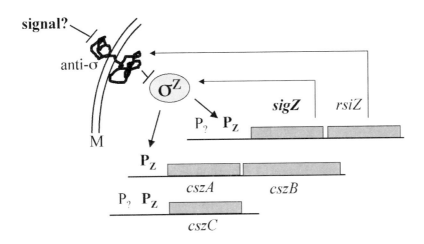

Figure 1   Properties of a generic ECF σ factor regulon. A typical bacterial ECF σ factor, here designated (arbitrarily) as σ^Z, is cotranscribed together with a downstream regulatory gene, *rsiZ* (regulator of sigZ; could also be designated *rszA*). In most cases, RsiZ functions as a specific anti-σ factor by forming an inactive complex with σ^Z. Usually, the anti-σ factor is located in the cytoplasmic membrane (M). Upon interaction with an extracytoplasmic signal (e.g. a protein or small molecule), σ^Z is released and is free to bind to core RNAP and direct transcription initiation from its specific promoter sites. As illustrated here, the *sigZ* operon is autoregulated: σ^Z activates transcription from the P_Z site. σ^Z also activates transcription of other genes (controlled by sigZ; *csz* genes), which may be distributed around the chromosome. In some cases, these genes may have other promoter elements ($P_?$) or may be additionally controlled by transcriptional repressors or activators. Thus, genes in the σ^Z regulon may also be part of other regulons.

The designation 'extracytoplasmic function' (or ECF) evolved from an analysis of the functions of the known examples of group 4 factors (Lonetto *et al.*, 1994). This phylogenetic cluster included regulators of a periplasmic stress and heat shock response (*E. coli* σ^E), iron transport (FecI in *E. coli*), a metal ion efflux system (CnrH in *Alcaligenes*), alginate secretion (AlgU/T in *P. aeruginosa*), and synthesis of membrane-localized carotenoids in *Myxococcus xanthus* (CarQ). The only unifying feature of these diverse physiological processes is that they all involve cell envelope functions (transport, secretion, extracytoplasmic stress). Hence, the name extracytoplasmic function (or ECF) was suggested for this family of σ factors. Even this broad generalization may be an oversimplification for this complex and rapidly growing family of regulators: at least one of the recently characterized ECF σ factors (*S. coelicolor* σ^R) controls a cytoplasmic stress response (see below).

In the last several years the complete genome sequences of dozens of bacteria have been determined. A survey of currently available genome sequences reveals a wide range in the numbers of ECF σ factors (Table 1): two in *E. coli*, seven in *B. subtilis*, 10 in *Mycobacterium tuberculosis*, and ~50 in *Streptomyces coelicolor*!

2.5. Group 5: The TxeR Subfamily

The discovery of the ECF subfamily of σ factors taught us that the biochemical identification of one or two regulators as σ factors can provide insight into the mechanism of action of a large family of related proteins. A similar story appears to be unfolding with the recent description of TxeR as a σ factor controlling toxin gene expression in *Clostridium difficile* (Mani and Dupuy, 2001). This regulatory protein functions biochemically as a σ factor despite the fact that the sequence of the protein bears little discernible resemblance to other members of the σ^{70} family. Addition of purified TxeR protein is sufficient for recognition of the *tox* promoters by either *E. coli* or *B. subtilis* core RNA polymerase (Mani and Dupuy, 2001). Since several other positive regulators of toxin and bacteriocin genes, including *C. tetani* TetR, *C. botulinum* BotR, and *C. perfingens* UviA, are related to TxeR (Marvaud *et al.*, 1998), it seems reasonable to suggest that these proteins are yet another distantly related group (herein designated group 5) of the σ^{70} family. A similarity between UviA and the overall structure of the σ^{70} family was noted earlier (Lonetto *et al.*, 1992). In fact, expression of UviA *in trans* activates a *C. difficile tox* promoter and expression of TxeR *in trans* activates the *C. perfringens bcn* promoter (a normal target of UviA; Garnier and Cole, 1988) (N. Mani and A.L. Sonenshein, personal communication). Similar functional complementarity has been seen for TetR and BotR *in vivo* (Marvaud *et al.*, 1998). Gel shift analyses show that TxeR enables *E. coli* core RNA polymerase to bind to DNA fragments containing the promoters of the *C. perfringens bcn* gene and the

Table 1 Survey of ECF σ factors in selected bacterial genomes[a].

Organism	Genome size (Mb)	ORF	Total σ factors (estimated)	ECF σ factors (estimated)	Year of completion
Streptomyces coelicolor A3(2)	8.3	7846	~65	~50	preliminary
Pseudomonas aeruginosa PAO1	6.3	5570	24	19	2000
Mesorhizobium loti	7	>7000	23	>10	2000
Caulobacter crescentus	4.0	3767	17	13	2001
Bacillus halodurans C-125	4.2	4066	19	11	2000
Mycobacterium tuberculosis H37Rv	4.4	3924	13	10	1998
Bacillus cereus	5.5	5477	18	8	preliminary
Bacillus subtilis 168	4.2	4100	17	7	1997
Porphyromonas gingivalis	2.3	2226	7	5	2001
Enterococcus faecalis V583	3.2	3334	6	2	2001
Magnetococcus MC-1	4.6	4528	7	2	preliminary
Vibrio cholerae El Tor N16961	4.0	3885	8	3	2000
Mycobacterium leprae TN	3.2	1604	4	2	2001
Escherichia coli K-12 MG1655	4.67	4288	7	2	1997
Haemophilus influenza	1.83	1740	5	2	1995
Pasteurella multocida, Pm70	2.26	2014	4	2	2001
Synechocystis sp.PCC 6803	3.57	3168	8	2	1996
Deinococcus radiodurans R1	3.28	3187	2	1	1999
Escherichia coli O157:H7	5.5	5361	5	1	2001
Lactococcus lactis IL1403	2.36	2266	2	1	2001
Neisseria meningitidis MC58	2.2	2158	4	1	2000
Staphylococcus aureus Mu50	2.8	2593	4	1	2001
Streptococcus pyogenes M1	1.8	1752	4	1	2001
Treponema pallidum Nichols	1.14	1041	5	1	1998
Thermotoga maritima MSB8	1.8	1877	4	1	1999
Aquifex aeolicus VF5	1.55	1544	4	0	1998
Borrelia burgdorferi B31	0.9	8430	3	0	1997
Campylobacter jejuni NCTC11168	1.64	1654	3	0	2001
Chlamydia trachomatis serovar D	1.04	896	3	0	1998
Helicobacter pylori 26695	1.67	1590	3	0	1997
Xylella fastidiosa CVC	2.7	2904	3	0	2000
Rickettsia prowazekii Madrid E	1.1	834	2	0	1998
Streptococcus pneumoniae	2.16	2236	1	0	2001
Buchnera sp. APS	0.64	564	1	0	2000
Mycoplasma genitalium G-37	0.58	470	1	0	1995
Mycoplasma pneumoniae M129	0.82	679	1	0	1996
Ureaplasma urealyticum serovar 3	0.75	3370	1	0	2000

[a] Genome sequences were analysed using the comprehensive microbial resource (Peterson *et al.*, 2001) available at the TIGR web site (www.tigr.org) and by using genome specific web sites and resources when available.

C. botulinum boNT gene (B. Dupuy, personal communication), suggesting a functional conservation between the respective positive regulators. *In vitro* transcription experiments show that, in the presence of TxeR, *E. coli* or *B. subtilis* core enzyme stimulates transcription from *bcn* promoters (N. Mani and A.L. Sonenshein, personal communication). It remains to be shown, however, that UviA, BotR and TetR function as bonafide σ factors *in vivo*.

3. FUNCTIONS OF ECF σ FACTORS

3.1. Strategies for Assigning Function

While many of the founding members of the ECF σ factor subfamily were discovered first as genetic regulators of known function, the advent of genomic sequencing has led to the inevitable discovery of numerous ECF σ factor genes for which functions can not easily be predicted. By far, the most dramatic example of this challenge is the recent discovery of ~50 ECF σ factors encoded by *S. coelicolor* genome (Paget *et al.*, 2002). Thus, we are faced with an increasingly familiar problem in functional genomics: how can we determine the function of regulators identified by genome sequencing?

In many ways, this is a familiar problem for those studying σ factors in Gram-positive bacteria. In the late 1970s, studies of *B. subtilis* RNAP revealed multiple associated σ factors with, *a priori*, no clear indication of function. The role for many of these factors was gradually elucidated by construction of mutant strains (using 'reverse genetics'), analyses of *in vitro* and *in vivo* transcription selectivity, and the identification of target genes (Haldenwang, 1995). These studies, for example, led to the identification of σ factors controlling flagellar motility (σ^D), the sporulation cascade (σ^E, σ^F, σ^G, σ^K), and the general stress response (σ^B). Similarly, biochemical fractionation studies in *S. coelicolor* revealed a large number of alternative σ factors with distinct *in vitro* transcription selectivity (Buttner *et al.*, 1988; Buttner, 1989; Chater *et al.*, 1989). Assigning physiological roles to these various factors represents a formidable challenge.

Several strategies can be envisioned to determine the physiological roles of newly described ECF σ factors. First, mutant strains lacking one or more ECF σ may have phenotypes that will provide clues to function. Second, target genes can be identified and, by understanding their functions, we may be able to predict the phenotype conferred by a σ factor mutation. Third, physical stimuli or genetic changes that act to induce expression of each σ factor regulon can be identified and used to infer possible function. For example, overexpression of a particular ECF σ regulon by mutation of the cognate anti-σ factor may reveal a more dramatic phenotype than mutation of the σ factor gene itself. Examples of these approaches, and their advantages and limitations, are summarized in Table 2.

3.2. *Escherichia coli*

E. coli contains seven σ factors, including two members of the ECF subfamily: σ^E and σ^{FecI}. The σ^E regulon is activated in response to 'periplasmic stress' or extreme heat shock and controls the expression of proteases and folding catalysts active in the periplasm. σ^{FecI} controls the expression of the ferric-citrate uptake system in response to the presence in the periplasm of ferric citrate (Angerer *et al.*, 1995; Braun, 1997). For more detailed discussions of the σ^E regulon and its role, the reader is referred to other recent reviews (Missiakas and Raina, 1998; Ravio and Silhavy, 2001).

3.2.1. σ^E

The activity of σ^E was originally discovered by Erickson and Gross (1989) as a holoenzyme form ($E\sigma^{24}$) necessary for transcription of the group 3 heat shock σ, σ^{32}, at very high temperatures (50°C) due to activation of a new promoter site. A similar promoter controls the heat-inducible periplasmic protease DegP (HtrA) (Lipinska *et al.*, 1988). This led to the hypothesis that *E. coli* contained a second heat shock regulon, activated by a new σ factor. This early work predated the isolation of the gene encoding σ^E (Raina *et al.*, 1995; Rouviere *et al.*, 1995), so it was not clear that σ^E would in fact become a founding member for the ECF subfamily.

Extensive work on the σ^E regulon in *E. coli* has led to a detailed model for the activation of this system in response to periplasmic stress (Ravio and Silhavy, 2001). Periplasmic stress can be elicited in several ways including the overexpression of outer membrane proteins or by the production of misfolded proteins in the periplasm (Missiakas and Raina, 1998). The latter condition can also be elicited genetically by mutations in folding catalysts. The importance of this stress response is underscored by the finding that σ^E is essential in *E. coli* (De Las Penas *et al.*, 1997a).

Many of the key regulators of σ^E activity (regulators of SigE; *rse* genes) are cotranscribed with *rpoE* to form an operon: *rpoE rseA rseB rseC* (De Las Penas *et al.*, 1997b; Missiakas *et al.*, 1997). The RseA membrane protein functions stoichiometrically as an anti-σ factor and its action is enhanced by the periplasmic RseB protein (Fig. 2). The role of the RseC protein is not yet clear, but it has been implicated in thiamine biosynthesis (Beck *et al.*, 1997). Activation of the σ^E regulon occurs when σ^E is released from RseA inhibition: a phenomenon accompanied by the degradation of RseA by Hho (DegS) protease (Ades *et al.*, 1999; Alba *et al.*, 2001).

The first two defined targets for σ^E, the *rpoH* and *htrA* genes, have identical promoter consensus elements (Table 3). Indeed, the similarity between these promoters and the *S. coelicolor dagA* P2 site contributed to the original

Table 2 Strategies for defining σ factor function.

Strategy	Requirements	Approach	Limitations	Advantages	Example
Genetics	• Genetic system for making directed mutations	Characterize phenotype of mutants lacking σ factor	Phenotype may not be obvious due to overlapping recognition by other σ factors or multiple promoters for target operons	Most direct approach to determining physiological role of a given σ factor	*E. coli* σ^E *S. coelicolor* σ^R *M. tuberculosis* σ^E
Define regulon by: (a) Consensus search	• Genome sequence • Some knowledge of promoter consensus	Search genome for sequences resembling consensus site for σ recognition	Only promoters matching the predefined consensus will be found	Promoters can be found even if the target operons have multiple promoter sites	*B. subtilis* σ^W *S. coelicolor* σ^R
(b) Transcriptional profiling	• Genome sequence • Appropriate mutant strains • DNA microarray	Define differences between RNA populations in WT and mutant strains	The σ regulon may not be active under the growth conditions of the experiment. Operons with multiple promoter sites may be missed	Very rapidly surveys the entire genome for transcriptional effects of a σ factor mutation. Best if coupled with knowledge of conditions that induce the regulon in question (e.g. *in vivo* expression of inducible σ factor)	*B. subtilis* σ^W *M. tuberculosis* σ^E and σ^H

Table 2 Cont.

Strategy	Requirements	Approach	Limitations	Advantages	Example
(c) ROMA	• Genomic DNA • Purified RNAP and σ factor • DNA microarray	Compare differences between RNA population synthesized *in vitro* with RNAP lacking and containing the σ factor in question	Only promoters active *in vitro* are recovered; requires purified transcription factors	Rapidly surveys the entire genome for direct targets of σ factor action; complementary to transcriptional profiling	*B. subtilis* σ^W and σ^X
(d) Proteomics	• 2-D gel system • N-terminal sequencing or mass-spectrometry for protein identification	Compare differences of proteins synthesized in WT and mutant strains	The σ regulon may not be active under the growth conditions used; operons with multiple promoter sites may be missed; can not distinguish direct or indirect effects of the σ factor	Very rapidly surveys the entire proteome for effects of a σ factor mutation	*E. coli* σ^E
Identification of conditions or mutations that activate the regulon	• Reporter fusion or other method to monitor σ activity	Screen for either physical stimuli or genetic changes that activate gene expression	May be difficult to identify stimuli; effects of mutations may be indirect	Does not require genomic sequence information	*E. coli* σ^E *B. subtilis* σ^W and σ^X *M. tuberculosis* σ^E and σ^H *S. coelicolor* σ^R

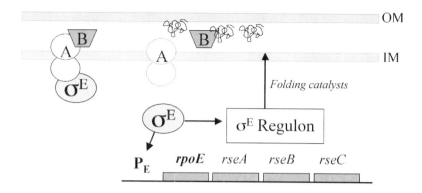

Figure 2 The *E. coli* σ^E regulon. The *E. coli* σ^E gene (*rpoE*) is cotranscribed with three regulator of *sigE* (*rse*) genes from an autoregulatory, σ^E-dependent promoter. RseA functions as an anti-σ factor and is located in the inner membrane (IM). The inhibitory action of RseA is stimulated by the RseB periplasmic protein (complex on the left). The accumulation of misfolded proteins in the periplasm (squiggles) is postulated to sequester RseB and expose RseA to degradation by the DegS(Hho) protease. As a result, σ^E is released and can activate transcription of its regulon. Several of the proteins in the σ^E regulon are known or postulated folding catalysts or proteases that can act to remove misfolded proteins (Table 4).

recognition of the ECF subfamily of regulators. Subsequent work identified another very similar sequence preceding the *rpoE* operon that functions as an autoregulatory site.

Recently, the σ^E regulon of *E. coli* has been defined using a genetic strategy to identify 20 promoter regions that are up-regulated in response to σ^E overexpression (Table 3). These genes include the known targets of σ^E (including *rpoH*, *htrA*, and the *rpoE* operon itself) together with genes encoding proteins associated with either the inner or outer membrane and involved in functions such as lipopolysaccharide biogenesis and protein folding (Dartigalongue *et al.*, 2001; Table 4). As seems common in the characterization of regulons controlled by ECF σ factors, many of the target genes encode proteins of unknown function with a predicted location in the cell envelope. Note that in this work the authors have designated these unknown function genes as *ecf* (extracytoplasmic function). While this is reasonable in *E. coli*, which has only one other ECF σ factor, this practice would clearly lead to confusion in other organisms and a designation such as *cse* (controlled by *sigE*) is preferred (see Fig. 1). Even this may lead to confusion, as issues of regulon overlap prevent many genes from being uniquely assigned to a single regulon.

Inspection of the regulatory regions of many genes under σ^E control reveals candidate promoters similar to those that have been biochemically

Table 3 Promoters of the *E. coli* σ^E regulon (adapted from Dartigalongue *et al.*, 2001).

Gene	-35/spacer region	-10	+1
htrA	**GAACTT**CAGGCTATAAAACGAA	**TCTGA**	AGAACA<u>C</u>
*rpoH*P3	**GAACTT**GTGGATAAAATCACGG	**TCTGA**	TAAAAC<u>A</u>
*rpoE*P2	**GAACTT**TACAAAAACGAGACAC	**TCTAA**	CCCTTTAG<u>C</u>
*cutC*P3	**GAA**TTAGGTTTTCCTGGACT	**GGTGA**	CGGGCG<u>T</u>
dsbC	**GC**ATCACCCGCGGGCGTGATG	**TCTGA**	AAAGA<u>A</u>
*fkpA*P2	TAA**TTT**AAACAAAAAGAG	**TCTGA**	AAATAG<u>A</u>
*htrM*P4	**GA**CATTCGTGTCTGAGATTGTC	**TCTGA**	CTCCATA<u>A</u>
*lpxD*P2	**AAA**AGACATCACTGCCGACGT	**ACTGA**	AACAGG<u>T</u>
mdoG	**GAAC**GATACCGGGATTCTGGTTG	**TCGGA**	ATGGCTGG<u>T</u>
nlpB	**GC**CA**TT**ACACAACAAACTATT	**GTCGA**	ACCCAATCC<u>A</u>
ostA	**GC**ATATATTCCCCAAATCGACAC	**ACGGA**	TAT<u>C</u>
*rpoD*P3	**AAACT**GTCGATGTGGGACGATATA	**GCAGA**	TAAGAA<u>T</u>
skp	TAA**GAT**CGCCGGGCCACGCAAAGA	**ACTGC**	ACCCTCCGG<u>T</u>
ecfA	**AAA**A**TT**GCACGCGGGATGTTCTG	**GCTGA**	TGCTGCTT<u>A</u>
ecfD	GG**TTT**TCGGACACCGTTGCAGTGA	**GCTGT**	ACTCGTTG<u>T</u>
ecfE	TAA**TT**CCAGGACACGGTGGTATT	**TTAGA**	TCGTATTG<u>A</u>
*ecfF*P1	**GAACTT**TTCGACGTTTGGTGGG	**ACTAA**	GAAAGCAT<u>A</u>
*ecfF*P2	**AAACT**GCCTGATGCGCTACGCT	**TATCA**	GGCCTGGA<u>A</u>
ecfG	C**AATTT**GACGGGCGTAAAG	**TTTGA**	AGCAGTG<u>G</u>
ecfH	**GC**ATTCAGGAAATGCATATGCT	**GCTGA**	CGGTAAAT<u>T</u>
ecfI	CGG**CT**ACGATGTAAAAATGGG	**TCTGG**	AAATGA<u>A</u>
ecfJ	A**AACT**CAGACCCAAGTGG	**TCGGA**	TCACC<u>T</u>
ecfK	**AA**A**GAT**CAAGGGCGGACCGGTA	**TCCGA**	GCGGGTTCAAGAC<u>T</u>
*ecfL*P2	**GAA**T**TT**ATGTTTTTGAATGCT	**TCTTA**	TCTCAC<u>G</u>

characterized (Table 3). Interestingly, many of the proposed −35 and −10 elements do not display strong similarity to the presumed consensus, suggesting that σ^E may have a relaxed promoter selectivity compared with other ECF σ factors. One can speculate that the presence of only two ECF σ factors in *E. coli* has allowed σ^E promoters to tolerate deviations from consensus while retaining recognition by the σ^E holoenzyme. In contrast, in organisms with a great many more ECF σ factors, deviations from consensus can rapidly switch a target promoter from control by one ECF σ to a regulon controlled by a related, but functionally distinct paralogue (e.g. Qiu and Helmann, 2001). This can of course be advantageous, and many target promoters do belong to more than one regulon. However, if it is disadvantageous this will act to restrict the sequences of promoters within each regulon and could, in principle, account for the high degree of sequence conservation noted among promoters in some regulons (e.g. *B. subtilis* σ^W and *S. coelicolor* σ^R; see below).

Table 4   The *E. coli* σ^E Regulon (adapted from Dartigalongue *et al.*, 2001).

Gene name	Function (category)
	Transcriptional factors/Regulatory genes
rpoE	σ^E ECF sigma factor (autoregulation)
rpoH	σ^{32} heat shock factor (group 3)
rpoD	σ^{70} primary sigma factor
rseA	Negative regulator of σ^E (inner membrane)
rseB	Negative regulator of σ^E (periplasm)
rseC	? Positive regulator for σ^E (inner membrane)
	Periplasmic folding factors
skp	Outer membrane protein assembly/folding
dsbC	Thiol:disulphide oxidoreductase
fkpA	Peptidyl prolyl isomerase
surA	Peptidyl prolyl isomerase
	Proteases
htrA (degP)	Serine protease (periplasmic)
ecfE (yaeL)	Putative carboxypeptidase (inner membrane)
	LPS biogenesis
htrM (rfaD)	Lipopolysaccharide biosynthesis
lpxD lpxA	Lipid A biosynthesis
ecfA (f288)	Lipopolysaccharide biogenesis?
	Sensory proteins
mdoG	Synthesis of membrane-derived oligosaccharide
cutC	Copper sensing
	Unknown functions
nlpB	Lipoprotein
ecfD (yfiO)	Putative lipoprotein (similar to ComL)
ecfF (yggN)	Putative periplasmic protein
ecfG (htrG)	Putative inner membrane protein
ecfH (yraP)	Putative lipoprotein (similar to OsmY)
ecfI (yidQ)	Putative inner membrane protein
ecfJ (ytfJ)	Putative periplasmic protein
ecfK (UP0)	Putative outer membrane protein
ecfL (yqjA)	Putative inner membrane protein

3.2.2. σ^{FecI}

The *fecIR* genes encode regulators affecting transcription of the *fecA* operon which encodes a specific ferric citrate uptake system (Braun, 1997). The *fecA* operon is transcribed only under Fe-limiting growth conditions due to repression by the iron-sensing ferric uptake regulator (Fur) protein (Angerer and Braun, 1998). However, iron limitation alone is not sufficient to induce expression: the *fecA* operon is activated by the presence of ferric citrate (Zimmermann *et al.*, 1984). This activation requires the FecI and FecR proteins to signal the presence of the substrate for transport, ferric citrate. The σ^{FecI} regulon appears to consist of this single target operon. Unlike many other ECF σ factors, σ^{FecI} does not autoregulate its own synthesis (Braun, 1997).

The ability of ferric citrate to activate the transcription of the appropriate uptake genes provides an elegant example of *trans*-membrane signalling. The signal transduction mechanism probably involves a direct interaction between ferric citrate bound to the outer membrane FecA protein and the periplasmic domain of the FecR regulatory protein (Enz *et al.*, 2000). This interaction leads to the release of σ^{FecI}, bound to the cytoplasmic domain of FecR, and the resulting free σ factor then activates transcription of the *fecA* operon.

Genetic analyses indicate that FecR plays a positive regulatory role since *fecR* mutants are not able to activate *fecA* transcription efficiently (Ochs *et al.*, 1995). Expression of the first 81 amino acids of FecR, encoding just the cytoplasmic N-terminal domain, is sufficient for full activation of a *fecA* reporter fusion, but this expression no longer requires ferric citrate (Welz and Braun, 1998). These results support a model in which the cytoplasmic N terminus of the transmembrane FecR protein interacts with σ^{FecI} to convert it from an inactive to an active σ factor (Braun, 1997). However, the nature of this activation event has proved elusive. One could imagine that the activation of σ^{FecI} involves, for example, a post-translational modification. In other systems, σ factors are synthesized as an inactive pro-protein that is activated by proteolysis (Kroos *et al.*, 1999). The activity of other regulators (although not to date σ factors) can be controlled by reversible phosphorylation or other types of covalent modification. No evidence has been presented for any such changes in the FecR:σ^{FecI} system.

An alternative model can be envisioned that reconciles the apparent positive regulatory role of FecR with the role of σ^{FecI} as a σ factor for RNA polymerase. The free σ factor may be unstable, perhaps due to proteolytic turnover in the cell, and formation of a FecR:σ^{FecI} complex may stabilize the σ against degradation (Stiefel *et al.*, 2001). Then, upon release from the complex upon exposure to ferric citrate, the σ^{FecI} protein may bind RNAP and catalyse transcription initiation. By this model, FecR would function both as an anti-σ factor, and in a positive role to stabilize the otherwise unstable σ factor (Fig. 3). A similar scenario may pertain to the *P. fluorescens* ECF σ PrtI, which is regulated by a transmembrane 'activator' protein, PrtR (Burger *et al.*, 2000).

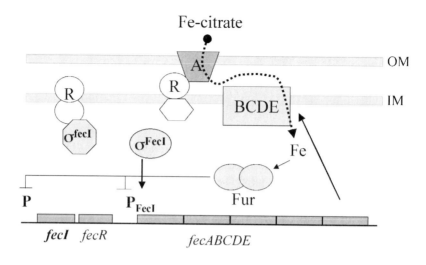

Figure 3   The *E. coli* σ^{FecI} regulon. The *fec* system is responsible for the inducible tran-scription of the ferric citrate transporter. σ^{fecI} is encoded in an operon with an inner membrane (IM) regulator protein, FecR. The *fecIR* operon is repressed, under iron-replete conditions, by the ferric uptake regulator (Fur) protein. When iron is limiting, the *fecIR* operon is expressed and there is presumably a modest amount of read through transcription into the downstream *fecA* operon. In the absence of ferric citrate, FecR and FecI are colo-calized to the inner membrane and σ^{FecI} is inactive (complex on the left). The FecA transporter, localized to the outer membrane (OM), signals the presence of ferric citrate by a direct protein–protein interaction with FecR and this, in turn, leads to the 'activation' of σ^{FecI} (complex on the right; see text). The free σ^{FecI} then binds to RNAP to activate expres-sion of the *fecABCDE* operon encoding both the outer and inner-membrane components of the ferric citrate transport system. As intracellular iron needs are met, the Fur protein will repress any further expression of this transporter and its regulatory proteins.

3.3. *Bacillus subtilis*

B. subtilis possesses seven ECF σ paralogues that were all initially identified during the international genome sequencing effort (Kunst *et al.*, 1997). None of these seven loci corresponds to previously identified genes, making it unlikely that they are essential regulators of any of the most well-studied processes in this organism, such as endospore formation, genetic competence, or the heat shock and general stress responses.

To begin to investigate the roles of the various ECF σ factors in *B. subtilis*, we and others have sought to determine mutant phenotypes for strains lacking each σ, identify target genes for each σ, and identify conditions leading to the activation of each σ factor regulon. Most studies to date have concentrated on three of these factors: σ^X, σ^W and σ^M. The recent discovery of genes encoding

11 ECF σ factors in *B. halodurans* and at least eight in *B. cereus* underscores the importance of these regulators in the bacilli. Remarkably, of the 11 ECF σ factors in *B. halodurans* only one is an obvious orthologue of a *B. subtilis* factor: σ^W (Takami *et al.*, 2000).

3.3.1. σ^X

The first ECF σ factor sequenced in *B. subtilis*, and the first to attract experimental scrutiny, was σ^X. The *sigX* gene was postulated to encode a σ factor based on its similarity to the newly described ECF subfamily in 1994 (Lonetto *et al.*, 1994). This inference was confirmed when the protein product of the *sigX* gene was overproduced and purified and found to have σ factor activity: addition of σ^X to RNAP leads to the specific recognition of a distinct autoregulatory site (P_X) not recognized by RNAP containing the major vegetative σ, σ^A (Huang *et al.*, 1997).

3.3.1.1. sigX Mutants are Slightly More Sensitive to Heat and Oxidative Stress. The function of this regulator was not immediately apparent as *sigX* mutants do not display gross phenotypic abnormalities. The only differences relative to wild-type detected in an initial survey were increased sensitivity to heat and oxidative stress (Huang *et al.*, 1997). However, these properties could well be the indirect effect of any number of changes in cell physiology. Since expression of *sigX* is itself not heat inducible, and heat shock regulation has been carefully investigated in *B. subtilis* without the identification of any link to *sigX* (Price, 2000; Hecker and Volker, 2001), it seems unlikely that σ^X is a central regulator of the heat shock response. Similarly, σ^X does not control transcription of any of the known antioxidant enzymes and oxidative stress responses, again suggesting an indirect effect of the *sigX* mutation on resistance.

3.3.1.2. σ^X is Not an Orthologue of σ^{FecI}. Since σ^X is related to the *E. coli* σ^{FecI} protein, Brutsche and Braun (1997) postulated that perhaps it also controlled iron uptake functions in *B. subtilis*. However, *B. subtilis* failed to use ferric citrate as an iron source and a *sigX* mutant was not affected in any known ferri-siderophore uptake systems. Surprisingly, expression of *sigX* in *E. coli* was found to partially complement a *fecI* mutation, suggesting that σ^X might be able to activate transcription of the *fec* transport genes (Brutsche and Braun, 1997). It has not been established whether or not this activation involves recognition of the same promoter sequence recognized by σ^{FecI}. In sum, the available evidence suggests that σ^X and σ^{FecI} are homologues (evolutionarily related), but not orthologues (they control distinct functions). Indeed, the only known orthologous alternative σ factors in *E. coli* and *B. subtilis* are the flagellar regulators σ^F and σ^D, respectively (Chen and Helmann, 1992).

Table 5 *Bacillus subtilis* σ^X regulon – promoters largely dependent on σ^X

Gene[a]	Operon size	Promoter site(s)[b]	Function[c]	Promoter sequence[d]	5' UTR[e]
sigX	2	P_A, P_X	Sigma factor	AAT**GTAAC**TTTTCAAGCTATTCATA**CGAC**AAAAAGT<u>G</u>AACG	17
csbB	1	P_X, P_B	Glucosyltransferase	ATT**GTAAC**AAAAAACAG–GTTTAAA**CGAC**TTTAAAAAA<u>A</u>GGA	80
lytR	1	P_X, P_A	Autolysin regulation	AAT**GAAAC**TTTTTTTTTA–TAAAAAA**CGAC**TATTTTAGG<u>A</u>TTT	84
rapD	1	P_X, P_A	Regulation	AAT**GTAAC**CAACTGTCAATGAGAGCC**GTC**AAAAGTTA<u>T</u>GATA	45
pssA	3	P_A, P_X	Phospholipid synthesis	CC**TGTAAC**GCTATTCGA–TCACTAT**CGTC**AAATAATATA<u>G</u>AT	34
dltA	5	(P_D?), P_X	D-alanylation of teichoic acids	AAT**GAAAC**TTTTTGAGC–ATCTGAT**CGTC**AAATAATC<u>A</u>TGTG	204
Total	13				
			Consensus	t**GtaAC** (15–16)	**CGwC**

[a] The first gene in the operon is listed.

[b] Known promoter sites preceding the operon: P_A is σ^A-dependent, P_X is σ^X-dependent, etc.

[c] The known or predicted function of the encoded gene products is summarized.

[d] The –35 and –10 elements are in bold. The transcriptional start site is underlined where known.

[e] The 5'-untranslated region (UTR) is the distance from the transcription start to the translation start for the first gene in the operon.

Table 6 ECF-type promoters dependent on multiple σ factors.

Gene[a]	Operon size	Function[b]	Promoter sequence[c]	5′ UTR[d]	σ factors[e]
abh	1	Regulation	CGG**GAAAC**TTTTTCAAAGTTTCATT**CGTC**TACGATATA$\underline{T}$ATTGA	62	σ^X, σ^W
divIC	2?	Cell division	TTT**GAAAC**TTCTTCCTGTGAAAATGC**GTC**TAACTTT$\underline{T}$AGACG	113	σ^X, σ^W
ywbN	1	Unknown	TA**CAAAAC**AAATGATCAGTCCTATAC**GTC**TTATGAT$\underline{A}$AATTA	105	σ^W, σ^X
yrhH	1	Methylase	AT**TGAAAC**ATTTTTCAATACATTGCC**CGTC**TAGTTGG$\underline{T}$ACCTT	87	σ^W, σ^X
yjbC	2?	Unknown	AAT**GAAAC**CATGGCGGAAGTTCGCAC**GTC**TTTATAGA$\underline{T}$GTAA	76	σ^W, σ^X, σ^M
pbpX	1	PBP	TT**GACAAC**TTTTTTAGGGCTTTATT**CGTC**TAACAAAACGTGT	40	σ^X, σ^W
yceC	6	Tellurium resistance	TAC**GAAAC**TTTGATATAATAACAAAC**GTA**TATATTAGTAATT	80	σ^W, σ^X
yqeZ	3	Protease	AAT**GAAAC**CCTTGATACATTTGTTAC**GTA**TGAAGAGAAGCA	38	σ^W, (σ^V)
ydcB	2	Holo-ACP synthase	TAT**GAGAC**AAACCATTTAAATAGTAC**GTA**CGCAAAGGAGGTT	13	σ^W, σ^X
ywoA	1	Bacteriocin transport	TTT**GAAAC**TTTTCATGAGTAAGATTA**GTC**TACTAAATATAAA	22	σ^X, σ^M, (σ^W)

a The first gene in the operon is listed.

b The known or predicted function of the encoded gene products is summarized.

c The −35 and −10 elements are in bold. The transcriptional start site is underlined where known.

d The 5′-untranslated region (UTR) is the distance from the transcription start to the translation start for the first gene in the operon.

e ECF σ factors known or suspected (in parentheses) to recognize the promoter illustrated.

In the course of their studies, Brutsche and Braun demonstrated that the gene immediately downstream of *sigX* encodes a negative regulator of *sigX* activity, designated *rsiX*. After overproduction in *E. coli*, they found that σ^X could direct transcription *in vitro* from its autoregulatory promoter site, P_X, but that when overexpressed with the negative regulator RsiX, the resulting σ^X:RsiX complex was inactive. Moreover, σ^X protein fractionated with the cell membrane when overexpressed with RsiX, whereas σ^X alone is a soluble protein. While their studies failed to provide a link between σ^X and iron utilization, they nevertheless demonstrated the anti-σ activity of RsiX and confirmed the predicted membrane localization of this regulatory factor (Brutsche and Braun, 1997).

3.3.1.3. Characterization of the σ^X Regulon by Promoter Consensus Search. We reasoned that by defining the promoter selectivity of σ^X we might be able to identify target genes and thereby assign a function to σ^X. To define the sequence determinants for σ^X-dependent recognition we took advantage of the fact that *sigX*, like many ECF σ factors, is transcribed (in part) from an autoregulatory promoter (P_X). Using reporter fusions containing only P_X, we performed saturation mutagenesis to define those bases in the -35 and -10 regions critical for promoter function. We then searched the *B. subtilis* genome for similar sequences preceding open reading frames. In total, more than a dozen candidate σ^X-dependent promoters were identified and tested for activity (Huang and Helmann, 1998). Of these, two were found to be exclusively recognized by σ^X *in vivo* (*csbB*, *lytR*) with the others exhibiting a variable level of residual transcription even in a *sigX* mutant strain. A current compilation of genes transcribed, in whole or in part, by the σ^X holoenzyme is presented in Tables 5 and 6.

Analysis of genes associated with σ^X-dependent promoters revealed that most have additional promoter sites. For example, *csbB* can be transcribed from either a σ^X- or a σ^B-dependent promoter while *lytR* is preceded by both σ^A- and σ^X-dependent sites (Huang and Helmann, 1998). Furthermore, even in a *sigX* mutant, transcripts could still be detected emanating from the sites recognized by σ^X, suggesting that other holoenzyme forms could also recognize these sequences. Because of these complexities, a *sigX* mutation may reduce, but is unlikely to eliminate, the expression of these target genes.

3.3.1.4. σ^X Controls Modifications of the Cell Envelope. We found σ^X-dependent promoters preceding several genes that affect the composition or metabolism of the cell envelope, including *lytR* (a negative regulator of autolysin; Lazarevic *et al.*, 1992), *csbB* (a membrane-bound glucosyl transferase; Akbar and Price, 1996), *pbpX* (a pencillin-binding protein), the *dlt* operon (controlling the D-alanylation of TA; Perego *et al.*, 1995), and the *pssA* operon controlling phosphatidylethanolamine synthesis. In addition, σ^X

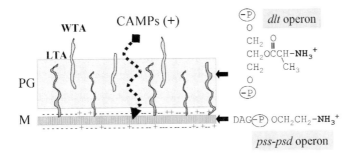

Figure 4 Roles of the *B. subtilis* σ^X protein in resistance to cationic antimicrobial peptides (CAMPs). The *B. subtilis* cell envelope includes both a cytoplasmic membrane (M) and a thick peptidoglycan layer (PG). Two of the operons controlled by σ^X are involved in modulating the net charge of the cell envelope. The *dlt* operon encodes proteins involved in the D-alanylation of both lipoteichoic acid (LTA) and wall teichoic acid (WTA) by esterification of the glycerol moieties with D-alanine. Since both LTA and WTA are glycerol-phosphate copolymers, the introduction of D-alanine esters reduces the net negative charge of the cell wall. Similarly, the cytoplasmic membrane contains an abundance of anionic phospholipids (indicated by the '–' signs) and the net charge of the membrane can be modulated by the incorporation of neutral constituents, such as glycolipids and the zwitterionic phosphatidyl ethanolamine (PE). The synthesis of PE requires the products of the *pss-psd* operon which is partially under σ^X control. The ability of cationic antimicrobial peptides (CAMPs) to penetrate the cell wall and permeabilize the membrane is reduced by the incorporation of these positively charged groups into the cell envelope.

contributes to the expression of *rapD*, a response regulator aspartate phosphatase of as yet unknown function (Reizer *et al.*, 1997; Perego, 1998). The σ^X regulon overlaps with regulons controlled by (at least) σ^D, σ^B and σ^W (Tables 5 and 6).

By defining the σ^X regulon, we were led to a model in which σ^X modifies the composition and properties of the cell envelope (Fig. 4). *B. subtilis* has a typical Gram-positive envelope containing a cytoplasmic membrane surrounded by a thick peptidoglycan (PG) layer. The cell wall is negatively charged, and functions in a manner analogous to the periplasm of Gram-negative bacteria, acting to bind and concentrate proteins, small molecules and ions near the cell (Merchante *et al.*, 1995; Pooley *et al.*, 1996). The activity of σ^X regulates the net charge of the cell wall by controlling transcription of the *dlt* operon and, in a parallel pathway, may modulate the net charge of the membrane by contributing to transcription of the *pssA* operon.

While it has been known for many years that teichoic acids are essential, their roles in cell physiology are not entirely clear. The σ^X-activated *dlt* operon controls the modification of teichoic acids by esterification with D-alanine (Perego *et al.*, 1995). *B. subtilis* contains both membrane-associated lipoteichoic acids (LTA) and wall teichoic acids (WTA) (Fischer, 1988). Both LTA and WTA are extensively substituted by esterification with sugars and D-alanine. The latter modification introduces free amino groups (NH_3^+) into the cell envelope, and thereby reduces the net negative charge (Perego et al., 1995).

Genetic studies indicate that *dlt* mutants have pleiotropic phenotypes. They often display altered patterns of autolysis (Wecke et al., 1996), perhaps due to alterations in autolysin binding to the cell wall, have alterations in adhesive properties (Clemans et al., 1999), carbohydrate metabolism (Spatafora et al., 1999), sensitivity to acid (Boyd et al., 2000), and may be affected in protein secretion and folding (Hyyrylainen et al., 2000).

Additional insight into the function of D-alanylation comes from the observation that *dlt* mutants in *Staphylococcus aureus* have a greatly increased sensitivity to cationic antimicrobial peptides (CAMPs) (Peschel et al., 1999). CAMPs are a broadly distributed family of peptides that kill bacterial cells (Hancock and Diamond, 2000; Hancock and Scott, 2000). Many are thought to act by accumulating within the cytoplasmic membrane to a critical concentration that allows the assembly of structures that permeabilize the cell. However, in other cases CAMPs also have effects on cell wall biosynthesis. For example, nisin and epidermin, two class I lantibiotics, interact with the lipid II PG synthesis intermediate (Brotz et al., 1998; Breukink et al., 1999). Inactivation of *dlt* also leads to altered methicillin resistance in *S. aureus* (Nakao et al., 2000).

The cell membrane also contains a net negative charge due to a preponderance of anionic phospholipids. However, phosphatidylethanolamine (PE), a neutral (zwitterionic) lipid, makes up as much as 50% of the membrane (Matsumoto et al., 1998). Since σ^X contributes to transcription of the *pssA* operon (which encodes both phosphatidyl serine synthase and phosphatidylserine decarboxylase; M. Cao and J.D. Helmann unpublished results), we predict that σ^X also regulates PE levels, and thus membrane net charge. Just as the *dlt* gene products lead to the incorporation of NH_3^+ groups into the cell wall, the *pss* operon products lead to the incorporation of NH_3^+ groups into the membrane.

These results lead to the hypothesis that modulation of surface charge, coordinated by σ^X, may function in resistance to CAMPs (Fig. 4). Using disk diffusion and MIC assays, we tested strains mutant in the *sigX, dlt* or *pssA* operons for sensitivity to a variety of CAMPs. The results, while not as dramatic as those reported for *S. aureus* (Peschel et al., 1999), demonstrate a two-to four-fold increase in sensitivity to various positively charged peptides in the *sigX* mutant, and in the *dlt pssA* double mutant strain. The *dlt* and *pssA* single mutants also had small, but reproducible, changes in sensitivity. These effects can be rationalized as a direct consequence of altered cell surface charge: D-alanylation acts to reduce the initial binding and accumulation of CAMPs near the cell membrane (Peschel *et al.*, 1999). Similarly, reduced surface charge and altered teichoic acids are associated with nisin resistance in the ruminal bacterium *Streptococcus bovis* (Mantovani and Russell, 2001). By analogy, an increase in PE content in the cytoplasmic membrane might also be predicted to increase CAMP resistance. Indeed, some *Listeria monocytogenes* strains selected for nisin resistance have increased PE contents in their membranes, although the genetic changes or mechanisms of resistance responsible for this effect are uncharacterized (Crandall and Montville, 1998).

3.3.1.5. Induction of the σ^X Regulon by Cell Wall Antibiotics. We have found that several antibiotics that target cell surface processes are strong inducers of the *sigX* operon and the σ^X regulon (M. Cao and J.D. Helmann submitted). The strongest inducers are inhibitors of PG biosynthesis and tunicamycin, a specific inhibitor of WTA synthesis (Pooley and Karamata, 2000). The genes for the biosynthetic enzymes controlling the synthesis of cell surface-associated polymers are generally not well characterized. However, recent studies have revealed a possible link between ECF σ factors and TA synthesis in the W23 strain of *B. subtilis*. When compared with the sequenced strain, 168, the W23 strain is found to carry a similar arrangement of TA biosynthesis genes organized into two divergent operons (the *tar* locus), with the additional presence of genes that specify synthesis of a ribitol-phosphate copolymer. The intergenic regulatory region for the W23 *tar* locus contains a 100 bp insertion relative to the sequenced *B. subtilis* 168 strain and this region carries an additional two promoter sites that resemble ECF σ recognition sequences, leading to the suggestion that ECF σ factors control TA biosynthesis in this strain (Minnig *et al.*, 2001).

3.3.2. σ^W

The σ^W regulon includes over 50 different genes activated by cell wall stress and is, to date, the most thoroughly studied ECF regulon in *B. subtilis* (Table 7). The *sigW* gene is contranscribed with an anti-σ factor gene, *rsiW*, from a single transcription start site (Huang *et al.*, 1998). This operon is positively autoregulated from a σ^W-dependent promoter, P_W, which is similar in sequence to P_X. However, there is no crosstalk: P_W is dependent on σ^W *in vivo* and is not well recognized by σ^X *in vitro* and, conversely, P_X is dependent on σ^X *in vivo* and is not well recognized by σ^W in *in vitro* transcription reactions (Huang *et al.*, 1998).

3.3.2.1. Overlap Between the σ^X and σ^W regulons. As discussed above, we initially sought to define the σ^X regulon by searching the *B. subtilis* genome for sequences similar to the autoregulatory site, P_X. In parallel with these studies,

Table 7 σ^W Regulon – promoter sites largely or exclusively dependent on σ^W.

Gene[a]	Operon size	Function[b]	Promoter sequence[c]	5′ UTR[d]
sigW	2	sigma	AAAATT**GAAAC**CTTTTGAAA-CGAAGCT**CGTA**TACATACAGACC	26
ydbS	2		AAGAAT**GAAAC**CTTTCTGTA-AAAGAGAC**GTA**TAAATAAC**G**ACG	24
yeaA	3		CTTTAT**GAAAC**CTTTGGCCC-TATTTATC**GTA**TTACGTA**A**AAAC	24
yxjI	1		GAGCCT**GAAAC**CTTTTCGCC-ACCTATCC**GTA**ATTTCATAC**A**AG	25
ysdB	1		AAAAGT**GAAAC**CTTTTTCTA-TGCTTTTC**GTA**TTACATCAG**A**TC	30
pbpE	2	PBP4*	ATATTT**GAAAC**GTTAGTAGG-TTAGTAAC**GTA**CAGAGATAT**G**GG	13
yjoB	1		TGGGAT**GAAAC**AAAATGCTA-TGTCAATC**GTA**TATATAAC**G**TTC	27
yteI	2	protease	TGAAGT**GAAAC**ATTTTTCAT-ATTGAATC**GTA**TAATGAGA**G**AGA	24
yknW	4	transport	AAACAT**GAAAC**TTTTTGATA-TCCTTCCC**GTA**CTATTTGTT**A**GA	23
ywrE	1		TTTTAT**GAAAC**GTTTTTCCT-TTTTCTTC**GTA**TAAAAGTAG**A**TT	15
yobJ	1		TTATAT**GAAAC**CTTTTTTAT-TTTAGAAC**GTA**TTAAAAGT**A**AAT	26
yfhL	2		ATGCAT**GAAAC**ATTTCTTCT-TTCTGCAC**GTA**ACAATGAGA**A**GG	15
yuaF	3		AATTTT**GAAAC**TTTTCCCGA-GGTGTCTC**GTA**TAAATGGTA**A**CG	44
yxzE	1		TGAAAT**GAAAC**CGGTCAGCG-TTTCATCC**GTA**TAACAGAT**A**TGG	24
yvlA	4		GAATTT**GAAAC**CTGAAGAGA-TTTTAAAC**GTA**TAAAATAAGT**A**AA	53
fosB	1	fosfomycin^R	CTGTAT**GAAAC**TTTCTTATGAAAAAAGTC**GTA**TATGTGGAT**G**AT	27
xpaC	>2		GAAGAT**GAAAC**TTGTTTAAG-GATTGAAC**GTA**GTAGATAATA**AT**	39
ythP	2	transport	GTTAAA**GAAAC**TTTTTTTAT-TCTATTTC**GTA**GTAAATTTT**GGA**	15
yoaF	1		AATAAT**GAAAC**CCGGAGTATGCCAAGCCC**GTA**TAACATAACATC	23
yoaG	1		ATTTTT**GAAAC**CTTTTATTAGCGCATTGCG**GTA**TAGCACGTTAGG	87
yqeZ	3	protease	AAAAAT**GAAAC**CTTTGATACATTTGTTAC**GTA**TGAAGAGAAGGC	38

Table 7 *Cont.*

Gene[a]	Operon size	Function[b]	Promoter sequence[c]	5′ UTR[d]
ybfO	1	detoxification	GAAAAG**GAAAC**TTTTTTCTATATCTATCTC**GTA**ATGACTAGAGAA	28
yozO	1		AATAT**TGAAAC**TTTTTTCTCTATATGTGC**GTA**TTACTTGATGAC	167
pspA (*ydjF*)	4 ?	phage shock protein	AAAAG**TGAAAC**TTTTAACGATAATAAATA**GTA**TATGTAACAAGG	35
		consensus	**tGAAAC** (16–17) **CGTA**	

[a] The first gene in the operon is listed.

[b] The known or predicted function of the encoded gene products is summarized.

[c] The −35 and −10 elements are in bold. The transcriptional start site is underlined where known.

[d] The 5′-untranslated region (UTR) is the distance from the transcription start to the translation start for the first gene in the operon.

we also began an investigation of a second ECF σ factor, σ^W. Serendipitously, these two lines of investigations were complementary: several of the promoters originally identified as candidate σ^X-dependent sites turned out also to be targets for σ^W. The first clue to this overlap came from primer extension mapping of transcripts corresponding to putative σ^X promoters using RNA preparations from various strains. For example, the transcripts corresponding to the *ywbN* and *yrhH* genes were easily detected in the *sigX* mutant sample, but not in wild-type or in a *sigX sigW* double mutant (Huang and Helmann, 1998). This suggested that these genes might be transcribed primarily by σ^W, and further suggested that the σ^W regulon might be expressed at a higher level in *sigX* mutant cells. Support for this model emerged when we purified the σ^W protein and demonstrated that *in vitro* both σ^X and σ^W could programme RNAP to recognize the *ywbN* and *yrhH* promoter elements.

The overlap between the σ^X and σ^W regulons appears to result from the similar, but nonidentical sequence recognition properties of the corresponding holoenzymes. As illustrated in Table 8, this difference can ultimately be traced to the –10 element. Based on a comparison of promoter sites exclusively recognized by σ^X (including the P_X autoregulatory site), exclusively recognized by σ^W (including the P_W autoregulatory site), and those recognized by both holoenzymes, we proposed a simple model for sequence discrimination (Huang *et al.*, 1998). According to this model, σ^X recognizes –10 elements with sequence CGaC, σ^W recognizes CGTa, and both can recognize CGTC (lower case reflects a noncritical base for recognition). To test this model, we mutated P_X and P_W and tested the effects on *in vitro* recognition by the σ^X and σ^W holoenzymes. Thus, we converted the σ^X-dependent site P_X (CGAC) into a σ^W-dependent promoter by mutation of two bases in the -10 element (CGTA). Conversely, we changed P_W (CGTA) into a promoter recognized by

Table 8   Mutations that switch promoter selectivity (Qiu and Helmann, 2001).

Promoter[a]	–35	–10	Activity
X	tgtaat**GtAAC**ttttcaagctattcata**CGAC**aaa		X >>> W[b]
X1	tgtaat**GtAAC**ttttcaagctattcata**CGTA**aaa		W >>> X[b]
X2	tgtaat**GtAAC**ttttcaagctattcata**CGTA**Taa		W >>> X
X3	tgtaat**GtAAC**ttttcaagctattcata**CGTC**aaa		X >>> W
W	aaaat**TGAAAC**cttttgaa-acgaagct**CGTA**tac		W >>> X
W2	aaaat**TGAAAC**cttttgaa-acgaagct**CGTC**tac		X > W
W3	aaaat**TGAAAC**cttttgaa-acgaagct**CGAC**tac		X >> W

[a] The sequence of the wild-type P_X (X) and three variants (X1, X2 and X3) is shown. The sequence of the wild-type (W) and two variants are also shown for P_W.

[b] Expression of X1 is σ^W-dependent *in vitro*, but it has very little activity *in vivo*.

either σ^X or σ^W with one base change (CGTC), and into an exclusively σ^X-dependent promoter with two base changes (CGAC) (Qiu and Helmann, 2001).

Despite the overlap between the σ^X and σ^W regulons, these two systems differ in several respects. First, σ^X regulated genes are usually turned on during late-logarithmic phase, while σ^W-dependent genes are not activated until early stationary phase under laboratory growth conditions (Huang *et al.*, 1997, 1998). Second, these two regulons respond to distinct but overlapping sets of chemical signals. For example, both regulons are induced by antibiotics active on the cell wall, but with differing efficiencies: vancomycin strongly induced the σ^W regulon while tunicamycin selectively induces the σ^X regulon (M. Cao and J.D. Helmann, submitted). Other cell-active antibiotics induce both regulons. The σ^W regulon is also strongly induced by alkali stress (Wiegert *et al.*, 2001), whereas the σ^X regulon is not. Finally, in a genetic analysis of transposon mutations that up-regulated either the σ^X or the σ^W regulon, all but one of the identified mutations activated only one of the two regulons (Turner and Helmann, 2000).

3.3.2.2. Defining the σ^W Regulon: (a) Promoter Consensus Searches. As a first approach to defining the σ^W regulon, we searched the genome for sites similar to the autoregulatory site, P_W. We were startled to find 16 perfect matches in positions suitable to function as promoter elements. Even more remarkable, all 16 promoters are largely, if not exclusively, σ^W-dependent *in vivo* (Huang *et al.*, 1999). Thus, unlike σ^X, σ^W appears to be required for the expression of its target genes. From this initial study, we concluded that σ^W controls a large regulon of at least 35 genes. By relaxing our search criteria to accommodate a 1 bp alteration in the spacer length, we identified five additional σ^W promoters including one preceding the *fosB* fosfomycin resistance gene (Cao *et al.*, 2001). While the consensus search approach was quite successful in identifying genes controlled by σ^W, this approach has several limitations (Table 2). Most obvious, any promoters that differ from the arbitrarily defined consensus sequence will be missed. One example is the *pspA(ydjF)* gene (Wiegert *et al.*, 2001), which is σ^W-dependent but differs in what was presumed to be an invariant base in the −10 region (Table 7).

3.3.2.3. Defining the σ^W Regulon: (b) Transcriptional Profiling. To complement the consensus search approach, we also defined the σ^W regulon by transcriptional profiling. A comparison of total RNA from wild-type and *sigW* mutant cells confirms the σ^W-dependence of many known targets and suggests several additional likely targets (Cao *et al.,* 2002). One limitation of this approach is that the level of expression of σ^W-dependent genes is quite low in the wild-type cells, so some operons are likely to be missed.

In general, transcriptional profiling experiments provide a much more powerful tool to defining regulons if strong inducing conditions can be identified.

The finding that the σ^W regulon is induced when cells are shifted to alkaline pH provides one such condition (Wiegert *et al.*, 2001). The alkaline shock stimulon includes 49 genes (of ~80 total) whose up-regulation is dependent on σ^W, either directly or indirectly. Significantly, we have observed that σ^W-dependent genes are the most strongly induced members of the vancomycin stimulon and this regulation requires the RsiW anti-σ factor (M. Cao *et al.*, submitted). Taken together, the consensus search and transcriptional profiling studies identify ~50 genes under σ^W control.

3.3.2.4 Defining the σ^W Regulon: (c) ROMA: Run-off Transcription – Macroarray Analysis. One difficulty with transcriptional profiling studies is that it is difficult to distinguish direct from indirect effects. To address this issue, Min Cao developed a complementary *in vitro* technique to identify the subset of genes that are directly dependent on σ^W for their expression (Cao *et al.*, 2002). In this experiment, total genomic DNA is fragmented with restriction enzymes and then used as a template for *in vitro* transcription using core RNA polymerase with and without a large molar excess of σ^W. The resulting radiolabelled RNA populations are then hybridized to cDNA macroarrays (Sigma-GenoSys) to identify those genes proximal to promoters active *in vitro*. In the case of σ^W, 44 strong signals are produced in response to the addition of σ^W and at least 22 of these correspond to promoters active *in vivo* (Cao *et al.*, 2002). The rest may result from a relaxed specificity of the holoenzyme under these *in vitro* conditions.

While the ROMA approach is technically challenging, and requires access to purified σ and RNA polymerase as well as DNA arrays, it provides a very useful complement to conventional transcriptional profiling. The presence of a signal in the ROMA experiment suggests that any effects seen in an *in vivo* experiment are likely to be direct, rather than indirect. Moreover, because of low levels of expression of many ECF regulons *in vivo*, particularly if inducing conditions are not known, *in vivo* transcriptional profiling experiments may fail to detect target genes (e.g. Manganelli *et al.*, 2001a). This problem is further compounded by the presence of additional promoter sequences and/or overlapping recognition among ECF σ regulons.

3.3.2.5. σ^W Controls an 'Antibiosis' Regulon. Our studies suggest that the σ^W regulon functions in both the synthesis of, and the defence against, antimicrobial compounds (Huang *et al.*, 1999; Cao *et al.*, 2001). Hence, we refer to the σ^W regulon as an 'antibiosis regulon'. Similarly, the cell envelope modifications orchestrated by σ^X may also be an adaptive response to the presence of antimicrobial agents. Historically, the term 'antibiosis' was coined to describe the ability of one organism to inhibit the growth of another. Ultimately, this term gave rise to the now much more familiar term 'antibiotic', used to refer to the chemicals mediating the growth inhibition.

The role of σ^W in controlling an antibiosis regulon is supported by two key observations: first, this regulon is strongly induced by antibiotics inhibiting cell wall biosynthesis and second, many of the gene products controlled by σ^W have known or putative roles in detoxification or antibiotic synthesis. In light of this conclusion, it is interesting to consider the observation of Wiegert *et al.* (2001), that the σ^W regulon accounts for a large portion of the alkali shock stimulon. We suggest that under the stress conditions used (a shift to pH 8.9) the growth-limiting event was an inability of the cell to synthesize cell wall. Since cytoplasmic pH is narrowly regulated over a range of extracellular pH conditions, it makes sense that the first essential enzymes to become inhibited at high pH would be those that function outside the cytoplasmic membrane and these are likely to be those involved in cell wall synthesis. Thus, it is probably not a coincidence that alkali stress and vancomycin stress are inducing the same target genes. Moreover, it is important to note that resistance to alkali shock itself is not likely to be the defining physiological role for σ^W, since a *sigW* mutant strain is no more sensitive to alkali shock than wild-type cells, and none of the target genes of the σ^W regulon have an obvious connection with pH homeostasis.

While most genes controlled by σ^W are of uncertain function, in several cases functional predictions can be made. For example, σ^W controls the transcription of PbpE (a low molecular weight penicillin-binding protein), the FosB fosfomycin resistance enzyme (Cao *et al.*, 2001), and several enzymes with possible functions in detoxification. These include a bromoperoxidase and an uncharacterized epoxide hydrolase (Huang *et al.*, 1999). In addition, σ^W directs the expression of several small, hydrophobic peptides that resemble bacteriocin precursors (Jack *et al.*, 1995) and an ABC transporter with similarity to bacteriocin export systems (Quentin *et al.*, 1999). Indeed, we have shown that *sigW* mutants display decreased expression of one or more bacteriocins (A. Gaballa and J.D. Helmann, unpublished results).

The σ^W regulon also includes two genes encoding signal-peptide peptidase homologues (*sppA* and *yqeZ*). The SppA family includes membrane-bound peptidases that have been proposed to function in the turnover of signal peptides left in the membrane by the action of leader peptidase (Suzuki *et al.*, 1987). However, *sppA* mutations do not greatly affect secretion, although some effects were noted (Bolhuis *et al.*, 1999), and the regulation of these two homologues by σ^W suggests another possible role. We speculate that SppA and YqeZ may function to cleave bacteriocins and thereby prevent their accumulation within the membrane to toxic levels. This could be either a defence mechanism or an immunity mechanism. In support of this idea, an operon encoding an SppA homologue has been implicated in immunity to enterocin A, a bacteriocin from *Enterococcus faecalis* (O'Keeffe *et al.*, 1999).

3.3.2.6. Transition State Regulators. One group of proteins likely to affect the activity of σ^X and σ^W are the transition state regulators: AbrB, Abh and Spo0A

(Strauch and Hoch, 1993). Bacilli produce numerous antibiotics and Spo0A, the key regulator of sporulation, is required for their synthesis (Schaeffer, 1969; Marahiel *et al.*, 1993). However, this Spo0A effect can be bypassed by mutations in *abrB*. This is explained by the observation that activation of Spo0A leads to repression of the AbrB repressor and thereby leads to derepression of antibiotic synthesis.

Several lines of evidence suggest that AbrB, and other transition state regulators, also affect the σ^W and σ^X regulons (Fig. 5). For example, AbrB represses the σ^W-dependent *pbpE* gene (Strauch, 1995), and has recently been found to repress both the *sigW* operon and several other σ^W target genes (Qian and Strauch, personal communication). In addition, Abh (an AbrB homologue)

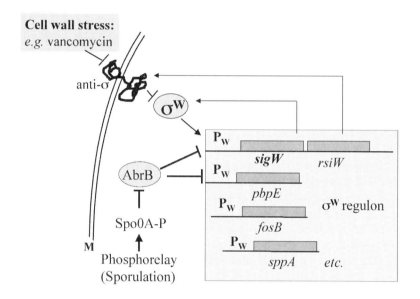

Figure 5   The *B. subtilis* σ^W regulon. In response to cell wall stress, such as that elicited by inhibitors of peptidoglycan biosynthesis, the σ^W regulon is activated. Activation of σ^W leads to expression of at least one new penicillin-binding protein (PbpE), a fosfomycin resistance protein (FosB), signal peptide peptidase (SppA) and a large number of other proteins (Table 7). Many of these proteins appear to be involved in either the detoxification of antimicrobials or the synthesis and secretion of bacteriocins. Both the *sigW* operon itself and several target genes are repressed by the 'transition state regulator' AbrB. AbrB functions to repress a large number of post-exponential functions (including antibiotic synthesis). When cells enter stationary phase under appropriate conditions the phosphorelay controlling the initiation of sporulation is activated and the resulting Spo0A-phosphate response regulator represses *abrB* expression, leading to derepression of these post-exponential functions. Thus, expression of the σ^W regulon is intimately connected to the control circuitry for sporulation and other post-exponential phase phenomena in *B. subtilis*.

is transcribed by σ^X (Huang and Helmann, 1998). The relationships between these transition state regulators, and the regulons controlled by σ^X and σ^W are currently under investigation.

3.3.3. σ^M

The *sigM* ECF σ factor was identified as a result of the *B. subtilis* genome sequencing project and the corresponding mutant attracted attention when it was found to have an apparent defect in spore outgrowth. Horsburgh and Moir (1999) have demonstrated that *sigM* is optimally expressed in early logarithmic phase cells from two promoter sites: one under σ^A control, and the second an autoregulatory site recognized by the σ^M holoenzyme. Expression of σ^M is up-regulated ~two-fold by growth in high salt, and the *sigM* mutant strain fails to grow in medium containing high levels of salt. This growth defect and the consequent formation of swollen and abnormally shaped cells, may be due to defects in cell wall biosynthesis, but biochemical analysis did not reveal gross differences in peptidoglycan structure in the mutant strain (Horsburgh and Moir, 1999).

The activity of σ^M, like many other ECF σ factors, is negatively regulated by two downstream genes. Using the pMUTIN plasmid vector, it was possible to regulate the level of expression of the downstream *yhdL* and *yhdK* genes and demonstrate that decreased expression of these genes leads to increased activity of σ^M. However, these genes cannot be disrupted, suggesting that the resulting up-regulation of the σ^M regulon impairs viability (Horsburgh and Moir, 1999).

In ongoing studies, the Moir and Hecker laboratories have used DNA macroarray and *lac* fusion analyses to identify genes that are up-regulated in response to the induction of σ^M synthesis (Thackray *et al.*, 2001). A preliminary assessment of the σ^M regulon reveals induction of the *yacK* and *yacL* genes (which may also be transcribed as part of the heat inducible *ctsR* operon), *radC(ysxA)*, *ydcF*, *ypbG*, *yjbC*, *yjbD* and *ywoA*. Interestingly, some of these genes are annotated as having a likely role in DNA repair. This is intriguing since it has been noted that desiccation, which is related to osmotic stress, may lead to DNA damage. Indeed, the extreme resilience of *Deinococcus radiodurans* against DNA damage is hypothesized to have evolved as a defence against desiccation stress (Mattimore and Battista, 1996).

Many of the newly identified members of the σ^M regulon are associated with candidate promoter elements resembling the well characterized *sigM* autoregulatory site. However, the *sigM* autoregulatory site has a distinctive -10 sequence, CGTG, not shared by most of these target operons. Most σ^M target genes have candidate -10 elements with sequence CGTC, similar to that noted above as being potentially recognized by either σ^X or σ^W. Indeed, two of the identified σ^M targets (*yjbC* and *ywoA*) are also known to be recognized by σ^W and/or σ^X (Table 6).

3.3.4. Other ECF σ Factors

The *B. subtilis* genome encodes four other ECF σ factors: σ^V, σ^Y, σ^Z and σ^{ylaC}. Although mutant strains have been generated for each of these factors, few clues have yet emerged as to their functions. At least for three of these σ factors, candidate autoregulatory promoters can be identified upstream of the σ factor operon. These sites have the characteristic ECF σ factor −35 element, with the conserved 'AAC' motif, followed by a candidate −10 element similar to that noted above for other ECF σ factors in *B. subtilis*. Remarkably, all of these sequences are very similar to each other. This raises several important questions: (1) are these in fact autoregulatory sites; (2) what is the promoter specificity for each σ and how are they distinct; (3) to what extent do the ECF σ factor regulons overlap; and (4) do any of the ECF σ factors regulate each other? Results from DNA microarray studies support the idea that these σ factors are all autoregulated and have led to lists of candidate target genes (Y. Fujita, personal communication). As noted above for the σ^M regulon, at least some of these target genes have been previously identified as targets of other ECF σ factors. If, as this suggests, these σ factors overlap in their promoter selectivity, why do they not contribute to gene expression of the promoters we have studied to date? The answer may be that most ECF σ factors are synthesized as part of a 'two component' regulatory system: the ECF σ and the cognate anti-σ. Under most conditions, these systems are essentially inactive. Only upon receiving the proper stimulus will σ be released and become active. Much additional work will be required to understand the sequence differences that determine which genes are subject to control by which σ factor and to define the extent of regulon overlap among these seven ECF σ factors.

3.4. *Streptomyces coelicolor*

Unravelling the complexities of the many ECF σ factor regulons in *S. coelicolor* is a truly daunting task. To date, most studies have focused on just three of the ~50 ECF σ factors in this organism: σ^E, σ^R and σ^{BldN} (Paget *et al.*, 2002). The σ^E regulon includes an operon involved in cell wall biosynthesis and mutants display an increased sensitivity to cell wall perturbants. The σ^R regulon responds to oxidative stress conditions that lead to the formation of disulphide bonds in the cytoplasm (more accurately referred to as disulphide stress). The σ^{BldN} regulator participates in the sporulation pathway and mutants in this σ are defective for the formation of aerial hyphae (bald phenotype). Analysis of these systems has been very productive: σ^E is unusual in that its expression is activated by a two-component regulatory system, studies of σ^R led to the discovery of a family of zinc-containing anti-σ factors (the ZAS family), while the σ^{BldN} system provides us with the first example of an ECF σ factor regulated by proteolytic processing from an inactive precursor.

3.4.1. σ^E

The biochemical activity referred to as σ^E was first detected when RNA polymerase fractions were analysed for the ability to recognize several promoter sites upstream of the *dagA* agarase gene (Buttner *et al.*, 1988). The fraction that activated the P2 promoter was found to contain a 20 kDa protein designated σ^E. As noted above, the cloning of the *sigE* gene was a key event in the original discovery of the ECF family of regulators (Lonetto *et al.*, 1994).

Isolation of the *sigE* gene allowed the construction of *sigE* null mutant strains. Strains lacking σ^E activity have several dramatic phenotypes, including the overproduction of the blue-pigmented antibiotic actinorhodin, altered colony morphology and poor sporulation. In addition, the mutants display an increased sensitivity to cell wall hydrolytic enzymes, including both muramidases that cleave the glycan backbone (such as lysozyme) and amidases that cleave the peptide cross-links. These phenotypes can all be suppressed by millimolar levels of Mg(II), which is known to stabilize the cell envelope (Paget *et al.*, 1999b).

The *sigE* gene is part of a four gene operon that also encodes a predicted membrane protein (CseA), a response regulator (CseB), and a membrane-bound histidine protein kinase (CseC) (Paget *et al.*, 1999a) (Fig. 6). Expression of this operon does not appear to be regulated by σ^E: instead, the major *sigE* operon promoter shows similarities to the promoter for *whiG*, which is under the control of an as yet uncharacterized form of holoenzyme. Expression of *sigE* requires activation by the CseB response regulator. Since this operon is expressed in culture, the signals perceived by the CseC sensor kinase are presumably present in the laboratory growth conditions used. The CseA protein appears to play a negative role in modulating expression of the *sigE* operon: an in-frame *cseA* deletion mutant has increased expression of the *sigE* operon. While the biochemical activity of CseA is not yet known, it could affect the CseC-CseB two component system that is required for activation of *sigE* operon transcription (Paget *et al.*, 2002).

The σ^E regulon has not yet been extensively characterized. To date, two promoter sites recognized by σ^E both *in vitro* and *in vivo* have been documented. The first, hrdDp1 contributes to the expression of the group 2 σ factor, HrdD. However, since the role of HrdD is as yet unknown (Paget *et al.*, 2002), the significance of this regulation is unclear. The second target operon includes 12 genes thought to specify cell wall glycan synthesis and has been designated *cwg*. While the precise role of the σ^E regulon has yet to be determined, an important clue comes from the recent observation that the *cwg*

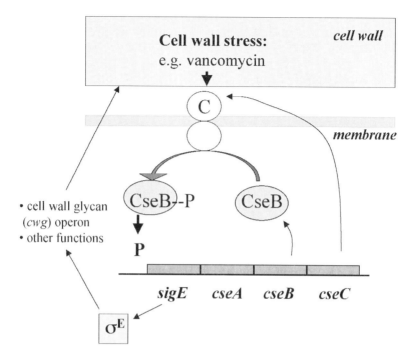

Figure 6 The *S. coelicolor* σ^E regulon. In response to cell wall stress (elicited by inhibitors of PG biosynthesis or by treatment with lysozyme), the CseC histidine protein kinase phosphorylates the CseB response regulator (note that in this case, *cse* stand for control *of* sigE). This protein, in turn, activates transcription of the *sigE* operon leading to the expression of σ^E protein. Note that there is no known anti-σ in this regulon. The σ^E regulon is not completely characterized, but includes at least one large operon involved in cell wall functions (the *cwg* operon). The functions of the Cwg proteins, and how they may act to restore cell wall homeostasis, are not yet understood.

genes are induced by vancomycin (Paget *et al.*, 2002). Thus, the σ^E regulon may function to coordinate responses to cell wall stress, much as described above for the σ^X and σ^W regulons in *B. subtilis*. The relationship between *cwg* operon expression and the phenotypes of a *sigE* mutant are currently under investigation (Paget *et al.*, 2002). Unexpectedly, the *sigE* mutant is unaffected in utilization of the *dagA* P2 promoter *in vivo*, despite the fact that it was transcription from this site that led to the original isolation of σ^E protein. This suggests that other σ factors can also recognize this site in addition to, or even instead of, σ^E.

While a *sigE* null mutant of *S. coelicolor* overproduces actinorhodin, in *S. antibioticus* σ^E plays a positive regulatory role in antibiotic biosynthesis (Jones *et al.*, 1997). Sequencing of the gene for phenoxazinone synthase, which catalyses the penultimate step in actinomycin biosynthesis, revealed a promoter sequence closely resembling known target promoters for σ^E. Purified σ^E holoenzyme indeed recognizes this site *in vitro*, but curiously, is dispensable *in vivo*. Nevertheless, a *sigE* mutant fails to produce actinomycin. These results suggest that σ^E does play an essential positive regulatory role in actinomycin synthesis, but this role must extend beyond recognition of the *phs* promoter region. In fact, the *sigE* null mutant has greatly diminished activity of the key biosynthetic enzyme actinomycin synthase I. Unlike the situation in *S. coelicolor*, the *S. antibioticus sigE* null mutant was unaffected in colony size or development. Nevertheless, the identity and arrangement of the neighbouring genes suggests that these are in fact orthologous σ factors.

3.4.2 . σ^R

Biochemical fractionation of purified RNA polymerase allowed Kang *et al.* (1997) to identify a 31-kDa σ that was enriched in stationary phase cells. When these fractions were incubated with various purified DNA templates, this σ was found to activate expression from a second promoter preceeding the *hrdD* gene, hrdDP2.

To begin to address the physiological role of σ^R, Paget *et al.* (1998) determined the amino-terminal sequence of the purified protein and designed primers to identify the corresponding gene. By constructing a null mutant, they were able to demonstrate that σ^R plays a key role in regulation of oxidative stress responses: the mutant strain is sensitive to both superoxide generators and to diamide, a chemical that oxidizes thiols in the cell cytoplasm leading to the formation of disulphides. Disulphide bonds are normally not present in proteins in the reducing environment of the cytoplasm and their formation can inhibit enzyme activity (Aslund and Beckwith, 1999). The resulting type of oxidative stress has been named 'disulphide stress'.

In response to oxidizing conditions, σ^R activates the transcription of the *trxBA* operon encoding both the thiol reductant thioredoxin and thioredoxin reductase (Paget *et al.*, 1998). Together, these proteins allow reducing equivalents, in the form of NADPH, to be efficiently used to reduce oxidized thiols (Carmel-Harel and Storz, 2000). In addition to the *trxBA* operon, σ^R also activates its own synthesis from one of two promoters that precede the *sigR* operon. These observations lead to a simple model (Fig. 7), in which disulphide stress activates σ^R-dependent transcription to restore the intracellular redox balance (Paget *et al.*, 1998).

The *sigR* operon contains two genes: *sigR* and *rsrA* (regulator of *sigR*). As noted for other ECF σ factors, the second gene in the operon encodes a negative regulator functioning as a specific anti-σ factor. However, RsrA is unusual

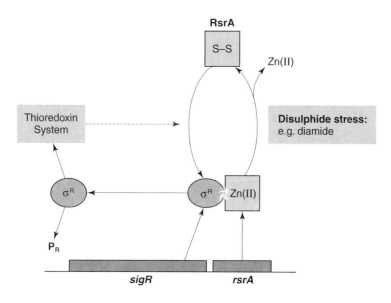

Figure 7   The *S. coelicolor* σ^R regulon. The *sigR* gene is cotranscribed with the regulator of *sigR* (*rsrA*) anti-σ factor. In the absence of stress, the cytoplasmic RsrA protein binds and inactivates σ^R. In response to diamide treatment, which oxidizes thiols in the cytoplasm, the RsrA zinc-metalloprotein is oxidized, leading to the release of σ^R. This oxidation is accompanied by the release of Zn(II) and the formation of at least one disulphide bond in RsrA. The σ^R regulon includes thioredoxin and thioredoxin reductase, together with numerous other genes (Table 9; see text). Thioredoxin acts to re-reduce oxidized thiols, including those in the RsrA anti-σ (dotted arrow), and thereby helps to restore intracellular redox balance.

in that it is a soluble, rather than a membrane-localized, protein. In a series of elegant collaborative experiments, the Roe and Buttner laboratories have established that RsrA is a small, zinc-containing protein that directly serves as the sensor of disulphide stress (Kang *et al.*, 1999; Paget *et al.*, 2001a). In its reduced form, RsrA forms a 1:1 complex with σ^R and prevents transcription initiation. When exposed to a thiol oxidant, such as diamide, at least one disulphide bond is formed in RsrA and the inhibition of σ^R is relieved. The bound Zn(II) ion is lost upon RsrA oxidation, suggesting that at least one of the cysteine residues that is oxidized also functions to coordinate Zn(II) ion. Altogether, RsrA contains seven Cys residues, but only three are required for activity. These appear to function as Zn(II) ligands and/or components of the thiol-disulphide redox switch.

   To further define the σ^R regulon, Paget *et al.* used a promoter consensus search strategy (Paget *et al.*, 2001b). By searching the *S. coelicolor* genome for similarities to the sequence GGAAT 18 bp GTT, they identified 34 candidate promoters located upstream of target genes. Remarkably, 30 of these are

Table 9   The σ^R Regulon (adapted from Paget *et al.*, 2001b).

σ^R target genes	σ^R target promoter sequence	(distance to translation start)
Subclass A	-35 -10	
8E4A.04c	GCAGCCG**GGAAT**GGGCGGGCCGGTCGTTC**GGTT**GCCGGGTTGAG	(+20)
E87.13	CGGGTGA**GGAAT**GGTGGACGGTCGGTGGC**GGTT**GCATCCACTGT	(+35)
folE	CAGGCCC**GGAAT**GTATGCCGCACCCCCCA**GGTT**CTAGCCTTGGG	(+54)
6G10.34c	CCGTCCC**GGAAT**GAATCCGCGGTCCCGCC**GGTT**GGAACCGTCGG	(+35)
rpmE	ACACCGG**GGAAT**AGTTCACCCCGCCCCCC**GGTT**TTGGGGGATGG	(+95)
E22.4	ATCCCGC**GGAAT**AGGTCACTATGGACCGTC**GTT**AGCACTCATCG	(+25)
E19A.11c	CGGACCG**GGAAT**CGATACGATGAGCCCAT**GGTT**CTCCTCGACGT	(+179)
7H2.09c	CGCACAC**GGAAT**AGCGGGTCCGGCGCGTGC**GTT**GATCCCTTGAA	(+20)
pepN	GGCCGCG**GGAAT**GGGTCGTGCCGTGTGGGC**GTT**CATGCGTCAAG	(+105)
7H2.11c	TCTCCCG**GGAAT**GCCCCACCCCGCAAGGGT**GTT**GTGACGTACGA	(+24)
sigR	ACGCCTG**GGAAT**GACCGGCGGACGGCCACT**GTT**GTGCCGTTCGT	(+214)
ssrA	GCCGCCG**GGAAT**AGGCTGGCACGCGTCCGT**GTT**GGTCACGTACG	(+62)
3D11.22	CGGCGGC**GGAAT**AGCCCGCCCGGCCCGGA**GGTT**GCTGCTGCTCA	(+53)
trxC	CGTGCGC**GGAAT**ACCCACCAGGTCGGCAGT**GTT**GTGCCGACGGC	(+34)
E20.23	TGGCCGG**GGAAT**CAGCGGTGGGTACCCGT**GGTT**TGGCAGGCAGA	(+43)
M1.15	CAGTAGT**GGAAT**GTTCAACCGTGCGGCGT**GGTT**GCAGGGCCCGA	(+13)
6A5.08	GTTCTCA**GGAAT**CTTGAAGTCAATCGGTGT**GTT**GTTGTGATGCG	(+242)
Subclass B		
6D7.18c	CGGAACG**GGAAT**CTTTACCGCCGCCCGGA**CGTT**GACCGGATGAC	(+51)
23B6.11c	CGCGACG**GGAAT**GGTCGTGCCGCCACGGA**CGTT**CACTCCGTGCG	(+52)
relA	GCCCGGA**GGAAT**CCGGCGCCCGCCCCAGG**CGTT**GCAGGGGTTGA	(+113)
4B5.01c	ACGGTCA**GGAAT**GCGTCCGGGCTCGTGATC**GTT**GACGATATGAG	(+52)
hrdD	CAGGTTG**GGAAT**TCTGTCCGGATTCCAGT**CGTT**GTTTCCATCGG	(+35)
hflX	GGCCCCG**GGAAT**CTCCCCGGTGGTTCGCA**CGTT**GTACACGATGT	(+50)
trxBA	CACAGCG**GGAAT**GCGCGCCCCGTAGGATC**CGTT**AGGTGGGGTGA	(+62)
cinA	CCGCCGG**GGAAT**CGCCGAACGCAGAGCGG**CGTT**GAAGGAGACGG	(+5)
E25.24c	CGTGAGG**GGAAT**CACTGATCGGCGTAGCGT**GTT**GAGTCGCAAGG	(+95)
P8.26c	AGCGTAG**GGAAT**GTTTGCAGAGGGCATCC**CGTT**GACCACTACGG	(+76)
E9.22	ATTCCTG**GGAAT**CATATTCCTTGGAATACG**GTT**GGCCCGAACAG	(+180)
guaB	TGATCGT**GGAAT**GAGCCACGTCCACTTCC**CGTT**AGCATGGACAT	(+34)
moeB	GTGTCTC**GGAAT**GAAAAAGGCCATGGCAGT**GTT**ACGTGGAGAAC	(+34)

functional sites that are induced by diamide in a σ^R-dependent manner (Table 9). Approximately one-half of the σ^R target genes are also transcribed from an additional, σ^R-independent promoter site. In addition, many of the σ^R promoters are still active even in a *sigR* mutant strain suggesting that another σ factor, presumably a member of the ECF subfamily, can also recognize these sites. The differences between the promoter sites recognized exclusively by σ^R (class A) and those with residual transcription (class B) are not entirely clear, but a correlation with the sequence of the −10 region is apparent: class A sites usually contain the sequence GGTT, while most class B sites contain CGTT (Table 9).

While these genes are unlikely to represent the entire σ^R regulon, they nevertheless provide a very informative overview of the types of functions that comprise this regulon. As expected, several of the regulated operons participate in thiol metabolism including genes likely to be involved in either cysteine biosynthesis or synthesis of the low molecular weight thiol mycothiol (functionally analogous to glutathione in many other bacteria; Newton *et al.*, 1996). Other genes in the σ^R regulon are likely to be involved in modulating translation during disulphide stress (Paget *et al.*, 2001b). These include *rpmE, relA* and *ssrA*. The *rpmE* gene encodes ribosomal protein L31 which contains a Cys-x-x-Cys motif. This leads to the speculation that this protein serves to sense redox stress and slow or pause translation until redox balance can be restored. Similarly, induction of RelA will lead to the synthesis of ppGpp which acts as a global modulator of growth rate. Finally, *ssrA* encodes a stable RNA that functions to rescue ribosomes stalled due to the lack of a termination codon (for example, at the 3′-end of incomplete mRNAs; reviewed in Karzai *et al.*, 2000). In addition to modulating translational capacity, induction of the σ^R regulon may also affect transcription: another σ^R target encodes a small RNA polymerase associated protein (RapA) of as yet unknown function. As discussed in more detail below, σ^R is likely to be the orthologue of *Mycobacterium tuberculosis* σ^H which also controls a large, diamide-inducible oxidative stress response (Manganelli *et al.*, 2001a; Raman *et al.*, 2001). It will be interesting to learn how the regulons controlled by these functionally similar systems are related, and how they might differ.

3.4.3. σ^{BldN}

The gene encoding σ^{BldN} was discovered in a screen for mutations that block morphological differentiation (Bibb *et al.*, 2000). Originally classified as a white mutant (*whiN*), these mutations blocked the formation of spores, but not aerial hyphae. In contrast, subsequent studies of null mutations in this locus demonstrated a failure to form aerial hyphae; thus, *whiN* was reclassified as a bald mutant, *bldN*.

When the *bldN* gene was sequenced it was found to encode a member of the ECF σ family (Bibb *et al.*, 2000). Subsequent work has demonstrated that *bldN* is expressed during development and this regulation involves, at least in part, regulation by a DNA-binding protein (BldD) that represses transcription during growth (Elliot *et al.*, 2001). Intriguingly, BldD also represses expression of another σ factor required for development, σ^{WhiG}.

The downstream targets of σ^{BldN} action are not yet well characterized. One promoter that depends on σ^{BldN} is the bldMp1 site that contributes to expression of the BldM response regulator. Several upstream regulators have also been defined. The expression of *bldN* depends on several other genetically defined *bld* loci, but their roles in regulating *bldN* are not yet well defined.

One of the most intriguing features of σ^{BldN} is the presence of a large (86 amino acid) N-terminal extension. It is proposed that σ^{BldN}, like some of the late σ factors regulating sporulation in *B. subtilis* (Kroos *et al.*, 1999), is synthesized as an inactive pro-protein that must be proteolytically processed to become active (Paget *et al.*, 2002). The factors and signals that control the processing of σ^{BldN} have yet to be elucidated.

The σ^{BldN} orthologue in *S. griseus* (σ^{AdsA}) is also implicated in sporulation (Yamazaki *et al.*, 2000). However, in this case we have a somewhat clearer picture of some of the upstream regulatory factors. In this organism expression of aerial hyphae and the antibiotic streptomycin is regulated in response to a γ-butyrolactone signalling molecule known as A-factor (Horinouchi, 1999). A specific A-factor-dependent transcriptional activator (AdpA) coordinates the responses of the cell to this global regulator. One of the direct targets for A-factor action is the *bldN* orthologue *adsA* gene (AdpA-dependent sigma factor), while another target is a gene (*strR*) that functions as an activator of the streptomycin biosynthesis gene cluster (Yamazaki *et al.*, 2000).

3.4.4. Other ECF Sigma Factors

Deconvoluting the myriad ECF-σ factor regulons in *S. coelicolor* is likely to keep many researchers busy for a long time to come. In addition to the three regulons described above, analyses of only two other ECF σ factors have been reported.

The first, σ^U, was identified during a screen for mutations affecting development (Gehring *et al.*, 2001). However, unlike *bldN*, in this case the developmentally affected mutant had a transposon insertion in a putative anti-σ gene, *rsuA*, and the developmental phenotype was a result of up-regulated activity of an ECF σ factor, rather than loss of activity. Indeed, a *sigU* mutant is able to differentiate normally, suggesting that the developmental phenotype in this case is a consequence of its uncontrolled activity, perhaps by competing for core enzyme with other σ factors required for development.

The second, σ^T, is associated with a downstream gene that encodes a putative anti-σ factor similar to RsrA, but likely to be membrane-associated (J-H. Roe, personal communication). Like σ^R, the σ^T regulon functions in defending the cell against oxidative stress; *sigT* mutants are also sensitive to diamide. This suggests that both of these regulons include functions that are essential for optimal resistance to diamide and a failure to express either one can lead to sensitivity. Thus, even though there may be regulon overlap, these two regulons are not redundant.

3.5. *Mycobacterium tuberculosis*

The re-emergence of tuberculosis as a clinically important disease during the last couple of decades has led to a resurgence of interest in both *M. tuberculosis* and its more rapidly growing relative, *M. smegmatis*. Genome sequencing reveals 10 ECF σ factors encoded in the *M. tuberculosis* genome (Cole *et al.*, 1998).

Using molecular beacons and real-time PCR as a tool to quantify mRNA levels, Manganelli *et al.* (1999) reported a survey of the expression of 10 σ factor genes, including seven members of the ECF subfamily, in response to various environmental stresses. During growth, *sigC* mRNA is even more abundant than that for the primary (group 1) σ factor gene, *sigA*. The mRNAs for *sigD, sigE* and *sigM* are also fairly abundant, while *sigF, sigH* and *sigI* messages are present at only low levels. The *sigG* mRNA could not be detected under these growth conditions. Both *sigE* and *sigH* were found to be induced by heat shock, and *sigE* was additionally induced by the detergent, sodium dodecyl sulphate (SDS). Both of these σ factors are also induced during phagocytosis by macrophages, suggesting that these stress responses may be important during infection (Graham and Clark-Curtiss, 1999; Jensen-Cain and Quinn, 2001). Most studies to date have focused on these two ECF σ factors.

3.5.1. σ^E

The first ECF σ factor to attract experimental scrutiny in *M. tuberculosis* was σ^E. This gene was originally identified as part of the *M. leprae* genome sequencing project and the corresponding gene was amplified from several Mycobacterial species (Wu *et al.*, 1997). The encoded σ^E proteins are >90% identical in *M. tuberculosis, M. smegmatis* and *M. avium*. Construction of a *sigE* null mutant in *M. smegmatis* revealed an increased sensitivity to oxidative stress, heat shock, low pH and detergent (SDS) stress (Wu *et al.*, 1997). Moreover, it was found that wild-type cells display an adaptive response to hydrogen peroxide that is lacking in a *sigE* mutant strain.

Studies of the role of σ^E in *M. tuberculosis* have established that it is also important in resistance to both heat, oxidative and detergent stresses, and the mutant strain is more sensitive to killing by macrophages (Manganelli *et al.*, 2001b). When the expression of various σ factors was analysed in the *sigE* mutant strain it was found that *sigE* mRNA was itself slightly elevated, suggesting that this gene may not be autoregulated. In addition, the level of mRNA for the group 2 σ factor, σ^B, was reduced about 10-fold.

To define the σ^E regulon further, DNA microarray studies were performed to compare the mRNA populations of wild-type and *sigE* mutant strains both during exponential growth and after imposition of detergent stress. Thirty-eight genes were expressed at levels at least two-fold lower in the *sigE* mutant than in the wild-type strain, including *sigB*. Of 62 genes induced by SDS in wild-type, 23 were not significantly induced in the *sigE* mutant strain. Of these 23, at least 10 operons (13 genes) are associated with putative promoter sites resembling other sites known to be recognized by ECF σ factors (Table 10).

Table 10 Putative promoters of the *M. tuberculosis* σ^E regulon[a] (adapted from Manganelli *et al.*, 2001b).

Gene	−35	−10	distance to start
σ^R consensus	GGGAAT-----------------cGTTg		
sigB	GGgaactcaaagtccggctttgtcGTTa		64
hsp	GGaacttaagcgtggtcgactcagGTTc		94
fadB2	GGacccgaattcgtcgaccaccgcGTTg		142
htpX	GGaaccagcggatacccctcaggcGTTg		77
Rv1057	GGcagtcggcgtcgcacagcacgGTTg		134
Rv1130	GGgccgctggtcattctccaactgGTTg		188
dinG	GGcacggcagctggtggcatccggGTTg		127
Rv2053c	GGaactcgggcctagcccgcagacGTTg		38
Rv2745c	GGaaccaaccccaccgccggcggcGTTc		62

[a] Genes shown to be induced by SDS in a σ^E-dependent manner were analysed for the presence of possible promoter elements resembling the *S. coelicolor* σ^R consensus (line 1).

3.5.2. σ^H

The *sigH* gene was originally identified as a paralogue of *sigE* by searching the *M. tuberculosis* genome sequence (Fernandes *et al.*, 1999). This gene is conserved in the faster growing species, *M. smegmatis*, and encodes a protein with 89% identity to *M. tuberculosis* σ^H. Analysis of *sigH* transcription by primer extension start site mapping identifies two promoters that are active under heat shock conditions (Fernandes *et al.*, 1999), consistent with the results of RT-PCR measurements of RNA levels in stressed cells of *M. tuberculosis* (Manganelli *et al.*, 1999). Construction of a *sigH* mutant in *M. smegmatis* revealed no significant difference from wild-type in the survival of a variety of stress conditions including heat shock, cold shock, acid stress and hydrogen peroxide treatment. However, the *sigH* mutant is significantly more sensitive to organic hydroperoxides. Reasoning that σ^H and σ^E might be partially redundant in function, Fernandes *et al.* (1999) created a *sigH sigE* double mutant. This strain was more sensitive to heat shock and organic peroxide stress than either single mutant.

In the closely related species *M. tuberculosis,* a *sigH* null mutant is more sensitive than wild-type to a variety of stresses including heat shock, hydrogen peroxide, organic peroxide and diamide, but not to superoxide generators. However, unlike *sigE* (Manganelli *et al.*, 2001b), disruption of *sigH* does not adversely affect the ability of *M. tuberculosis* to survive and multiply inside macrophages (Manganelli *et al.*, 2001a).

Since σ^H is the closest homologue of *S. coelicolor* σ^R it seems likely that these two σ factors may be functional orthologues. To test this hypothesis, microarray analyses were used to measure global mRNA profiles of wild-type and *sigH* null mutant strains after exposure to the thiol oxidizing agent diamide (Manganelli *et al.*, 2001a). Of the 48 strongly induced genes in the wild-type, 39 were no longer induced in the *sigH* null mutant. Therefore, σ^H is a key regulator of the diamide (disulphide stress) stimulon in *M. tuberculosis*, as it is in *S. coelicolor*. In contrast, comparison of the mRNA profiles in non-stressed cells failed to reveal a significant effect of the *sigH* mutation, indicating that this regulon is not induced during logarithmic growth of non-stressed cells (Manganelli *et al.*, 2001a).

The σ^H regulon includes at least two heat shock proteins (Hsp and ClpB) and several transcription factors including σ^B and σ^E as well as σ^H itself. Interestingly, the *sigB* promoter recognized by σ^H is the same as that controlled by σ^E, demonstrating that these two σ factors have overlapping promoter selectivity (Manganelli *et al.*, 2001a; Raman *et al.*, 2001). In addition, σ^H controls an operon encoding thioredoxin and thioredoxin reductase, as well as a set of genes involved in cysteine biosynthesis (Manganelli *et al.*, 2001a; Raman *et al.*,

Table 11 Putative promoters of the *M. tuberculosis* σ^H regulon[a] (Manganelli *et al.*, 2001a).

Gene	−35 spacer −10
σ^R *consensus*	**GGGAAT**-17-**cGTTg**
hsp	gt**GGAAc**-18-**gGTTc**t
Rv0355c	gc**GGAAc**-21-t**GTTg**g
clpB	gc**GGAAc**-17-**cGTTg**a
Rv0991c	gg**GGAA**t-17-**gGTTc**t
Rv1334	cc**GGAA**t-17-t**GTT**tc
trxB	ct**GGAAc**-18-**cGTTg**a
Rv2466c	cg**GGAAc**-17-**gGTTg**t
Rv2699c	cg**GGc**A**c**-17-**cGTTc**g
sigB	t**gGGAAc**-17-**cGTT**aa
Rv3054c	t**cGGAA**t-17-**gGTTg**g
Rv3463	cc**GGAA**t-18-**gGTT**ag
moeZ	g**gGGAAg**-18-**cGTTg**a
sigH	ct**GGAA**a-18-**gGTTg**a
trxB2	cg**GGAAc**-17-**cGTTc**g

[a] Genes shown to be induced by diamide in a σ^H-dependent manner were analysed for the presence of possible promoter elements resembling the *S. coelicolor* σ^R consensus (line 1).

2001). Control of these latter functions by σ^H is consistent with their control in *S. coelicolor* by σ^R and further emphasizes that these are orthologous systems.

Inspection of the promoter regions for the 27 putative transcription units induced by diamide stress in a σ^H-dependent manner reveals candidate ECF-type promoter elements for 14 operons (Table 11). At least one of these sites (*sigB*) has been validated experimentally (Manganelli *et al.*, 2001a; Raman *et al.*, 2001). Thus, the σ^H regulon probably includes these 14 operons as direct targets and a roughly equal number that are either controlled by promoters not easily identified by sequence inspection or which are indirect targets for σ^H (Manganelli *et al.*, 2001a). It is intriguing that the σ^E- and σ^H-dependent promoters described to date (Tables 10 and 11) appear to be very similar in sequence. The basis for promoter discrimination between these two σ factors is not yet clear.

3.6. *Pseudomonas aeruginosa*

ECF σ factors have been extensively studied in *Pseudomonas* spp. where they participate in various iron uptake pathways, alginate secretion and the expression of virulence factors. The best characterized systems are the *P. aeruginosa* σ^E (also known as AlgU or AlgT) regulator of alginate biosynthesis (Hughes and Mathee, 1998) and σ^{PvdS}, a regulator of pyoverdine siderophore biosynthesis (Vasil and Ochsner, 1999).

3.6.1. σ^E

In patients with cystic fibrosis, the chronic colonization of the lungs by *P. aeruginosa* is a major factor in the progression and ultimate lethality of the disease. Pathogenic isolates of *P. aeruginosa* from cystic fibrosis patients are typically mucoid in appearance and produce an abundant exopolysaccharide known as alginate. Genetic investigations of the molecular changes leading to mucoidy identified several genetic loci (*muc* genes) that, in retrospect, are negative regulators of the σ^E factor that activates transcription of the alginate biosynthesis genes. The gene encoding σ^E, *algU(algT)*, is cotranscribed with the *mucABCD* genes. The MucA protein functions directly as an anti-σ factor for σ^E while the MucB protein is located in the periplasm (Hughes and Mathee, 1998). This system is analogous to the regulation of σ^E in *E. coli* by the RseA and RseB proteins. Indeed, expression of *E. coli* σ^E in *P. aeruginosa* can complement an *algU* mutation and restore mucoidy (Yu *et al.*, 1995). Alginate is also synthesized by a variety of other Pseudomonads, including the plant pathogen *P. syringae* (Keith and Bender, 1999). For a detailed review of this system, see Hughes and Mathee (1998).

3.6.2. σ^{PvdS} and its Relatives

Pseudomonads contain a family of ECF σ factors that are all regulated by the Fur protein in response to iron limitation and activate the expression of siderophore biosynthesis and/or transport genes (Leoni *et al.*, 2000). The best characterized member of this family of regulators is σ^{PvdS}, which is required for the synthesis of the siderophore pyoverdine (reviewed in Vasil and Ochsner, 1999). The σ^{PvdS} regulator is highly similar to σ^{PbrA} (89% identity) and σ^{PfrI} (85% identity) which are required to activate pseudobactin biosynthesis in *P. fluorescens* and *P. putida*, respectively (Leoni *et al.*, 2000). A cluster of more distantly related Fur-regulated ECF σ factors includes σ^{PupI} and *E. coli* σ^{FecI} (Leoni *et al.*, 2000), both of which direct expression of siderophore receptors and appear to require activation by their cognate membrane-bound regulator (Koster *et al.*, 1993).

The σ^{PvdS} regulon is likely to include at least six target promoters including genes required for pyoverdine synthesis, a positive activator of toxin production, and an extracellular protease (Ochsner *et al.,* 1997; Wilson *et al.*, 2001). The coordinate regulation of these genes by the σ^{PvdS} holoenzyme accounts for the observation that these genes all have a conserved DNA sequence motif in their regulatory regions (Rombel *et al.*, 1995). The σ^{PvdS} protein has been purified and shown to program recognition of its cognate target sites *in vitro* (Wilson and Lamont, 2000).

Although *P. aeruginosa* encodes at least eight ECF σ factors only two have been studied in detail. However, it is worth noting that cycle selection experiments using purified Fur protein identified candidate Fur-binding sequences proximal to three additional genes encoding ECF σ factors (Ochsner and Vasil, 1996). It remains to be determined whether these ECF σ factors are involved in the regulation of iron uptake functions. However, at least one of these σ factors (σ^{FiuI}; Genbank entry AF051691) is located adjacent to a putative siderophore receptor gene consistent with the idea that it may regulate iron transport.

The *P. fluorescens* siderophore, pseudobactin, is transcriptionally regulated by a close homologue of σ^{PvdS} designated σ^{PbrA} (Sexton *et al.*, 1995). Like σ^{PvdS}, the expression of σ^{PbrA} is regulated by Fur in response to iron levels (Sexton *et al.*, 1996). Interestingly, the expression of both siderophore and its receptor is reduced in mutants in siderophore biosynthesis, suggesting that pseudobactin (like ferric citrate in *E. coli*) may have a role in inducing expression of the corresponding uptake operon. However, this response does not require σ^{PbrA} but is instead mediated by an unidentified factor (Callanan *et al.*, 1996). An ECF σ factor, σ^{PrtI}, has also been implicated in production of protease in a *P. fluorescens* (Burger *et al.*, 2000).

Table 12 ECF σ factors and their regulons in various bacteria.

Organism	ECF σ	Known stimuli	Regulon	Proposed function	Regulation	References
Stress responses						
B. subtilis	σ^W	Antimicrobial compounds that inhibit cell wall synthesis; alkali pH	~50 proteins	Detoxification, antibiosis	RsiW (anti-σ)	(Huang et al., 1999; Turner and Helmann, 2000; Cao et al., 2001, 2002)
B. subtilis	σ^M	Salt stress		Survival in high concentrations of salt	YhdL–YhdK (anti-σ)	(Horsburgh and Moir, 1999)
Bartonella henselae	σ^E	Heat shock	HtrA			(Resto-Ruiz et al., 2000)
E. coli	σ^E	Misfolded proteins in periplasm; extreme heat shock	~20 proteins	Folding of proteins in periplasm	RseA-RseB (anti-σ)	(Danese and Silhavy, 1997; De Las Penas et al., 1997a, b; Missiakas et al., 1997; Dartigalongue et al., 2001)
M. tuberculosis	σ^E	Heat shock, acidic pH, detergent, oxidative stress				(Wu et al., 1997)
M. tuberculosis	σ^H	Heat shock; disulphide stress, organic peroxides		Respond to heat shock and protease production		(Fernandes et al., 1999)
P. fluorescens LS107d2	σ^{PrtI}	Temperatures above 25°C	AprX (protease)		PrtR (anti-σ)	(Burger et al., 2000)
S. coelicolor	σ^R	Disulphide stress	~28 proteins (TrxB, TrxA, HrdD, et)	Maintains intracellular reducing environment	RsrA (anti-σ)	(Kang et al., 1999; Paget et al., 1998, 2001b)
Vibrio angustum	σ^E	High temperature				(Hild et al., 2000)

Table 12 *cont.*

Organism	ECF σ	Known stimuli	Regulon	Proposed function	Regulation	References
Metal transport						
B. bronchiseptica	σ^BupI	Siderophore	BfrZ (siderophore receptor)	Iron uptake	BupR (anti-σ)	(Pradel and Locht, 2001)
E. coli	σ^FecI	Iron starvation; ferric citrate	FecA, B, C, D, E	Ferric citrate transport	FecR (anti-σ)	(Angerer et al., 1995; Ochs et al., 1996)
P. aeruginosa	σ^PvdS	Iron starvation	RegA, PvdA, PvdD, PvdE, PvcA,B,C,D, PtxR, ToxA and PrpL	Regulates biosynthesis of pyoverdine, exotoxin A and PrpL (extracellular proteinase)	Repressed by Fur in iron-rich media	(Leoni et al., 2000; Wilson and Lamont, 2000; Wilson et al., 2001)
P. fluorescens M114	σ^PbrA	Iron starvation	PbuA (receptor), a casein protease	Iron uptake	Repressed by Fur in iron-rich media	(Sexton et al., 1995, 1996)
Rhizobium leguminosarum	σ^RpoI			Regulates siderophore production		(Yeoman et al., 1999)
Ralstonia eutropha CH34	σ^CnrH	Nickel, cobalt	CnrY, CnrX, CnrH, CnrC, CnrB, CnrA	Cobalt and nickel resistance		(Tibazarwa et al., 2000)
Virulence-related						
Azotobacter vinelandii	σ^AlgU		AlgD	Regulates alginate production and cyst formation	Repressed by MucA, MucC (anti-σ)	(Martinez-Salazar et al., 1996; Moreno et al., 1998; Nunez et al., 2000)

Table 12   cont.

Organism	ECF σ	Known stimuli	Regulon	Proposed function	Regulation	References
P. aeruginosa	σ^E		AlgD	Alginate production and tolerance to environmental stress	MucA-MucB	(Hershberger et al., 1995; Xie et al., 1996; Mathee et al., 1997; Rowen and Deretic, 2000)
P. syringae	σ^{AlgT}	Heat, osmotic stress, H_2O_2, paraquat and copper sulphate	AlgD	Alginate production and tolerance to environmental stress		(Keith and Bender, 1999)
Salmonella typhimurium	σ^E		HtrA, FkpA	Virulence and immunogenicity	RseA-RseB (possible anti-σ)	(Humphreys et al., 1999)
Erwinia amylovora Ea321	σ^{HrpL}	Nutrition (carbon, nitrogen), pH, temperature	HrpN, HrpI	Hypersensitive response, pathogenesis	HrpS	(Wei and Beer, 1995)
P. syringae	σ^{HrpL}	Nutrition (carbon, nitrogen), pH, osmotic stress	Hrp proteins, HrmA	Hypersensitive response, pathogenesis	Activated by HrpR-HrpS	(Xiao et al., 1994)
Cell surface-related						
B. subtilis	σ^X	Cell density signals and antimicrobial compounds	>10 proteins	Modification of cell envelope	RsiX (anti-σ)	(Huang et al., 1997; Huang and Helmann, 1998; Turner and Helmann, 2000)
S. coelicolor	σ^E	Cell wall active antibiotics; lysozyme	HrdD	Required for normal cell wall structure	CseB-CseC	(Paget et al., 1999a, b)
P. aeruginosa and P. fluorescens	σ^X		OprF	Regulate the expression of major outer membrane protein		(Brinkman et al., 1999)
Photobacterium species strain SS9	σ^E	Cold temperature and high pressure		Controls outer membrane protein synthesis		(Chi and Bartlett, 1995)

Table 12   cont.

Organism	ECF σ	Known stimuli	Regulon	Proposed function	Regulation	References
M. xanthus	σ^{RpoE1}			Motility		(Ward et al., 1998)
Xanthomonas campestris	σ^{RfaY}			Lipopolysaccharide biosynthesis		(Dow et al., 1995)
Developmental and morphological changes						
Caulobacter crescentus	σ^T, σ^U	G1–S phase transition		Regulates early S-phase gene expression	Repressed by CtrA	(Laub et al., 2000)
S. coelicolor A3(2)	σ^{BldN}		BldM	Aerial mycelium formation	Regulated by BldG, H	(Bibb et al., 2000)
S. griseus	σ^{AdsA}	A-factor (γ-butyrolactone)		Aerial hyphae formation	Activated by AdpA	(Yamazaki et al., 2000)
Others						
M. xanthus	σ^{CarQ}	Light	CarC, CarS	Regulation of carotenoid biosynthesis	CarR (anti-σ)	(Gorham et al., 1996; Martinez-Argudo et al., 1998)
Rhodobacter sphaeroides	σ^E	A signal from tetrapyrrole biosynthesis	CycA	Regulation of cytochrome c2 biosynthesis	ChrR (anti-σ)	(Newman et al., 1999)
S. antibioticus	σ^E			Actinomycin production		(Jones et al., 1997)
Starkeya novella	σ^E		SorAB	Thiosulphate oxidation		(Kappler et al., 2001)

3.7. ECF σ Factors in Other Organisms

ECF σ factors are widely distributed among bacteria and offer a convenient mechanism for coordinating gene expression with extracellular signals (Table 12). In general, bacteria with greater metabolic or developmental complexity tend to have larger genomes compared with highly specialized organisms that may have undergone reductive genome evolution (Table 11). There is a clear correlation between genome size and the proportion of the genome devoted to regulatory functions: in bacterial genomes of 0.5 to 2 Mb in size, <3% of the open reading frames encode likely regulatory proteins while this

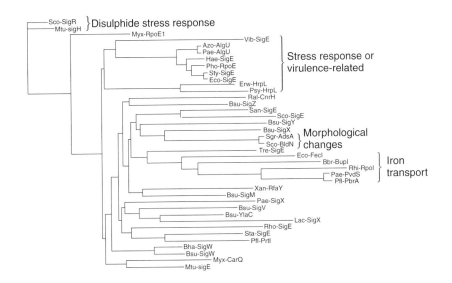

Figure 8 Phylogenetic relationships among the ECF subfamily σ factors. σ factors with similar functions tend to form clusters. Illustrated here are an iron transport cluster, a stress and virulence cluster, a disulphide stress cluster, and a cluster involved in morphological changes in Streptomycetes. Species abbreviations are: Azo, *Azotobacter vinelandii*; Bbr, *Bordetella bronchiseptica*; Bha, *Bacillus halodurans*; Bsu, *Bacillus subtilis*; Eco, *Escherichia coli*; Erw, *Erwinia amylovora*; Hae, *Haemophilus influenzae*; Lac, *Lactococcus lactis*; Mtu, *Mycobacterium tuberculosis*; Myx, *Myxococcus xanthus*; Pae, *Pseudomonas aeruginosa*; Pfl, *Pseudomonas fluorescens*; Pho, *Photobacterium* sp.; Psy, *Pseudomonas syringae*; Ral, *Ralstonia eutropha*; Rhi, *Rhizobium leguminosarum*; Rho, *Rhodobacter sphaeroides*; San, *Streptomyces antibioticus*; Sco, *Streptomyces coelicolor*; Sgr, *Streptomyces griseus*; Sta, *Starkeya novella*; Sty, *Salmonella typhimurium*; Tre, *Treponema pallidum*; Vib, *Vibrio angustum*; and Xan, *Xanthomonas campestris*. The genetic distances are based on comparisons of amino acid sequences and were calculated by the FITCH program from the BioEdit package (http://jwbrown.mbio.ncsu.edu/BioEdit/bioedit.html).

number increases to >8% in the 6.3 Mb *P. aeruginosa* genome (Stover *et al.*, 2000). Similarly, *Caulobacter crescentus*, which has a complex dimorphic lifestyle, encodes an estimated 100 or more two-component signal transduction proteins, as well as 13 ECF σ factors in its 4 Mb genome (Nierman *et al.*, 2001). The expansion, presumably by gene duplication and divergence, of the ECF σ factors in many of the more complex bacterial genomes apparently occurs in preference to other classes of σ factors. Whereas many simple bacterial genomes encode few σ factors, and may not encode any ECF σ factors, in the largest genomes sequenced to date the majority of the σ factors are of the ECF class.

Inspection of the phylogenetic relatedness among various ECF σ factors reveals several clusters of proteins with related functions (Fig. 8). For example, there is a cluster of proteins that regulate iron transport conserved in various Pseudomonads, there is a close relationship between the disulphide stress σ factors in *S. coelicolor* and *M. tuberculosis*, and there are many homologues of *E. coli* σ^E that may regulate periplasmic or heat stress responses similar to that controlled by σ^E. This correlation of sequence with function is reminiscent of that observed among group 3 σ factors (Lonetto *et al.*, 1992) and suggests that alternative σ factors of the ECF subfamily arose early in many of these lineages.

4. RECURRING THEMES IN THE STUDY OF ECF σ FACTORS

As more ECF σ factors have been investigated, several common regulatory features have emerged. While none of these features is universal, in most cases ECF σ factors coordinate transcriptional responses to extracellular signals, the signal transduction pathway often involves a membrane-localized anti-σ factor, and ECF σ factors often positively regulate their own synthesis. In addition, numerous examples have now been documented of overlapping promoter recognition and potential functional redundancy.

As defined originally for *E. coli* σ^E and the closely related *P. aeruginosa* σ^E protein, ECF σ factors are often cotranscribed with one or more negative regulators (Hughes and Mathee, 1998). The immediately adjacent gene often encodes an anti-σ factor, typically membrane-localized, that binds and stoichiometrically inactivates the ECF σ factor. The transmembrane localization of the anti-σ factor allows the cell to control the activity of an intracellular transcription factor in response to signals present outside the cytoplasmic membrane. Examples include the induction of the *E. coli* σ^E regulon in response to unfolded proteins in the periplasm (Ravio and Silhavy, 2001) and the induction of several iron transport systems in response to the presence of the cognate ferri-siderophore complex (Braun, 1997). In this respect, ECF σ factors together with their *trans*-membrane anti-σ factors are functionally analogous to classical two-component regulatory systems in which a transmembrane sensory protein (a histidine protein kinase) controls the activity of a cytoplasmic response regulator.

In general, ECF σ factors are often quite divergent in sequence not only from other σ factors, but from each other. Sequence analysis alone is sufficient to predict that a protein functions as an ECF σ factor, but only in rare cases does it allow one to predict function (hopefully this will change as more and more ECF σ factors become characterized). For example, the role of *M. tuberculosis* σ^H in controlling a disulphide stress regulon was a reasonable hypothesis based on the close similarity to σ^R (Manganelli *et al.*, 2001a; Raman *et al.*, 2001). Our ability to make functional inferences based on the sequences of known and putative anti-σ factors is even more limited. One of the best studied anti-σ factors is RsrA, the redox-sensing regulator of *S. coelicolor* σ^R. This protein contains a bound zinc ion and is the prototype for an emerging family of Zn-containing anti-σ factors (the ZAS subfamily; Paget *et al.*, 2001a). However, many anti-σ factors have sequences suggestive of a metal-binding site and many of these regulators do not respond to oxidation. Therefore, the mere presence of a ZAS-type anti-σ factor, linked to a σ factor gene is not a strong predictor of function.

A second recurring feature of ECF σ factors is positive autoregulation: most genes encoding ECF σ factors are preceded by a promoter recognized by the corresponding σ. This presumably serves to amplify the signal produced by release of active σ from the inactive σ-anti-σ complex. Despite their limited sequence identity, promoters recognized by ECF subfamily σ factors often share characteristic sequence features, including the common occurrence of an 'AAC' motif in the −35 consensus region. Inspection of the DNA sequence upstream of ECF σ factor genes for possible autoregulatory sites can provide a preliminary indication of sequence selectivity for newly identified σ factors of this class.

A third emerging theme in the analysis of ECF σ factor regulons is regulon overlap. Particularly in those organisms containing many ECF σ factors, some promoter sites appear to function as targets for regulation by more than one ECF σ factor. The overlap between the *B. subtilis* σ^X and σ^W regulons appears to be due, at least in part, to overlapping recognition properties in the −10 consensus element (Qiu and Helmann, 2001). There is also limited overlap between the *M. tuberculosis* σ^E and σ^H regulons: both σ factors recognize the same promoter for the *sigB* gene (Manganelli *et al.*, 2001a; Raman *et al.*, 2001). Despite the presence of regulon overlap, it is often possible to observe phenotypic consequences of mutations in ECF σ factor genes. This lack of functional redundancy may be due, at least in part, to the fact that so many regulons are likely to be silent under normal growth conditions. In other words, the potential functional redundancies among ECF σ factors is masked by the presence of anti-σ factors that prevent the expression of many of these regulons unless appropriate inducing conditions are present.

Finally, we can consider prospects and strategies for understanding the roles of ECF σ factors in the post-genomic era. Clearly, transcriptional profiling will continue to play an important role in defining ECF σ factor regulons, particularly if conditions are employed that activate the regulon in question. In the absence of detailed knowledge of the physiological inducers, induction can be artificially elicited by either deletion or repression of the cognate anti-σ or by overexpression of the σ factor itself.

A complementary approach is provided by the computer-based identification of candidate target genes using the 'consensus-search' method. This has worked quite well in those cases where the recognition sequence seems quite highly conserved (e.g. *B. subtilis* σ^W and *S. coelicolor* σ^R), but would not work very well if the consensus is not well defined (e.g. *E. coli* σ^E). Analysis of the σ^W regulon provides an instructive example of the bioinformatics of promoter recognition. In the whole genome there are 27 perfect matches to the P_W autoregulatory sequence (Huang *et al.*, 1999), TGAAAC N_{16} CGTA, and 16 of these sites (all of those positioned upstream of reading frames) function as promoters (Huang *et al.*, 1999). What about the other 11 sites? Most of these are in the middle of transcription units and/or are inappropriately oriented to serve an obvious role as promoters. It seems likely that many of these sites are 'false-positives'. How does the σ^W holoenzyme distinguish the functional promoters from these other sites? We hypothesize that sequences in addition to the -35 and -10 elements are also important in promoter recognition. Candidates for such discriminatory elements include extensions of the -35 or -10 sequence elements (e.g. σ^W promoters frequently have a T-rich segment adjacent to the -35 element and a -10 element of CGTAta; Table 7) and the upstream promoter (UP element; Estrem *et al.*, 1998) region (Huang *et al.*, 1999). It is likely that similar considerations will pertain to other attempts to identify target genes using consensus-search methods. Clearly, we are still just beginning in our attempts to use computer-assisted analyses to extract biological information from genome sequences.

ACKNOWLEDGEMENTS

I thank Min Cao for her invaluable assistance with the Tables and Figures. I also thank Mark Buttner, Mark Paget, Nagraj Mani, Issar Smith, Anne Moir, Wolfgang Schumann, Jeff Errington and Yasutaro Fujita for generously sharing their unpublished data. All work on ECF σ factors in the Helmann laboratory is funded by a grant from the NIH (GM47446).

REFERENCES

Ades, S.E., Connolly, L.E., Alba, B.M. and Gross, C.A. (1999) The *Escherichia coli* sigma(E)-dependent extracytoplasmic stress response is controlled by the regulated proteolysis of an anti-sigma factor. *Genes Dev.* **13**, 2449–2461.

Akbar, S. and Price, C.W. (1996) Isolation and characterization of *csbB*, a gene controlled by *Bacillus subtilis* general stress transcription factor σ^B. *Gene* **177**, 123–128.

Alba, B.M., Zhong, H.J., Pelayo, J.C. and Gross, C.A. (2001) *degS* (*hhoB*) is an essential *Escherichia coli* gene whose indispensable function is to provide sigma activity. *Mol. Microbiol.* **40**, 1323–1333.

Angerer, A. and Braun, V. (1998) Iron regulates transcription of the *Escherichia coli* ferric citrate transport genes directly and through the transcription initiation proteins. *Arch. Microbiol.* **169**, 483–490.

Angerer, A., Enz, S., Ochs, M. and Braun, V. (1995) Transcriptional regulation of ferric citrate transport in *Escherichia coli* K-12. FecI belongs to a new subfamily of sigma-70-type factors that respond to extracytoplasmic stimuli. *Mol. Microbiol.* **18**, 163–174.

Aslund, F. and Beckwith, J. (1999) Bridge over troubled waters: sensing stress by disulfide bond formation. *Cell* **96**, 751–753.

Beck, B.J., Connolly, L.E., De Las Penas, A. and Downs, D.M. (1997) Evidence that *rseC*, a gene in the *rpoE* cluster, has a role in thiamine synthesis in *Salmonella typhimurium*. *J. Bacteriol.* **179**, 6504–6508.

Becker, G. and Hengge-Aronis, R. (2001) What makes an *Escherichia coli* promoter σ^S dependent? Role of the −13/−14 nucleotide promoter positions and region 2.5 of σ^S. *Mol. Microbiol.* **39**, 1153–1165.

Bibb, M.J., Molle, V. and Buttner, M.J. (2000) Sigma(BldN), an extracytoplasmic function RNA polymerase sigma factor required for aerial mycelium formation in *Streptomyces coelicolor* A3(2). *J. Bacteriol.* **182**, 4606–4616.

Bolhuis, A., Matzen, A., Hyyrylainen, H.L., Kontinen, V.P., Meima, R., Chapuis, J., Venema, G., Bron, S., Freudl, R. and van Dijl, J.M. (1999) Signal peptide peptidase- and ClpP-like proteins of *Bacillus subtilis* required for efficient translocation and processing of secretory proteins. *J. Biol. Chem.* **274**, 24585–24592.

Boyd, D.A., Cvitkovitch, D.G., Bleiweis, A.S., Kiriukhin, M.Y., Debabov, D.V., Neuhaus, F.C. and Hamilton, I.R. (2000) Defects in D-alanyl-lipoteichoic acid synthesis in *Streptococcus mutans* results in acid sensitivity. *J. Bacteriol.* **182**, 6055–6065.

Braun, V. (1997) Surface signaling: novel transcription initiation mechanism starting from the cell surface. *Arch. Microbiol.* **167**, 325–331.

Breukink, E., Wiedemann, I., van Kraaij, C., Kuipers, O.P., Sahl, H. and de Kruijff, B. (1999) Use of the cell wall precursor lipid II by a pore-forming peptide antibiotic. *Science* **286**, 2361–2364.

Brinkman, F.S., Schoofs, G., Hancock, R.E. and De Mot, R. (1999) Influence of a putative ECF sigma factor on expression of the major outer membrane protein, OprF, in *Pseudomonas aeruginosa* and *Pseudomonas fluorescens*. *J. Bacteriol.* **181**, 4746–4754.

Brotz, H., Josten, M., Wiedemann, I., Schneider, U., Gotz, F., Bierbaum, G. and Sahl, H.G. (1998) Role of lipid-bound peptidoglycan precursors in the formation of pores by nisin, epidermin and other lantibiotics. *Mol. Microbiol.* **30**, 317–327.

Brutsche, S. and Braun, V. (1997) SigX of *Bacillus subtilis* replaces the ECF sigma factor FecI of *Escherichia coli* and is inhibited by RsiX. *Molec. Gen. Genet.* **256**, 416–425.

Burger, M., Woods, R.G., McCarthy, C. and Beacham, I.R. (2000) Temperature regulation of protease in *Pseudomonas fluorescens* LS107d2 by an ECF sigma factor and a transmembrane activator. *Microbiology* **146**, 3149–3155.

Buttner, M.J. (1989) RNA polymerase heterogeneity in *Streptomyces coelicolor* A3(2). *Mol. Microbiol.* **3**, 1653–1659.

Buttner, M.J. and Lewis, C.G. (1992) Construction and characterization of *Streptomyces coelicolor* A3(2) mutants that are multiply deficient in the nonessential hrd-encoded RNA polymerase sigma factors. *J. Bacteriol.* **174**, 5165–5167.

Buttner, M.J., Smith, A.M. and Bibb, M.J. (1988) At least three different RNA polymerase holoenzymes direct transcription of the agarase gene (*dagA*) of *Streptomyces coelicolor* A3(2). *Cell* **52**, 599–607.

Callanan, M., Sexton, R., Dowling, D.N. and O'Gara, F. (1996) Regulation of the iron uptake genes in *Pseudomonas fluorescens* M114 by pseudobactin M114: the *prbA* sigma factor gene does not mediate the siderophore regulatory response. *FEMS Microbiol. Lett.* **144**, 61–66.

Cao, M., Bernat, B.A., Wang, Z., Armstrong, R.N. and Helmann, J.D. (2001) FosB, a cysteine-dependent fosfomycin resistance protein under the control of sigma(W), an extracytoplasmic-function sigma factor in *Bacillus subtilis*. *J. Bacteriol.* **183**, 2380–2383.

Cao, M., Kobel, P.A., Morshedi, M.M., Wu, M.F., Paddon, C. and Helmann, J.D. (2002) Defining the *Bacillus subtilis* σ^W regulon: a comparative analysis of promoter consensus search, run-off transcription/macroarray analysis (ROMA), and transcriptional profiling approaches. *J. Mol. Biol.* **316**, 443–457.

Carmel-Harel, O. and Storz, G. (2000) Roles of the glutathione- and thioredoxin-dependent reduction systems in the *Escherichia coli* and *Saccharomyces cerevisiae* responses to oxidative stress. *Annu. Rev. Microbiol.* **54**, 439–461.

Caslake, L.F., Gruber, T.M. and Bryant, D.A. (1997) Expression of two alternative sigma factors of *Synechococcus* sp. strain PCC 7002 is modulated by carbon and nitrogen stress. *Microbiology* **143**, 3807–3818.

Chater, K.F., Bruton, C.J., Plaskitt, K.A., Buttner, M.J., Mendez, C. and Helmann, J.D. (1989) The developmental fate of *S. coelicolor* hyphae depends upon a gene product homologous with the motility sigma factor of *B. subtilis*. *Cell* **59**, 133–143.

Chen, Y.F. and Helmann, J.D. (1992) Restoration of motility to an *Escherichia coli fliA* flagellar mutant by a *Bacillus subtilis* sigma factor. *Proc. Natl Acad. Sci. USA* **89**, 5123–5127.

Chi, E. and Bartlett, D.H. (1995) An *rpoE*-like locus controls outer membrane protein synthesis and growth at cold temperatures and high pressures in the deep-sea bacterium *Photobacterium* sp. strain SS9. *Mol. Microbiol.* **17**, 713–726.

Clemans, D.L., Kolenbrander, P.E., Debabov, D.V., Zhang, Q., Lunsford, R.D., Sakone, H., Whittaker, C.J., Heaton, M.P. and Neuhaus, F.C. (1999) Insertional inactivation of genes responsible for the D-alanylation of lipoteichoic acid in *Streptococcus gordonii* DL1 (Challis) affects intrageneric coaggregations. *Infect. Immun.* **67**, 2464–2474.

Cole, S.T., Brosch, R., Parkhill, J., Garnier, T., Churcher, C., Harris, D., Gordon, S.V., Eiglmeier, K., Gas, S., Barry, C.E. et al. (1998) Deciphering the biology of *Mycobacterium tuberculosis* from the complete genome sequence. *Nature* **393**, 537–544.

Crandall, A.D. and Montville, T.J. (1998) Nisin resistance in *Listeria monocytogenes* ATCC 700302 is a complex phenotype. *Appl. Environ. Microbiol.* **64**, 231–237.

Danese, P.N. and Silhavy, T.J. (1997) The sigma(E) and the Cpx signal transduction systems control the synthesis of periplasmic protein-folding enzymes in *Escherichia coli*. *Genes Dev.* **11**, 1183–1193.

Dartigalongue, C., Missiakas, D. and Raina, S. (2001) Characterization of the *Escherichia coli* sigma E regulon. *J. Biol. Chem.* **276**, 20866–20875.

De Las Penas, A., Connolly, L. and Gross, C.A. (1997a) SigmaE is an essential sigma factor in *Escherichia coli*. *J. Bacteriol.* **179**, 6862–6864.

De Las Penas, A., Connolly, L. and Gross, C.A. (1997b) The sigma-E-mediated response to extracytoplasmic stress in *Escherichia coli* is transduced by RseA and RseB, two negative regulators of sigma-E. *Mol. Microbiol.* **24**, 373–385.

Dow, J.M., Osbourn, A.E., Wilson, T.J. and Daniels, M.J. (1995) A locus determining pathogenicity of *Xanthomonas campestris* is involved in lipopolysaccharide biosynthesis. *Mol. Plant. Microbe Interact.* **8**, 768–777.

Elliot, M.A., Bibb, M.J., Buttner, M.J. and Leskiw, B.K. (2001) BldD is a direct regulator of key developmental genes in *Streptomyces coelicolor* A3(2). *Mol. Microbiol.* **40**, 257–269.

Enz, S., Mahren, S., Stroeher, U.H. and Braun, V. (2000) Surface signaling in ferric citrate transport gene induction: interaction of the FecA, FecR and FecI regulatory proteins. *J. Bacteriol.* **182**, 637–646.

Erickson, J.W. and Gross, C.A. (1989) Identification of the σ^E subunit of *Escherichia coli* RNA polymerase: a second alternative σ factor involved in high temperature gene expression. *Genes Dev.* **3**, 1462–1471.

Estrem, S.T., Gaal, T., Ross, W. and Gourse, R.L. (1998) Identification of an UP element consensus sequence for bacterial promoters. *Proc. Natl Acad. Sci. USA*, **95**, 9761–9766.

Fernandes, N.D., Wu, Q.L., Kong, D., Puyang, X., Garg, S. and Husson, R.N. (1999) A mycobacterial extracytoplasmic sigma factor involved in survival following heat shock and oxidative stress. *J. Bacteriol.* **181**, 4266–4274.

Fischer, W. (1988) Physiology of lipoteichoic acids in bacteria. *Adv. Microb. Physiol.* **29**, 233–302.

Garnier, T. and Cole, S.T. (1988) Studies of UV-inducible promoters from *Clostridium perfringens in vivo* and *in vitro*. *Mol. Microbiol.* **2**, 607–614.

Gehring, A.M., Yoo, N.J. and Losick, R. (2001) A RNA polymerase sigma factor that blocks morphological differentiation by *Streptomyces coelicolor*. *J. Bacteriol.* **183**, 5991–5996.

Gorham, H.C., McGowan, S.J., Robson, P.R. and Hodgson, D.A. (1996) Light-induced carotenogenesis in *Myxococcus xanthus*: light-dependent membrane sequestration of ECF sigma factor CarQ by anti-sigma factor CarR. *Mol. Microbiol.* **19**, 171–186.

Graham, J.E. and Clark-Curtiss, J.E. (1999) Identification of *Mycobacterium tuberculosis* RNAs synthesized in response to phagocytosis by human macrophages by selective capture of transcribed sequences (SCOTS). *Proc. Natl Acad. Sci. USA* **96**, 11554–11559.

Gross, C.A., Lonetto, M. and Losick, R. (1992) Bacterial sigma factors. In: *Transcriptional Regulation* vol. 1 (S.L. McKnight and K.R. Yamamoto, eds), pp. 129–176. Cold Spring Harbor Press, Cold Spring Harbor, NY.

Gross, C.A., Chan, C., Dombroski, A., Gruber, T., Sharp, M., Tupy, J. and Young, B. (1998) The functional and regulatory roles of sigma factors in transcription. *Cold Spring Harb. Symp. Quant. Biol.* **63**, 141–155.

Grossman, A.D., Erickson, J.W. and Gross, C.A. (1984) The *htpR* gene product of *E. coli* is a sigma factor for heat-shock promoters. *Cell* **38**, 383–390.

Haldenwang, W.G. (1995) The sigma factors of *Bacillus subtilis*. *Microbiol. Rev.* **59**, 1–30.

Hancock, R.E. and Diamond, G. (2000) The role of cationic antimicrobial peptides in innate host defences. *Trends Microbiol.* **8**, 402–410.

Hancock, R.E. and Scott, M.G. (2000) The role of antimicrobial peptides in animal defenses. *Proc. Natl Acad. Sci. USA* **97**, 8856–8861.

Hecker, M. and Volker, U. (2001) General stress response of *Bacillus subtilis* and other bacteria. *Adv. Microb. Physiol.* **44**, 35–91.

Helmann, J.D. (1991) Alternative sigma factors and the regulation of flagellar gene expression. *Mol. Microbiol.* **5**, 2875–2882.

Helmann, J.D. (1994) Bacterial sigma factors. In: *Transcription: Mechanisms and*

Regulation. vol. 3 (R.C. Conaway and J. Conaway, eds), pp. 1–17. Raven Press, New York.

Helman, J.D. (1999) Anti-sigma factors. *Curr. Opin. Microbiol.* **2**, 135–141.

Helmann, J.D. and Chamberlin, M.J. (1988) Structure and function of bacterial σ factors. *Annu. Rev. Biochem.* **57**, 839–872.

Helmann, J.D. and Moran, C.P., Jr. (2002) RNA polymerase and sigma factors. In: *Bacillus subtilis and Its Relatives: From Genes to Cells.* (A.L. Sonenshein and R. Losick, eds), ASM Press, Washington D.C.

Hengge-Aronis, R. (1999) Interplay of global regulators and cell physiology in the general stress response of *Escherichia coli. Curr. Opin. Microbiol.* **2**, 148–152.

Hengge-Aronis, R. (2000) The general stress response in *Escherichia coli.* In: *Bacterial Stress Responses.* (G. Storz and R. Hengge-Aronis, eds), pp. 161–178. ASM Press, Washington, D.C.

Hershberger, C.D., Ye, R.W., Parsek, M.R., Xie, Z.D. and Chakrabarty, A.M. (1995) The *algT* (*algU*) gene of *Pseudomonas aeruginosa,* a key regulator involved in alginate biosynthesis, encodes an alternative sigma factor (sigma E). *Proc. Natl Acad. Sci. USA* **92**, 7941–7945.

Hild, E., Takayama, K., Olsson, R.M. and Kjelleberg, S. (2000) Evidence for a role of *rpoE* in stressed and unstressed cells of marine *Vibrio angustum* strain S14. *J. Bacteriol.* **182**, 6964–6974.

Horinouchi, S. (1999) γ-Butyrolactones that control secondary metabolism and differentiation in *Streptomyces.* In: *Cell–Cell Signaling in Bacteria.* (G.M. Dunny and S.C. Winans, eds) pp. 193–207. ASM Press, Washington, D.C.

Horsburgh, M.J. and Moir, A. (1999) Sigma M, an ECF RNA polymerase sigma factor of *Bacillus subtilis* 168, is essential for growth and survival in high concentrations of salt. *Mol. Microbiol.* **32**, 41–50.

Huang, X. and Helmann, J.D. (1998) Identification of target promoters for the *Bacillus subtilis* σ^X factor using a consensus-directed search. *J Mol. Biol.* **279**, 165–173.

Huang, X., Decatur, A., Sorokin, A. and Helmann, J.D. (1997) The *Bacillus subtilis* σ^X protein is an extracytoplasmic function sigma factor contributing to the survival of high temperature stress. *J. Bacteriol.* **179**, 2915–2921.

Huang, X., Fredrick, K.L. and Helmann, J.D. (1998) Promoter recognition by *Bacillus subtilis* σ^W: Autoregulation and partial overlap with the σ^X regulon. *J. Bacteriol.* **180**, 3765–3770.

Huang, X., Gaballa, A., Cao, M. and Helmann, J.D. (1999) Identification of target promoters for the *Bacillus subtilis* extracytoplasmic function σ factor, σ^W. *Mol. Microbiol.* **31**, 361–371.

Hughes, K.T. and Mathee, K. (1998) The anti-sigma factors. *Annu. Rev. Microbiol.* **52**, 231–286.

Humphreys, S., Stevenson, A., Bacon, A., Weinhardt, A.B. and Roberts, M. (1999) The alternative sigma factor, sigmaE, is critically important for the virulence of *Salmonella typhimurium. Infect. Immun.* **67**, 1560–1568.

Hyyrylainen, H.L., Vitikainen, M., Thwaite, J., Wu, H., Sarvas, M., Harwood, C.R., Kontinen, V.P. and Stephenson, K. (2000) D-Alanine substitution of teichoic acids as a modulator of protein folding and stability at the cytoplasmic membrane/cell wall interface of *Bacillus subtilis. J. Biol. Chem.* **275**, 26696–26703.

Jack, R.W., Tagg, J.R. and Ray, B. (1995) Bacteriocins of Gram-positive bacteria. *Microbiol. Rev.* **59**, 171–200.

Jensen-Cain, D.M. and Quinn, F.D. (2001) Differential expression of *sigE* by *Mycobacterium tuberculosis* during intracellular growth. *Microb. Pathog.* **30**, 271–278.

Jones, G.H., Paget, M.S.B., Chamberlin, L. and Buttner, M.J. (1997) Sigma-E is required for the production of the antibiotic actinomycin in *Streptomyces antibioticus. Mol. Microbiol.* **23**, 169–178.

Kang, J.G., Hahn, M.Y., Ishihama, A. and Roe, J.H. (1997) Identification of sigma factors for growth phase-related promoter selectivity of RNA polymerases from *Streptomyces coelicolor* A3(2). *Nucleic Acids Res.* **25**, 2566–2573.

Kang, J.G., Paget, M.S., Seok, Y.J., Hahn, M.Y., Bae, J.B., Hahn, J.S., Kleanthous, C., Buttner, M.J. and Roe, J.H. (1999) RsrA, an anti-sigma factor regulated by redox change. *EMBO J.* **18**, 4292–4298.

Kappler, U., Friedrich, C.G., Truper, H.G. and Dahl, C. (2001) Evidence for two pathways of thiosulfate oxidation in *Starkeya novella* (formerly *Thiobacillus novellus). Arch. Microbiol.* **175**, 102–111.

Karzai, A.W., Roche, E.D. and Sauer, R.T. (2000) The SsrA-SmpB system for protein tagging, directed degradation and ribosome rescue. *Nat. Struct. Biol.* **7**, 449–455.

Keith, L.M. and Bender, C.L. (1999) AlgT (sigma22) controls alginate production and tolerance to environmental stress in *Pseudomonas syringae. J. Bacteriol.* **181**, 7176–7184.

Koster, M., van de Vossenberg, J., Leong, J. and Weisbeek, P.J. (1993) Identification and characterization of the *pupB* gene encoding an inducible ferric-pseudobactin receptor of *Pseudomonas putida* WCS358. *Mol. Microbiol.* **8**, 591–601.

Kroos, L., Zhang, B., Ichikawa, H. and Yu, Y.T. (1999) Control of sigma factor activity during *Bacillus subtilis* sporulation. *Mol. Microbiol.* **31**, 1285–1294.

Kunst, F., Ogasawara, N., Moszer, I., Albertini, A.M., Alloni, G., Azevedo, V., Bertero, M.G., Bessieres, P., Bolotin, A., Borchert, S., *et al.* (1997) The complete genome sequence of the Gram-positive bacterium *Bacillus subtilis. Nature* **390**, 249–256.

Laub, M.T., McAdams, H.H., Feldblyum, T., Fraser, C.M. and Shapiro, L. (2000) Global analysis of the genetic network controlling a bacterial cell cycle. *Science* **290**, 2144–2148.

Lazarevic, V., Margot, P., Soldo, B. and Karamata, D. (1992) Sequencing and analysis of the *Bacillus subtilis lytRABC* divergon: a regulatory unit encompassing the structural genes of the N-acetylmuramoyl-L-alanine amidase and its modifier. *J. Gen. Microbiol.* **138:** 1949–1961.

Lee, S.J. and Gralla, J.D. (2001) Sigma38 (*rpoS*) RNA polymerase promoter engagement via-10 region nucleotides. *J. Biol. Chem.* **276**, 30046–30071.

Leoni, L., Orsi, N., de Lorenzo, V. and Visca, P. (2000) Functional analysis of PvdS, an iron starvation sigma factor of *Pseudomonas aeruginosa. J. Bacteriol.* **182**, 1481–1491.

Lipinska, B., Sharma, S. and Georgopoulos, C. (1988) Sequence analysis and regulation of the *htrA* gene of *Escherichia coli*: a sigma 32-independent mechanism of heat-inducible transcription. *Nucleic Acids Res.* **16**, 10053–10067.

Lonetto, M., Gribskov, M. and Gross, C.A. (1992) The σ^{70} family: Sequence conservation and evolutionary relationships. *J. Bacteriol.* **174**, 3843–3849.

Lonetto, M.A., Brown, K.L., Rudd, K.E. and Buttner, M.J. (1994) Analysis of the *Streptomyces coelicolor sigE* gene reveals the existence of a subfamily of eubacterial σ factors involved in the regulation of extracytoplasmic functions. *Proc. Natl Acad. Sci. USA* **91**, 7573–7577.

Manganelli, R., Dubnau, E., Tyagi, S., Kramer, F.R. and Smith, I. (1999) Differential expression of 10 sigma factor genes in *Mycobacterium tuberculosis. Mol. Microbiol.* **31**, 715–724.

Manganelli, R., Voskuil, M.I., Schoolnik, G.K., Gomez, M. and Smith, I. (2001a) Role of the extracytoplasmic-function σ factor σ^{H} in *Mycobacterium tuberculosis* global gene expression. *Mol. Microbiol.* (in press).

Manganelli, R., Voskuil, M.I., Schoolnik, G.K. and Smith, I. (2001b) The *Mycobacterium tuberculosis* ECF sigma factor σ^E: role in global gene expression and survival in macrophages. *Mol. Microbiol.* **41**, 423–437.

Mani, N. and Dupuy, B. (2001) Regulation of toxin synthesis in *Clostridium difficile* by an alternative RNA polymerase sigma factor. *Proc. Natl Acad. Sci. USA* **98**, 5844–5849.

Mantovani, H.C. and Russell, J.B. (2001) Nisin resistance of *Streptococcus bovis*. *Appl. Environ. Microbiol.* **67**, 808–813.

Marahiel, M.A., Nakano, M.M. and Zuber, P. (1993) Regulation of peptide antibiotic production in *Bacillus*. *Mol. Microbiol.* **7**, 631–636.

Martin, D.W., Holloway, B.W. and Deretic, V. (1993) Characterization of a locus determining the mucoid status of *Pseudomonas aeruginosa*: AlgU shows sequence similarities with a *Bacillus* sigma factor. *J. Bacteriol.* **175**, 1153–1164.

Martinez-Argudo, I., Ruiz-Vazquez, R.M. and Murillo, F.J. (1998) The structure of an ECF-sigma-dependent, light-inducible promoter from the bacterium *Myxococcus xanthus*. *Mol. Microbiol.* **30**, 883–893.

Martinez-Salazar, J.M., Moreno, S., Najera, R., Boucher, J.C., Espin, G., Soberon-Chavez, G. and Deretic, V. (1996) Characterization of the genes coding for the putative sigma factor AlgU and its regulators MucA, MucB, MucC, and MucD in *Azotobacter vinelandii* and evaluation of their roles in alginate biosynthesis. *J. Bacteriol.* **178**, 1800–1808.

Marvaud, J.C., Eisel, U., Binz, T., Niemann, H. and Popoff, M.R. (1998) TetR is a positive regulator of the tetanus toxin gene in *Clostridium tetani* and is homologous to botR. *Infect. Immun.* **66**, 5698–5702.

Mathee, K., McPherson, C.J. and Ohman, D.E. (1997) Posttranslational control of the *algT* (*algU*)-encoded sigma22 for expression of the alginate regulon in *Pseudomonas aeruginosa* and localization of its antagonist proteins MucA and MucB (AlgN). *J. Bacteriol.* **179**, 3711–3720.

Matsumoto, K., Okada, M., Horikoshi, Y., Matsuzaki, H., Kishi, T., Itaya, M. and Shibuya, I. (1998) Cloning, sequencing, and disruption of the *Bacillus subtilis psd* gene coding for phosphatidylserine decarboxylase. *J. Bacteriol.* **180**, 100–106.

Mattimore, V. and Battista, J.R. (1996) Radioresistance of *Deinococcus radiodurans*: functions necessary to survive ionizing radiation are also necessary to survive prolonged desiccation. *J. Bacteriol.* **178**, 633–637.

Merchante, R., Pooley, H.M. and Karamata, D. (1995) A periplasm in *Bacillus subtilis*. *J. Bacteriol.* **177**, 6176–6183.

Minnig, K., Beggah, S., Kehl, S. and Mauel, C. (2001) Growth-phase dependent regulation of teichoic acid biosynthesis in *Bacillus subtilis* – a comparison of strains 168 and W23. *Proc. 11th Int. Conf. Bacilli: Functional Genomics of Gram-Positive Microorganisms,* Abstr. P105.

Missiakas, D. and Raina, S. (1998) The extracytoplasmic function sigma factors: role and regulation. *Mol. Microbiol.* **28**, 1059–1066.

Missiakas, D., Mayer, M.P., Lemaire, M., Georgopoulos, C. and Raina, S. (1997) Modulation of the *Escherichia coli* sigma-E (RpoE) heat-shock transcription-factor activity by the RseA, RseB and RseC proteins. *Mol. Microbiol.* **24**, 355–371.

Moreno, S., Najera, R., Guzman, J., Soberon-Chavez, G. and Espin, G. (1998) Role of alternative sigma factor *algU* in encystment of *Azotobacter vinelandii*. *J. Bacteriol.* **180**, 2766–2769.

Muro-Pastor, A.M., Herrero, A. and Flores, E. (2001) Nitrogen-regulated group 2 sigma factor from *Synechocystis* sp. strain PCC 6803 involved in survival under nitrogen stress. *J. Bacteriol.* **183**, 1090–1095.

Nakao, A., Imai, S. and Takano, T. (2000) Transposon-mediated insertional mutagenesis of the D-alanyl-lipoteichoic acid (*dlt*) operon raises methicillin resistance in *Staphylococcus aureus*. *Res. Microbiol.* **151**, 823–829.

Newman, J.D., Falkowski, M.J., Schilke, B.A., Anthony, L.C. and Donohue, T.J. (1999) The *Rhodobacter sphaeroides* ECF sigma factor, sigma(E), and the target promoters *cycA* P3 and *rpoE* P1. *J. Mol. Biol.* **294**, 307–320.

Newton, G.L., Arnold, K., Price, M.S., Sherrill, C., Delcardayre, S.B., Aharonowitz, Y., Cohen, G., Davies, J., Fahey, R.C. and Davis, C. (1996) Distribution of thiols in microorganisms: mycothiol is a major thiol in most actinomycetes. *J. Bacteriol.* **178**, 1990–1995.

Nierman, W.C., Feldblyum, T.V., Laub, M.T., Paulsen, I.T., Nelson, K.E., Eisen, J., Heidelberg, J.F., Alley, M.R., Ohta, N., Maddock, J.R., *et al.* (2001) Complete genome sequence of *Caulobacter crescentus*. *Proc. Natl Acad. Sci. USA* **98**, 4136–4141.

Nunez, C., Leon, R., Guzman, J., Espin, G. and Soberon-Chavez, G. (2000) Role of *Azotobacter vinelandii mucA* and *mucC* gene products in alginate production. *J. Bacteriol.* **182**, 6550–6556.

Ochs, M., Veitinger, S., Kim, I., Welz, D., Angerer, A. and Braun, V. (1995) Regulation of citrate-dependent iron transport of *Escherichia coli*: FecR is required for transcription activation by FecI. *Mol. Microbiol.* **15**, 119–132.

Ochs, M., Angerer, A., Enz, S. and Braun, V. (1996) Surface signaling in transcriptional regulation of the ferric citrate transport system of *Escherichia coli*: mutational analysis of the alternative sigma factor FecI supports its essential role in *fec* transport gene transcription. *Mol. Gen. Genet.* **250**, 455–465.

Ochsner, U.A. and Vasil, M.L. (1996) Gene repression by the ferric uptake regulator in *Pseudomonas aeruginosa*: cycle selection of iron-regulated genes. *Proc. Natl Acad. Sci. USA* **93**, 4409–4414.

Ochsner, U.A., Johnson, Z., Lamont, I.L., Cunliffe, H.E. and Vasil, M.L. (1997) Exotoxin A production in *Pseudomonas aeruginosa* requires the iron-regulated *pvdS* gene encoding an alternative sigma factor. *Mol. Microbiol.* **21**, 1019–1028.

O'Keeffe, T., Hill, C. and Ross, R.P. (1999) Characterization and heterologous expression of the genes encoding enterocin A production, immunity, and regulation in *Enterococcus faecium* DPC1146. *Appl. Environ. Microbiol.* **65**, 1506–1515.

Paget, M.S., Kang, J.G., Roe, J.H. and Buttner, M.J. (1998) sigmaR, an RNA polymerase sigma factor that modulates expression of the thioredoxin system in response to oxidative stress in *Streptomyces coelicolor* A3(2). *EMBO J.* **17**, 5776–5782.

Paget, M.S., Leibovitz, E. and Buttner, M.J. (1999a) A putative two-component signal transduction system regulates sigmaE, a sigma factor required for normal cell wall integrity in *Streptomyces coelicolor* A3(2). *Mol. Microbiol.* **33**, 97–107.

Paget, M.S.B., Chamberlin, L., Atrih, A., Foster, S.J. and Buttner, M.J. (1999b) Evidence that the extracytoplasmic function sigma factor sigmaE is required for normal cell wall structure in *Streptomyces coelicolor* A3(2). *J. Bacteriol.* **181**, 204–211.

Paget, M.S., Bae, J.B., Hahn, M.Y., Li, W., Kleanthous, C., Roe, J.H. and Buttner, M.J. (2001a) Mutational analysis of RsrA, a zinc-binding anti-sigma factor with a thiol-disulphide redox switch. *Mol. Microbiol.* **39**, 1036–1047.

Paget, M.S., Molle, V., Cohen, G., Aharanowitz, Y. and Buttner, M.J. (2001b) Defining the disulphide stress response in *Streptomyces coelicolor* A3(2): identification of the SigR regulon. *Mol. Microbiol.* **42**, 1007–1020.

Paget, M., Hong, H.-J., Bibb, M. and Buttner, M.J. (2002) The ECF sigma factors of *Streptomyces coelicolor* A3(2). In: *Switches, Signals, Regulons & Cascades: Control of Bacterial Gene Expression: Society for General Microbiology Symposium, Volume No. 61*. (C.M. Thomas and D.A. Hodgson, eds). Cambridge University Press, Cambridge.

Perego, M. (1998) Kinase-phosphatase competition regulates *Bacillus subtilis* development. *Trends Microbiol.* **6**, 366–370.

Perego, M., Glaser, P., Minutello, A., Strauch, M.A., Leopold, K. and Fischer, W. (1995) Incorporation of D-alanine into lipoteichoic acid and wall teichoic acid in *Bacillus subtilis*. Identification of genes and regulation. *J. Biol. Chem.* **270**, 15598–15606.

Peschel, A., Otto, M., Jack, R.W., Kalbacher, H., Jung, G. and Gotz, F. (1999) Inactivation of the *dlt* operon in *Staphylococcus aureus* confers sensitivity to defensins, protegrins, and other antimicrobial peptides. *J. Biol. Chem.* **274**, 8405–8410.

Peterson, J.D., Umayam, L.A., Dickinson, T., Hickey, E.K. and White, O. (2001) The comprehensive microbial resource. *Nucleic Acids Res.* **29**, 123–125.

Pooley, H.M. and Karamata, D. (2000) Incorporation of [2-^{3}H] glycerol into cell surface components of *Bacillus subtilis* 168 and thermosensitive mutants affected in wall teichoic acid synthesis: effect of tunicamycin. *Microbiology* **146**, 797–805.

Pooley, H.M., Merchante, R. and Karamata, D. (1996) Overall protein content and induced enzyme components of the periplasm of *Bacillus subtilis*. *Microb. Drug Resist.* **2**, 9–15.

Pradel, E. and Locht, C. (2001) Expression of the putative siderophore receptor gene *bfrZ* is controlled by the extracytoplasmic-function sigma factor BupI in *Bordetella bronchiseptica*. *J. Bacteriol.* **183**, 2910–2917.

Price, C.W. (2000) Protective function and regulation of the general stress response in *Bacillus subtilis* and related gram-positive bacteria. In: *Bacterial Stress Responses* (G. Storz and R. Hengge-Aronis, eds), pp. 179–197. ASM Press, Washington, D.C.

Qiu, J. and Helmann, J.D. (2001) The −10 region is a key promoter specificity determinant for the *Bacillus subtilis* extracytoplasmic-function sigma factors sigma(X) and sigma(W). *J. Bacteriol.* **183**, 1921–1927.

Quentin, Y., Fichant, G. and Denizot, F. (1999) Inventory, assembly and analysis of *Bacillus subtilis* ABC transport systems. *J. Mol. Biol.* **287**, 467–484.

Raina, S., Missiakas, D. and Georgopoulos, C. (1995) The *rpoE* gene encoding the sigma-E (sigma-24) heat shock sigma factor of *Escherichia coli*. *EMBO J.* **14**, 1043–1055.

Raman, S., Song, T., Puyang, X., Bardarov, S., Jacobs, W.R. Jr. and Husson, R.N. (2001) The alternative sigma factor SigH regulates major components of oxidative and heat stress responses in *Mycobacterium tuberculosis*. *J. Bacteriol.* **183**, 6119–6125.

Ravio, T. and Silhavy, T.J. (2001) Periplasmic stress and ECF sigma factors. *Annu. Rev. Microbiol.* **55**, 591–624.

Reizer, J., Reizer, A., Perego, M. and Saier, M.H. Jr. (1997) Characterization of a family of bacterial response regulator aspartyl-phosphate (RAP) phosphatases. *Microb. Comp. Genomics* **2**, 103–111.

Resto-Ruiz, S.I., Sweger, D., Widen, R.H., Valkov, N. and Anderson, B.E. (2000) Transcriptional activation of the *htrA* (high-temperature requirement A) gene from *Bartonella henselae*. *Infect. Immun.* **68**, 5970–5978.

Rombel, I.T., McMorran, B.J. and Lamont, I.L. (1995) Identification of a DNA sequence motif required for expression of iron-regulated genes in pseudomonads. *Mol. Gen. Genet.* **246**, 519–528.

Rouviere, P.E., De Las Penas, A., Mecsas, J., Lu, C.Z., Rudd, K.E. and Gross, C.A. (1995) *rpoE*, the gene encoding the second heat-shock sigma factor, sigma-E, in *Escherichia coli*. *EMBO J.* **14**, 1032–1042.

Rowen, D.W. and Deretic, V. (2000) Membrane-to-cytosol redistribution of ECF sigma factor AlgU and conversion to mucoidy in *Pseudomonas aeruginosa* isolates from cystic fibrosis patients. *Mol. Microbiol.* **36**, 314–327.

Schaeffer, P. (1969) Sporulation and the production of antibiotics, exoenzymes, and exotoxins. *Bacteriol. Rev.* **33**, 48–71.

Sexton, R., Gill, P.R. Jr., Callanan, M.J., O'Sullivan, D.J., Dowling, D.N. and O'Gara, F. (1995) Iron-responsive gene expression in *Pseudomonas fluorescens* M114: cloning and characterization of a transcription-activating factor, PbrA. *Mol. Microbiol.* **15**, 297–306.

Sexton, R., Gill, P.R. Jr., Dowling, D.N. and O'Gara, F. (1996) Transcriptional regulation of the iron-responsive sigma factor gene *pbrA*. *Mol. Gen. Genet.* **250**, 50–58.

Spatafora, G.A., Sheets, M., June, R., Luyimbazi, D., Howard, K., Hulbert, R., Barnard, D., el Janne, M. and Hudson, M.C. (1999) Regulated expression of the *Streptococcus mutans dlt* genes correlates with intracellular polysaccharide accumulation. *J. Bacteriol.* **181**, 2363–2372.

Stiefel, A., Mahren, S., Ochs, M., Schindler, P.T., Enz, S. and Braun, V. (2001) Control of the ferric citrate transport system of *Escherichia coli*: mutations in region 2.1 of the FecI extracytoplasmic-function sigma factor suppress mutations in the FecR transmembrane regulatory protein. *J. Bacteriol.* **183**, 162–170.

Stover, C.K., Pham, X.Q., Erwin, A.L., Mizoguchi, S.D., Warrener, P., Hickey, M.J., Brinkman, F.S., Hufnagle, W.O., Kowalik, D.J., Lagrou, M., *et al.* (2000) Complete genome sequence of *Pseudomonas aeruginosa* PA01, an opportunistic pathogen. *Nature* **406**, 959–964.

Strauch, M.A. (1995) Delineation of AbrB-binding sites on the *Bacillus subtilis spo0H, kinB, ftsAZ,* and *pbpE* promoters and use of a derived homology to identify a previously unsuspected binding site in the *bsuBI* methylase promoter. *J. Bacteriol.* **177**, 6999–7002.

Strauch, M.A. and Hoch, J.A. (1993) Transition-state regulators: sentinels of *Bacillus subtilis* post-exponential gene expression. *Mol. Microbiol.* **7**, 337–342.

Suzuki, T., Itoh, A., Ichihara, S. and Mizushima, S. (1987) Characterization of the *sppA* gene coding for protease IV, a signal peptide peptidase of *Escherichia coli*. *J. Bacteriol.* **169**, 2523–2528.

Takami, H., Nakasone, K., Takaki, Y., Maeno, G., Sasaki, R., Masui, N., Fuji, F., Hirama, C., Nakamura, Y., Ogasawara, N., Kuhara, S. and Horikoshi, K. (2000) Complete genome sequence of the alkaliphilic bacterium *Bacillus halodurans* and genomic sequence comparison with *Bacillus subtilis*. *Nucleic Acids Res.* **28**, 4317–4331.

Tatusov, R.L., Galperin, M.Y., Natale, D.A. and Koonin, E.V. (2000) The COG database: a tool for genome-scale analysis of protein functions and evolution. *Nucleic Acids Res.* **28**, 33–36.

Thackray, P.D., Houston, C.W., Horsburgh, M.J., Trewhitt, S.E.J., Zuber, U., Hecker, M. and Moir, A. (2001) SigM, an ECF sigma factor of *Bacillus subtilis* involved in response to salt stress. *Proc. 11th Int. Conf. Bacilli: Functional Genomics of Gram-Positive Microorganisms.*

Tibazarwa, C., Wuertz, S., Mergeay, M., Wyns, L. and van Der Lelie, D. (2000) Regulation of the *cnr* cobalt and nickel resistance determinant of *Ralstonia eutropha (Alcaligenes eutrophus)* CH34. *J. Bacteriol.* **182**, 1399–1409.

Turner, M.S. and Helmann, J.D. (2000) Mutations in multidrug efflux homologs, sugar isomerases, and antimicrobial biosynthesis genes differentially elevate activity of the sigma(X) and sigma(W) factors in *Bacillus subtilis*. *J. Bacteriol.* **182**, 5202–5210.

Vasil, M.L. and Ochsner, U.A. (1999) The response of *Pseudomonas aeruginosa* to iron: genetics, biochemistry and virulence. *Mol. Microbiol.* **34**, 399–413.

Ward, M.J., Lew, H., Treuner-Lange, A. and Zusman, D.R. (1998) Regulation of motility behavior in *Myxococcus xanthus* may require an extracytoplasmic-function sigma factor. *J. Bacteriol.* **180**, 5668–5675.

Wecke, J., Perego, M. and Fischer, W. (1996) D-alanine deprivation of *Bacillus subtilis* teichoic acids is without effect on cell growth and morphology but affects the autolytic activity. *Microb. Drug Resist.* **2**, 123–129.

Wei, Z.M. and Beer, S.V. (1995) *hrpL* activates *Erwinia amylovora hrp* gene transcription and is a member of the ECF subfamily of sigma factors. *J. Bacteriol.* **177**, 6201–6210.

Welz, D. and Braun, V. (1998) Ferric citrate transport of *Escherichia coli*: functional regions of the FecR transmembrane regulatory protein. *J. Bacteriol.* **180**, 2387–2394.

Wiegert, T., Homuth, G., Versteeg, S. and Schumann, W. (2001) Alkaline shock induces the *Bacillus subtilis* σ^W regulon. *Mol. Microbiol.* **41**, 59–71.

Wilson, M.J. and Lamont, I.L. (2000) Characterization of an ECF sigma factor protein from *Pseudomonas aeruginosa. Biochem. Biophys. Res. Commun.* **273**, 578–583.

Wilson, M.J., McMorran, B.J. and Lamont, I.L. (2001) Analysis of promoters recognized by PvdS, an extracytoplasmic-function sigma factor protein from *Pseudomonas aeruginosa. J. Bacteriol.* **183**, 2151–2155.

Wise, A., Brems, R., Ramakrishnan, V., and Villarejo, M. (1996) Sequences in the –35 region of *Escherichia coli rpoS*-dependent genes promote transcription by E sigma S. *J Bacteriol.* **178**, 2785–2793.

Wosten, M.M. (1998) Eubacterial sigma-factors. *FEMS Microbiol. Rev.* **22**, 127–150.

Wu, Q.L., Kong, D., Lam, K. and Husson, R.N. (1997) A mycobacterial extracytoplasmic function sigma factor involved in survival following stress. *J. Bacteriol.* **179**, 2922–2929.

Xiao, Y., Heu, S., Yi, J., Lu, Y. and Hutcheson, S.W. (1994) Identification of a putative alternate sigma factor and characterization of a multicomponent regulatory cascade controlling the expression of *Pseudomonas syringae* pv. syringae Pss61 hrp and *hrmA* genes. *J. Bacteriol.* **176**, 1025–1036.

Xie, Z.D., Hershberger, C.D., Shankar, S., Ye, R.W. and Chakrabarty, A.M. (1996) Sigma factor–anti-sigma factor interaction in alginate synthesis: inhibition of AlgT by MucA. *J. Bacteriol.* **178**, 4990–4996.

Yamazaki, H., Ohnishi, Y. and Horinouchi, S. (2000) An A-factor-dependent extracytoplasmic function sigma factor (σ^{AdsA}) that is essential for morphological development in *Streptomyces griseus. J. Bacteriol.* **182**, 4596–4605.

Yeoman, K.H., May, A.G., deLuca, N.G., Stuckey, D.B. and Johnston, A.W. (1999) A putative ECF sigma factor gene, *rpoI*, regulates siderophore production in *Rhizobium leguminosarum. Mol. Plant Microbe Interact.* **12**, 994–999.

Yu, H., Schurr, M.J. and Deretic, V. (1995) Functional equivalence of *Escherichia coli* sigma-E and *Pseudomonas aeruginosa* AlgU: *E. coli rpoE* restores mucoidy and reduces sensitivity to reactive oxygen intermediates in *algU* mutants of *P. aeruginosa. J. Bacteriol.* **177**, 3259–3268.

Zimmermann, L., Hantke, K. and Braun, V. (1984) Exogenous induction of the iron dicitrate transport system of *Escherichia coli* K-12. *J. Bacteriol.* **159**, 271–277.

How Oxygen Damages Microbes: Oxygen Tolerance and Obligate Anaerobiosis

James A. Imlay

Department of Microbiology, University of Illinois, Urbana, IL 61801, USA

ABSTRACT

The orbital structure of molecular oxygen constrains it to accept electrons one at a time, and its unfavourable univalent reduction potential ensures that it can do so only with low-potential redox partners. In *E. coli*, this restriction prevents oxygen from oxidizing structural molecules. Instead, it primarily oxidizes reduced flavins, a reaction that is harmful only in that it generates superoxide and hydrogen peroxide as products. These species are stronger oxidants than is oxygen itself. They can oxidize dehydratase iron-sulphur clusters and sulphydryls, respectively, and thereby inactivate enzymes that are dependent upon these functional groups. Hydrogen peroxide also oxidizes free iron, generating hydroxyl radicals. Because hydroxyl radicals react with virtually any biomolecules they encounter, their reactivity is broadly dissipated, and only their reactions with DNA are known to have an important physiological impact. *E. coli* elaborates scavenging and repair systems to minimize the impact of this adventitious chemistry; mutants that lack these defences grow poorly in aerobic habitats. Some of the growth deficits of these mutants cannot be easily ascribed to sulphydryl, cluster, or DNA damage, indicating that important aspects of oxidative stress still lack a biochemical explanation.

Obligate anaerobes cannot tolerate oxygen because they utilize metabolic schemes built around enzymes that react with oxidants. The reliance upon low-potential flavoproteins for anaerobic respiration probably causes substantial superoxide and hydrogen peroxide to be produced when anaerobes are exposed to air. These species then generate damage of the same type that they produce in aerotolerant bacteria. However, obligate anaerobes also utilize several classes of dioxygen-sensitive enzymes that are not needed by aerobes. These

ADVANCES IN MICROBIAL PHYSIOLOGY VOL 46
ISBN 0-12-027746-8

enzymes are used for processes that help maintain the redox balance during anaerobic fermentations. They catalyse reactions that are chemically difficult, and the reaction mechanisms require the solvent exposure of radicals or low-potential metal clusters that can react rapidly with oxygen. Recent work has uncovered adaptive strategies by which obligate anaerobes seek to minimize the damage done by superoxide and hydrogen peroxide. Their failure to divest themselves of enzymes that can be directly damaged by molecular oxygen suggests that evolution has not yet provided economical options to them.

ABBREVIATIONS

CuZnSOD   Cu,Zn-superoxide dismutase
Frd   Fumarate reductase

MnSOD   Mn-superoxide dismutase
Pdh   Pyruvate dehydrogenase
Pfl   Pyruvate:formate lyase
PfOR   Pyruvate:ferredoxin oxidoreductase
PUFA   Polyunsaturated fatty acid
SOD   Superoxide dismutase
SOR   Superoxide reductase
TBAR   Thiobarbituric acid-reactive species
TPP   Thiamine pyrophosphate

1. INTRODUCTION

A characteristic property of bacteria is that they cannot tolerate oxygen levels that substantially exceed those of their native habitats. Bacteria have traditionally been classified by whether they can grow in air-saturated media, and sensitivity to oxygen is likely to be a feature that has a substantial impact upon their ability to thrive in important environmental habitats. Yet the process by which oxygen damages bacteria, and the strategies by which bacteria defend themselves against it, are understood only in part. In the past 30 years, the details of oxidative damage have been most fully revealed in *Escherichia coli*. This organism provides several technical advantages, but the most significant is that one can most easily discover what goes wrong in an organism if one understands in detail how it is supposed to work. At this point it is worth considering the extent to which the lessons gleaned from *E. coli* apply to those microbes which are unable to surmount the stress that oxygen creates. The goal of this review is to place in a chemical and physiological context those questions that remain.

2. THE REDOX PROPERTIES OF MOLECULAR OXYGEN

Molecular oxygen is an unusual chemical. Its thermodynamic potential as an oxidant ($+ 0.818$ V, O_2/H_2O) is remarkable, but the molecular–orbital arrangement of its electrons ensures that its potential can only be realized through the agency of transition-metal catalysts. Molecular oxygen is a triplet species, with two spin-aligned, unpaired electrons. For this reason, potential electron donors that are singlet species – that is, which have their electrons spin-paired – cannot form productive orbital overlaps with the half-filled orbitals of oxygen and therefore cannot concertedly transfer two electrons to it (Naqui and Chance, 1986). This is the kinetic restriction that prevents the spontaneous combustion of biological molecules in aerobic environments.

$$O_2 \xrightarrow{-0.16\ V} O_2^- \xrightarrow{+0.94\ V} H_2O_2 \xrightarrow{+0.38\ V} HO\cdot + H_2O \xrightarrow{+2.33\ V} H_2O$$

Figure 1   Univalent reduction potentials of oxygen species. The redox potential cited here for the O_2/O_2^- couple is the one that is most widely used in the biological literature and is based upon a standard condition of 1 M dissolved oxygen (Wood, 1987). Most papers in the chemical literature cite –0.33 V, using a standard condition of water in equilibrium with 1 atmosphere of oxygen.

However, triplet species are not forbidden from participating in univalent electron-transfer reactions. Therefore molecular oxygen can be reduced by one electron, to superoxide, by any electron donor with the driving force to do so. However, the redox potential of the O_2/O_2^- redox couple (–0.16 V) is low, which means that it can oxidize only those molecules with similarly low potentials (Fig. 1). This is a critical constraint: most biomolecules are themselves averse to univalent oxidation because of the loss of the pairing energy that it would entail. As one example, the redox potential of the $NADH^{\cdot+}/NADH$ couple is so high – +0.93 V (Carlson et al., 1984) – that the reaction rate of NADH with oxygen is vanishingly small. This is true as well of the main structural molecules from which cells are constructed: molecular oxygen does not directly oxidize carbohydrates, nucleic acids, lipids or amino acids at rates that would have repercussions for the viability of the organism. Indeed, the apparent inertness of biomaterials confounded initial attempts to understand the toxic actions of oxygen.

3. THE UBIQUITY OF SUPEROXIDE DISMUTASE AND CATALASE

The existence of superoxide dismutase (SOD) was not anticipated; the enzyme was serendipitously discovered by Joe McCord and Irwin Fridovich as a contaminating activity that blocked electron movement between xanthine oxidase and cytochrome c (McCord and Fridovich, 1969). Superoxide was revealed to be the mediator of the electron transfer, and McCord and Fridovich realized that the biological purpose of the enzyme must be to suppress levels of superoxide inside the cell. They soon established that this enzyme is present in most aerotolerant organisms, but is absent or scant in obligately anaerobic bacteria (McCord et al., 1971). This correlation suggested that the sensitivity of the latter organisms to oxygen might be due to their inability to scavenge superoxide. Because a similar biotic distribution was observed for catalase, a scavenger of hydrogen peroxide, the attention of workers interested in oxygen toxicity quickly turned to these species. The McCord–Fridovich hypothesis raised two questions that still drive research in this field: if superoxide and hydrogen peroxide are the agents of oxygen toxicity, how are they formed? And what do they damage?

In 1986, Carlioz and Touati reported the phenotype of mutants of *E. coli* which lacked superoxide dismutase (Carlioz and Touati, 1986). These mutants exhibited defects in catabolism, biosynthesis and DNA replication that could be complemented by heterologous SODs (Natvig *et al.*, 1987), thereby proving that superoxide is formed inside *E. coli* and, if not scavenged, will damage critical biomolecules.

4. HOW ARE SUPEROXIDE AND HYDROGEN PEROXIDE FORMED INSIDE AEROBIC CELLS?

The availability of SOD mutants in *E. coli* gave further momentum to studies of oxidative stress in this bacterium. To find the sources of superoxide, physiological reductants were added to cell extracts, and superoxide evolution was detected by its ability to reduce cytochrome *c* (Imlay and Fridovich, 1991). This approach has identified a number of *E. coli* enzymes that, when reduced, can transfer electrons to oxygen at a significant rate: NADH dehydrogenase II, succinate dehydrogenase, fumarate reductase, and sulphite reductase (Imlay, 1995; Messner and Imlay, 1999). What these enzymes have in common is that they are members of electron-transport chains and that they utilize redox moieties that are adept at univalent electron-transfer reactions – the kind of reactions that are required for the generation of superoxide. Interestingly, in each case it is a solvent-exposed flavin that reacts adventitiously with oxygen.

The reactivity of flavins with oxygen is due to the stability of the flavosemiquinone that is formed upon univalent electron transfer (Fig. 2) (Muller, 1987; Massey, 1994). This feature is important to the normal catalytic cycle of these enzymes, during which the flavins must donate or accept electrons from obligate univalent redox cofactors, such as metal centres. Thus, the active sites of these enzymes are configured to stabilize the flavosemiquinone, which enhances the autoxidizability of the dihydroflavin. Most sites of superoxide production found thus far in mammalian systems have also been the flavins of redox enzymes (Massey *et al.*, 1969; Fridovich, 1970). The rate of superoxide production with these enzymes is first-order in oxygen concentration (K.R. Messner and J.A. Imlay, unpublished data), confirming that the reaction is wholly adventitious and indicating that the degree of oxidative stress that an organism experiences is likely to be proportionate to the oxygen concentration of its immediate environment.

Comparison of flavoproteins reveals that they react with oxygen at widely disparate rates (Table 1). The reasons for these differences have been

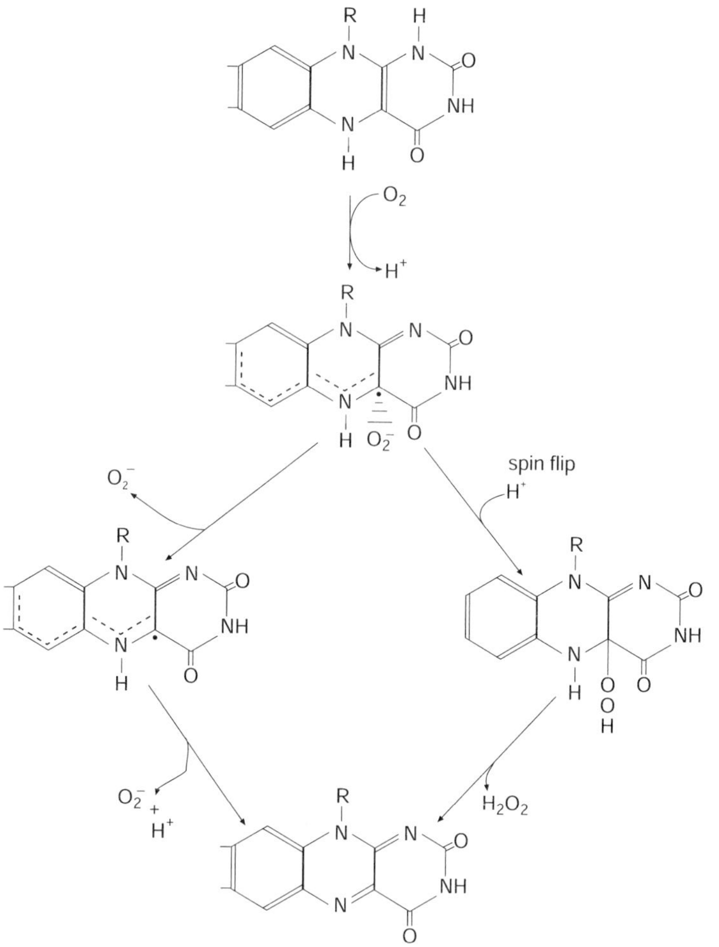

Figure 2 Oxidation of dihydroflavin by molecular oxygen. The initial electron trans-fer (top) generates flavosemiquinone and superoxide. Diffusion of superoxide out of the active site (left) leads to the formation of two molecules of superoxide. In contrast, spin inversion and recombination of radicals (right) generates a peroxyl adduct that dissociates as hydrogen peroxide (Massey, 1994). Flavoproteins that react adventitiously with oxygen produce superoxide and hydrogen peroxide in different ratios.

elucidated in a few cases. Not surprisingly, enzymes whose flavins have a high electron density and substantial solvent exposure are most likely to react with oxygen. The univalent redox potential of the flavin is dependent upon protein context, and it is likely that this feature is also critical, although this prediction has not been directly tested. Interestingly, although most

Table 1 Turnover numbers of selected autoxidizing flavoenzymes.

Enzyme	O_2^-/min	H_2O_2/min	Oxidizing moiety
Flavodoxin[a]	0.055	NR	Flavin
Glutathione reductase[a]	0.83	NR	Flavin
Lipoyl dehydrogenase[a]	2.4	NR	Flavin
Ferredoxin:NADP$^+$ oxidoreductase[a]	3.3	NR	Flavin
Old Yellow enzyme[a]	4.9	NR	Flavin
Sulphite reductase (per flavoprotein)[b]	2.0	7.5	Flavin
Succinate dehydrogenase[c]	12	<1	Flavin
NADH dehydrogenase II[b]	15	130	Flavin
Xanthine 'oxidase'[d]	290	800	Flavin
Fumarate reductase[c]	310	60	Flavin
Neutrophilic NADPH oxidase[e]	3900	NR	Cytochrome b_{558}

NR, not reported.
Turnover numbers vary with substrate concentration; the maximum reported rates are listed here.
[a] Massey *et al.*, 1969;
[b] Messner and Imlay, 1999;
[c] K.R. Messner and J.A. Imlay, unpublished data;
[d] Fridovich, 1970;
[e] Miki *et al.*, 1992.
NADPH oxidase is provided for comparison.

electron-transport chains also contain quinones, iron-sulphur clusters, and haems, all of which are facile univalent reductants, these moieties seem not to react substantially with oxygen (Messner and Imlay, 1999). Cytochrome oxidase, of course, is a special case, and although superoxide and hydrogen peroxide might be construed to be formal intermediates in the catalytic cycle of this enzyme, neither species is released from the active site in detectable amounts (Minghetti and Gennis, 1988).

Although autoxidizing flavoproteins are often initially identified by their ability to generate superoxide, in most cases the predominant product of flavin oxidation is actually hydrogen peroxide (Messner and Imlay, 1999). At first blush this appears to violate the spin restriction that forbids divalent electron transfers to oxygen, but studies of flavoprotein oxidases have resolved this dilemma (Anderson, 1982; Muller, 1987). These oxidases include enzymes whose physiological function appears to be the divalent transfer of electrons to molecular oxygen. The current model (Fig. 2) is that the reduced flavins initially transfer a single electron to molecular oxygen, forming flavosemiquinone and superoxide. Before superoxide escapes the active-site solvent cage, either it or the flavosemiquinone undergoes a spin inversion, allowing productive orbital overlap between the two radical electrons and the formation of a peroxy adduct at the C4 carbon of the isoalloxazine moiety. Subsequent heterolytic cleavage releases hydrogen peroxide. It seems likely that a similar series of events happens in flavoenzymes whose reaction with oxygen is adventitious. In those enzymes, the recombination of superoxide and flavosemiquinone radicals is a less-efficient event, and some superoxide escapes the active site before the adduct can be formed. Among the closely studied enzymes, the yield of superoxide accounts varies between 5% and 80% of the oxygen product (e.g. Messner and Imlay, 1999).

The only enzyme that is known to generate substantial superoxide other than by flavin oxidation is the respiratory cytochrome bc_1 complex, which is found in many bacteria and in eukaryotic mitochondria, but not in E. coli. In the normal course of its catalytic cycle, an unstable ubisemiquinone is formed on the periplasmic face of the enzyme. The radical nature of oxygen allows it to react rapidly with other radicals, and so oxygen intercepts up to 1% of the ubisemiquinone intermediate (Boveris and Cadenas, 1982). Superoxide formation is markedly accelerated by inhibitors that block normal electron movement from the semiquinone to its cytochrome b partner. The superoxide that is formed is detected on the periplasmic face of the membrane and therefore seems most likely to pose an oxidative threat for biomolecules in the bacterial periplasm.

5. HOW MUCH SUPEROXIDE AND HYDROGEN PEROXIDE DO CELLS EXPERIENCE?

The number of flavoproteins in E. coli is likely to number less than 50, and only a handful can be counted among the most abundant proteins in any single growth circumstance. For example, during aerobic growth on glucose, only the NADH dehydrogenases, succinate dehydrogenase, the pyruvate and oxoglutarate complexes, glutathione reductase, and alkyl hydroperoxide reductase are flavoproteins that handle a large electron flux. Of these, NADH dehydrogenase II has, by an order of magnitude, the highest autoxidation turnover number (Messner and Imlay, 1999). Thus it is possible that in this circumstance, this one enzyme produces the majority of endogenous superoxide and hydrogen peroxide – in effect, that it establishes the degree of oxidative stress with which these cells must cope. In other environments, this is not true: when anaerobic cells are shifted into air, fumarate reductase, which is induced under anaerobic conditions, is exposed to air and surpasses NdhII as an oxidant source (Imlay, 1995). Thus the amount of superoxide that is formed inside a cell can be determined largely by the reactivities and titres of its one or two most autoxidizable enzymes. For this reason, the degree of oxidative stress depends upon the enzyme profile of the bacterium. It may vary markedly from one organism to another, and within a single organism from one growth circumstance to another. This feature of oxidative stress is likely to be an important factor in establishing the sensitivity of different bacteria to oxygen.

Methods do not exist that can measure the amount of superoxide inside bacteria. The rate of superoxide production has been suggested to be approximately 5 μM/s^{-1} (Imlay, 1995). This estimate is based upon extrapolations of the rates at which enzymes generate superoxide *in vitro*. Given the high level of SOD inside cells, the predicted level of superoxide is only about 10^{-10} M, less than one molecule per cell. The same approach indicated hydrogen peroxide yields of 5–10 μM s^{-1}. More recently, the formation of hydrogen peroxide (H_2O_2) was monitored *in vivo* using mutants which lack catalases and peroxidases. It was produced at a rate of 14 μM s^{-1} during exponential growth in glucose medium (Seaver and Imlay, 2001a).

H_2O_2 accumulation inside *E. coli* is limited primarily by alkylhydroperoxide reductase, an NADH-driven peroxidase that is more efficient at low H_2O_2 concentrations than is catalase. Steady-state H_2O_2 concentrations have been estimated to be about 10^{-7} M in glucose medium (Seaver and Imlay, 2001b). Importantly, H_2O_2 is also generated in culture medium when growth substrates react with oxygen. This H_2O_2 then diffuses into cells, often at a rate that exceeds the rate of internal production. Thus in laboratory cultures, and perhaps in nature, H_2O_2 stress may be primarily created by the environment rather than by endogenous metabolism.

6. HOW DOES SUPEROXIDE DAMAGE CELLS?

While molecular oxygen can abstract electrons from the flavins and semi-quinones of redox enzymes, these reactions are themselves not harmful, other than by fractionally short-circuiting the electron transport pathway. (Less than 1% of the electrons that flow through the *E. coli* chain are leaked to oxygen.) The harm arises from the fact that superoxide and hydrogen peroxide are both more reactive than is oxygen itself, and these species can efficiently damage biomolecules with which oxygen reacts slowly or not at all.

6.1. Damage to Iron-sulphur Clusters

Much of the early controversy over the role of SOD arose from scepticism that superoxide could react with biomolecules. The redox potentials of superoxide (Fig. 1) suggest that it is both a weak reductant and a moderate oxidant. Its activity as an oxidant, however, is expected to be kinetically restricted because, at physiological pH, superoxide is anionic and therefore a poor electrophile. In fact, *in vitro* experiments established that superoxide does not react detectably with amino acids, common carbohydrate metabolites, or nucleic acids (Bielski and Richter, 1977; Fitzsimons, 1979; Sawyer and Valentine, 1981; Fee, 1982). However, Carlioz and Touati's demonstration that SOD mutants exhibit specific phenotypes indicated that there must be something that superoxide can damage. The breakthrough was Olen Brown's observation that hyperbaric oxygen creates auxotrophies in *E. coli* for branched-chain, aromatic, and sulphur-containing amino acids (Boehme *et al.*, 1976; Brown and Seither, 1983) – the very same defects that were later observed under normoxia in SOD mutants. By feeding cell intermediates in the branched-chain pathway, Brown determined that metabolites downstream of dihydroxyacid dehydratase relieved the auxotrophy of hyperoxic cells, while those upstream of the enzyme did not. It was subsequently shown that this enzyme is inactive in SOD mutants and rapidly loses activity when it is exposed to superoxide *in vitro* (Kuo *et al.*, 1987; Brown *et al.*, 1995). Presumably Brown's hyperoxic treatment had also created superoxide stress, since the rate of superoxide formation is first-order in oxygen concentration.

Dihydroxyacid dehydratase belongs to a family of enzymes that utilize active-site [4Fe-4S] clusters to bind and dehydrate substrate (Flint and Allen, 1996) (Fig. 3). Three of the four iron atoms in the cluster are liganded

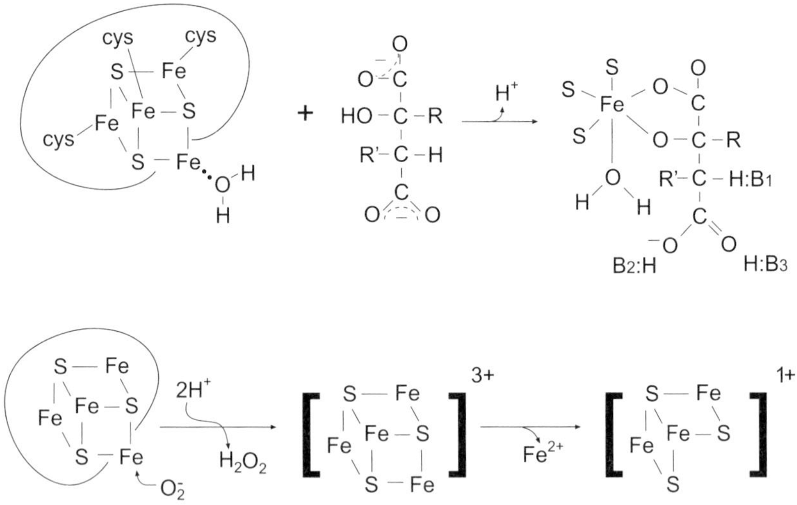

Figure 3 Reactivity of the [4Fe-4S] cluster of dehydratases. (Top) The solvent-exposed iron atom of aconitase moves to six-coordinate geometry when it binds substrate (Lauble *et al.*, 1992). Coordination by iron of the substrate hydroxyl group facilitates its removal. (Bottom) In the absence of substrate, the solvent-exposed cluster is oxidized by superoxide to an unstable state that releases iron.

by cysteinyl residues, while the fourth has an open coordination site occupied by water, which the hydroxyl group of substrate displaces. Through its action as a Lewis acid, the iron atom abstracts the hydroxide ion as part of the dehydration reaction. The vulnerability of this enzyme arises because of three unusual attributes: the cluster is exposed, the catalytic iron atom electrostatically attracts the superoxide anion, and the cluster is amenable to univalent oxidation. The oxidized cluster is unstable and decays, with loss of a ferrous atom, to a $[3Fe-4S]^{1+}$ cluster (Kuo *et al.*, 1987; Flint and Emptage, 1990):

(1) $[4Fe-4S]^{2+} + O_2^- + 2H^+ \rightarrow [4Fe-4S]^{3+} + H_2O_2$

(2) $[4Fe-4S]^{3+} \rightarrow [3Fe-4S]^{1+} + Fe^{2+}$

The resultant cluster is unable to coordinate substrate and activity is lost. This reaction series has been observed for several enzymes of this class, including aconitases A and B (Gardner and Fridovich, 1991b; S.M. Varghese and J.A. Imlay, unpublished data), fumarases A and B (Liochev and Fridovich, 1992), 6-phosphogluconate dehydratase (Gardner and Fridovich, 1991a), and serine dehydratase (Hofmeister *et al.*, 1994). Damage to aconitase and fumarase explains the inability of *E. coli* SOD mutants to catabolise non-fermentable carbon sources, such as succinate and fumarate. Yeast mutants that lack mitochondrial SOD exhibit the same defect, which has also been attributed to the inactivity of aconitase (van Loon *et al.*, 1986). Interestingly, yeast mutants that lack cytosolic SOD are lysine auxotrophs (Chang *et al.*, 1991), evidently because superoxide inactivates homoaconitase, an analogous enzyme in the lysine biosynthetic pathway (L-L. Liou, M. Wallace, J.S. Valentine and E.B. Gralla, unpublished results).

The second-order rate constant for cluster damage is very high: 10^6–10^7 M^{-1} s^{-1} for a range of [4Fe-4S] dehydratases (Hausladen and Fridovich, 1994; Flint and Allen, 1996). The rate constant for oxidation of these clusters by molecular oxygen is approximately six orders of magnitude lower, presumably reflecting the less favourable redox potential of that reaction.

If the concentration of superoxide in *E. coli* is 10^{-10} M, then the expected lifetime of a cluster is about 35 min (Gort and Imlay, 1998). Interestingly, *E. coli* has the capacity to repair damaged clusters. This is illustrated by the recovery of enzyme activity when the oxidative stress is removed, despite the presence of inhibitors of protein synthesis (Kuo *et al.*, 1987; Gardner and Fridovich, 1992; Keyer and Imlay, 1997; Gort and Imlay, 1998). Repair is reasonably fast, with a halftime of about 5 min, which ensures that, despite continual damage by endogenous superoxide, >90% of dehydratases are active in aerobic cells (Gardner and Fridovich, 1992; Gort and Imlay, 1998). The mechanism of repair is unknown.

It should be noted that in *E. coli* the iron-sulphur clusters of redox-active enzymes, such as succinate dehydrogenase, sulphite reductase and NADH dehydrogenase, are not damaged by superoxide (Imlay, 1995; Messner and Imlay, 1999). One reason may be that these clusters are buried within the protein or lipid bilayer, which anionic superoxide cannot penetrate. Alternatively, oxidation of the reduced forms of these clusters does not destabilize them, and over-oxidation of the oxidized clusters may be thermodynamically prohibited.

6.2. Other Damage Generated by Superoxide

The destruction of iron-sulphur clusters apparently also causes the hyper-mutagenesis that is evident in superoxide-stressed cells (Farr *et al.*, 1986): the released iron reacts with hydrogen peroxide, generating hydroxyl radicals that react with DNA (below) (Liochev and Fridovich, 1994; Keyer *et al.*, 1995; Keyer and Imlay, 1996). This raises the question of whether in fact all facets of superoxide toxicity are consequences, directly or not, of the oxidation of iron-sulphur clusters. It seems likely that this is not true. SOD-deficient mutants of *E. coli* cannot synthesize either cysteine or aromatic amino acids (Carlioz and Touati, 1986), and neither of these phenotypes is easily ascribed to cluster damage. The only cluster found in cysteine biosynthetic enzymes is that of sulphite reductase, and that enzyme apparently cannot be injured by superoxide (Messner and Imlay, 1999). Benov and Fridovich have noted that reduced sulphur species efflux from SOD mutants (Benov *et al.*, 1996). The loss of intermediates in the sulphur-assimilation pathway could explain the phenotype, although there is not yet any compelling explanation for the efflux.

Superoxide can also oxidize the dihydroxyethyl-thiamine intermediate of transketolase (Takabe *et al.*, 1980). The product is a thiaminyl radical that is likely to react rapidly with oxygen, thereby regenerating superoxide. A semi-stable ketohydroxyethyl adduct may be the other product. This adduct could be hydrolysed and released as glycolic acid. The dihydroxyethyl-thiamine intermediate, like the dehydratase clusters, fulfils the criteria for reactivity with superoxide: it is solvent-exposed, is neutral in charge, and can resonance-stabilize the thiaminyl radical that oxidation would produce. This reaction has been cited as the likely cause of the aromatic auxotrophy that superoxide imposes (Benov and Fridovich, 1997; Benov, 1999), since flux through transketolase is necessary for the production of erythrose-4-phosphate, a precursor of aromatic products. Indeed, the overproduction of transketolase relieves this phenotype.

These superoxide targets were detected because they confer biosynthetic deficits that can be clearly defined. It seems likely that other mechanisms of damage still await discovery.

7. MECHANISM OF OXIDATIVE DNA DAMAGE

DNA is the biomolecule most vulnerable to damage, because it is the least expendable. Substantial damage must occur to membranes before the permeability barrier is breached, and an enzymic activity must be almost completely depleted before growth defects emerge. In contrast, a single hit to DNA can be mutagenic or even lethal.

Soon after the discovery of SOD, Beauchamp and Fridovich used xanthine oxidase as an enzyme source to test whether superoxide could oxidize methional *in vitro*. The oxidation occurred, but interestingly, either SOD or catalase was sufficient to block the reaction (Beauchamp and Fridovich, 1970). The oxidation of methional had previously been used to detect hydroxyl radicals formed by gamma radiation, and these workers reasoned that the transfer of an electron from superoxide to hydrogen peroxide – itself formed by spontaneous superoxide dismutation – must create hydroxyl radicals. Further investigation determined that the direct electron transfer from O_2^- to H_2O_2 is too slow to occur at a detectable rate (Halliwell, 1976; McClune and Fee, 1976), and that in the xanthine oxidase system, the transfer was mediated by adventitious iron that contaminated the buffers (Halliwell, 1978; McCord and Day, 1978).

Neither superoxide nor hydrogen peroxide are themselves strong enough oxidizing agents to damage DNA, but the hydroxyl radical is as powerful an oxidant as can exist in an aqueous environment (since a more powerful oxidant would oxidize water, creating a hydroxyl radical). Hydroxyl radicals react at near diffusion-limited rates with virtually all organic molecules, including fatty acids, proteins and DNA. When DNA was exposed to the xanthine oxidase system, it was damaged and organic molecules that themselves react with hydroxyl radicals were able to competitively suppress the damage (Morgan *et al.*, 1976). Again, contaminating iron mediated the reaction (Brawn and Fridovich, 1981). In fact, until the reactivity of superoxide with iron-sulphur clusters was discovered, this series of reactions provided the most plausible explanation for the toxicity of superoxide.

(3) $\quad O_2^- + Fe^{3+} \rightarrow O_2 + Fe^{2+}$

(4) $\quad Fe^{2+} + H_2O_2 \rightarrow Fe^{3+} + OH^- + HO^.$

(5) $\quad HO^. + DNA \rightarrow H_2O + DNA\ lesion$

Experiments with intact cells supported several aspects of the scheme. First, the addition of hydrogen peroxide to cells accelerated DNA damage (Levin *et al.*, 1982; Imlay and Linn, 1986). Damage could be blocked by cell-permeable iron chelators (Imlay and Linn, 1988b) and accelerated by mutations that over-loaded cells with iron (Touati *et al.*, 1995). Finally, starvation slowed the rate of damage, indicating that metabolism must be the source of the electron that is transferred by iron to the peroxide (Imlay and Linn, 1986).

Reactions 2 (p. 121) and 3 (p. 123) are kinetically feasible. The Fenton reaction (2) has a second-order rate constant of 76 M^{-1} s^{-1} (Walling, 1975), so that the millimolar concentrations of H_2O_2 used by many experimenters will oxidize ferrous iron with a half-time of a few seconds. The rate constant for reaction of hydroxyl radical with DNA is approximately 10^9 M^{-1} s^{-1}, similar to the value for its reaction with other biomolecules (Hutchinson, 1985). Given the high concentration of potential hydroxyl-radical targets in a cell, HO· is likely to react very close to the site of the iron that formed it. DNA may be a frequent target, since both its phosphodiester backbone and its bases provide ligands that can tightly bind adventitious iron (Luo *et al.*, 1994; Rai *et al.*, 2001). The univalent oxidation of either bases or ribosyl residues creates DNA radicals, which often react rapidly with molecular oxygen; the ultimate damage product is not determined until the DNA or peroxyl radical is resolved by a second redox reaction. A wide spectrum of products is formed (Blakely *et al.*, 1990).

Thus reactions 2 and 3 of the scheme have substantial experimental support and are considered established. Reaction 1, however, is too slow to occur at a significant rate *in vivo*. The concentration of superoxide inside cells has been estimated to be 10^{-10} M (Imlay and Fridovich, 1991); with a second-order rate constant of only 10^5 M^{-1} s^{-1} (Bielski *et al.*, 1985), iron reduction by superoxide would have a half-time of >10 h (Keyer *et al.*, 1995). Kinetic studies of DNA damage indicate that iron is re-cycled *in vivo* much faster than that. More directly, neither overproduction of superoxide dismutase nor anaerobiosis protects DNA from exogenous H_2O_2, proving that superoxide is not an obligatory participant in the damage process.

Therefore other reductants must be responsible for reducing ferric to ferrous iron *in vivo*. Thiols (10^{-3} M), NAD(P)H (10^{-4} M), and free flavins (10^{-5} M) have all been considered (Winterbourn, 1979; Rowley and Halliwell, 1982); each can reduce free iron *in vitro*. It is thought that starved cells are resistant to DNA damage because these reductant pools are depleted (Imlay and Linn, 1986). Conversely, cells that are treated with cyanide are hypersensitive to peroxide, apparently because they accumulate unusually high concentrations of NADH (Imlay and Linn, 1988a). Mutations that eliminate an NADH-dependent flavin reductase activity greatly diminish the sensitivity of cyanide-treated cells, suggesting that electrons flow from NADH to free flavins to free iron (A.N. Woodmansee and J.A. Imlay, unpublished data). This result indicates that the rate of iron reduction can determine the rate of oxidative DNA damage, and that these rates can vary under different metabolic circumstances.

The origin and nature of the iron that catalyses the damage is unclear. Several studies have used electron paramagnetic resonance to visualize a small (10^{-5} M) pool of chelatable iron in the cytosol (Keyer and Imlay, 1996; Gort and Imlay, 1998). In *E. coli* cells that have not been exposed to peroxide, this iron is predominantly reduced (A.N. Woodmansee and J.A. Imlay, unpublished data). The amount of free iron is increased by regulatory mutations that elevate the rate of iron uptake beyond the capacity of the iron-storage proteins, and vulnerability to DNA damage is proportionately increased (Touati *et al.*, 1995; Keyer and Imlay, 1996). As mentioned, the amount of free iron is also high in superoxide-stressed cells. However, superoxide-mediated damage is not the source of the free iron in SOD-proficient cells, since SOD overproduction does not lower the iron level. An important challenge is to identify the source of this free iron. While it is formally possible that this iron merely represents iron that is brought into cells by the known transporters, it seems implausible that they simply dump iron inside, both because this act would expose cells to the threat of oxidative damage, and also because iron sticks to so many biomolecules that little might find its way to the apoproteins into which it must be incorporated. It is more attractive to imagine that specific iron trafficking mechanisms exist. If so, then the free iron represents iron that has escaped the trafficking system.

8. HOW DOES HYDROGEN PEROXIDE DAMAGE PROTEINS AND MEMBRANES?

Hydrogen peroxide can damage cells in ways that do not entail the formation of hydroxyl radicals. Most bacteria are killed when they are exposed to high doses of H_2O_2 (approximately 20 mM) for extended periods of time (approximately 20 min), and iron chelators do not protect them (Imlay and Linn, 1986). The dose–response curve exhibits a broad shoulder, as if many hits are required to cause death, and the 'dead' cells do not filament, suggesting that the DNA is not the target. Thus the killing kinetics are consistent with the many lesions that would be needed to permeabilize a membrane or to fully deplete the cell of a critical enzyme. To date, however, the target has not been identified. Because such a large H_2O_2 exposure is necessary for the killing effect, this type of injury may be significant only in unusual circumstances – for example, when bacteria are exposed to high levels of peroxide during the oxidative burst of phagocytes.

8.1. Oxidation of Protein Thiols

Much lower concentrations of H_2O_2, however, are sufficient to inhibit growth. The addition of about 50 micromolar H_2O_2 to the growth medium causes *E. coli* to suspend growth until the H_2O_2 is scavenged; at that point growth resumes. When catalase and alkylhydroperoxide reductase, the two main H_2O_2

scavengers inside *E. coli*, are eliminated by mutation, as little as 1 μM H_2O_2 is enough to slow the growth rate (Seaver and Imlay, 2001a). The injury that causes this growth inhibition has not been identified. One possibility is that the H_2O_2 oxidizes the cysteinyl residues of key enzymes. Millimolar quantities of H_2O_2 disrupt the metabolism of eukaryotic cells by oxidizing the active-site cysteine of glyceraldehyde-3-phosphate dehydrogenase (Vaidyanathan *et al.*, 1993; Schuppe-Koistinen *et al.*, 1994; Lind *et al.*, 1998). Other enzymes with key cysteinyl residues are also affected (Chen *et al.*, 1993; Meij *et al.*, 1994; Crow *et al.*, 1995; Caselli *et al.*, 1998; Denu and Tanner, 1998). *In vitro*, H_2O_2 oxidizes the thiolate anion of free cysteine to sulphenic acid, and this derivative reacts with additional thiols to form a disulphide bond:

(6)   Cys-SH $\leftrightarrow$ Cys-S$^-$ + H$^+$

(7)   Cys-S$^-$ + H_2O_2 $\rightarrow$ Cys-SOH + OH$^-$

(8)   Cys-SOH + Cys-SH $\rightarrow$ Cys-SS-Cys + H_2O

When both cysteine residues belong to the same protein, an intraprotein disulphide bond is formed. Such a reaction is responsible for the activation by H_2O_2 of the *E. coli* transcription factor OxyR and the yeast transcription factor Yap1.

However, puzzles remain. First, the rate constant for reaction 7 with free cysteine is only about 20 M^{-1} s^{-1} (Winterbourn and Metodiewa, 1999). That value is consistent with the rates at which the eukaryotic enzymes are inactivated by H_2O_2. However, it is hard to reconcile that constant with growth inhibition of *E. coli* by 1 μM H_2O_2, which should require 10 h to oxidize even half the cysteine residues of a given protein. Indeed, the general cellular thiol pool of *E. coli* is not substantially oxidized by even millimolar levels of H_2O_2 (Smirnova *et al.*, 1997). However, since cysteinyl residues of both OxyR and Ahp react readily with submicromolar concentrations of H_2O_2 (Ellis and Poole, 1997; Aslund *et al.*, 1999), there must exist some protein contexts that greatly accelerate thiol oxidations; perhaps they are also present in the putative H_2O_2-sensitive enzymes. Second, disulphide bonds are efficiently reduced by glutathione, forming mixed disulphides upon the first reaction and restoring reduced thiols upon the second. The oxidized glutathione is then reduced by glutathione reductase:

(9)   Cys-SS-Cys + GSH $\rightarrow$ Cys-S-SG + Cys-SH

(10) Cys-S-SG + GSH $\rightarrow$ Cys-SH + GSSG

(11) NADPH + H$^+$ + GSSG $\rightarrow$ NADP$^+$ + 2GSH

Thus cysteine oxidation can diminish enzyme activity only if it outpaces this disulphide reduction system. One possibility is that some thiols may be buried in active sites that exclude glutathione and therefore may be trapped in either the sulphenate or disulphide forms.

Recent studies of mammalian glyceraldehyde-3-phosphate dehydrogenase have indicated that after oxidation some isozymes of this enzyme are stabilized in the mixed-disulphide form (Lind *et al.*, 1998; Grant *et al.*, 1999). It has been speculated that this is an evolutionarily selected feature that protects the enzyme against the further oxidation of sulphenic acid to a sulphinate (Cys-SO$_2$H) species. Sulphinates cannot be re-reduced by known thiol-dependent reductive systems; therefore, the conversion of an active-site cysteine residue to a sulphinate species could irreversibly inactivate the enzyme.

Mutants of *E. coli* which lack the protective thiol-reducing systems accumulate disulphide-bonded proteins when the cells are cultured in air (Prinz *et al.*, 1997). Thus, at the moment, our understanding of thiol chemistry does not fully explain the vulnerability of these residues to oxidants.

8.2. Non-thiolate Protein Oxidations

Hydrogen peroxide also oxidizes iron-sulphur clusters, *in vitro* and *in vivo* (Flint *et al.*, 1993). The rate constant that was determined with selected dehydratases (approximately 10^2 M^{-1} s^{-1}) was substantially lower than that for superoxide and would indicate a cluster half-life of about 100 min in 1 µM H$_2$O$_2$. The activity of these enzymes has not yet been reported for cultures that are exposed to low levels of H$_2$O$_2$.

Finally, H$_2$O$_2$ can also create carbonyl and methionine sulphoxide adducts in proteins (Levine *et al.*, 1996; Dukan and Nystrom, 1999; Dukan *et al.*, 2000). Rate constants have not been reported, and the physiological significance of such lesions is not clear. Carbonylation occurs more readily with some proteins than others; the vulnerability of aconitase suggests that its metal centre may catalyse the reaction. The discovery that bacteria contain methionine sulphoxide reductases (Rahman *et al.*, 1992) seemed to imply that these lesions are produced in sufficient quantities to warrant a repair system. The reductase is not induced during H$_2$O$_2$ stress, but H$_2$O$_2$ especially inhibits the growth of mutants that lack it (Moskovitz *et al.*, 1995; John *et al.*, 2001).

8.3. Damage to Membranes

Lipid peroxidation has long been a hallmark of oxidative stress in mammalian systems (Gutteridge and Halliwell, 1990). Its significance in bacteria, however, is uncertain, since bacteria lack the polyunsaturated fatty acids that are easily

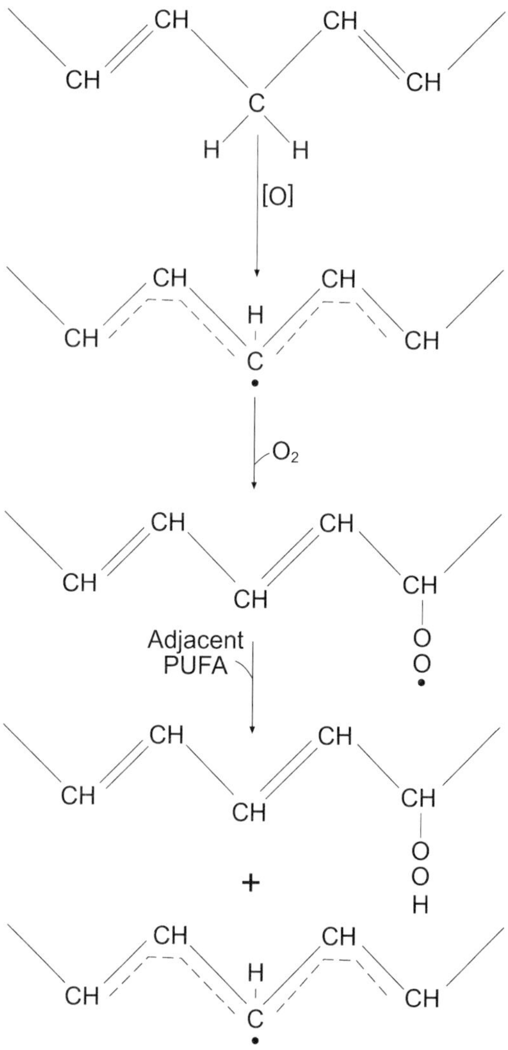

Figure 4 The current model of lipid peroxidation. The chain reaction is initiated by electron abstraction from a bis-allylic carbon atom. The oxidant [O] has not been identified, but may be a hydroxyl radical or iron-liganded molecular oxygen. The carbon-centred radical that is generated, is stabilized by delocalization over flanking carbon atoms. Oxygen addition is favoured at a distal carbon to maintain conjugation, producing a peroxyl radical that propagates the chain by oxidizing neighbouring polyunsaturated fatty acids (PUFAs). Because bacterial membranes lack polyunsaturated lipids, the univalent oxidation reaction is far less favourable, and lipid peroxidation is not known to occur.

oxidized. Mammalian lipids contain substantial linoleic acid. Its bis-allylic carbon atom is easily univalently oxidized, because conjugation to flanking double bonds stabilizes the radical product (Fig. 4). While thermodynamic considerations do not absolutely prohibit the oxidation of (mono-)allylic carbon atoms (Koppenol, 1990), which are present in bacterial monounsaturated fatty acids, these fatty acids were not affected by treatments *in vitro* that rapidly peroxidized polyunsaturated lipids (Bielski *et al.*, 1983). Further, linoleic, linolenic and arachidonic acids were oxidized in proportion to their content of bis-allylic carbon atoms, emphasizing their importance in the process. Still, workers have occasionally speculated that the debilitating effects of oxidants upon transport or energy processes are due to membrane damage (Farr *et al.*, 1988; Kohen and Chevion, 1988; Leven *et al.*, 1990). One study reported the accumulation of TBARs (thiobarbituric acid-reactive species), which are regarded as evidence of peroxidation, in *E. coli* strains that are deficient in alkyl hydroperoxide reductase (Ahp) or catalase and which contain high levels of H_2O_2 (Gonzalez-Flecha and Demple, 1997). More-focused studies will be needed to resolve this issue.

In summary, we are still in the early stages of identifying the types of damage that are created by physiological doses of H_2O_2. This is somewhat ironic, because genetic approaches have provided a comparatively advanced understanding of defences against it. The primary hurdle has been a technical one: because bacteria are so adept at scavenging H_2O_2, it has been difficult to set up systems in which they are exposed to low-but-constant doses. The construction of mutants that are devoid of scavenging activity may push this work forward, as happened when SOD mutants were created.

9. DOES MOLECULAR OXYGEN ITSELF DAMAGE BIOMOLECULES?

The research emphasis on superoxide and hydrogen peroxide may obscure the possibility that excess oxygen blocks the growth of bacteria, including *E. coli*, because molecular oxygen itself directly damages the cell. Because it is a radical species, molecular oxygen reacts rapidly with other radicals. Enzymes which contain free radicals are therefore potential targets of oxygen. In *E. coli*, pyruvate:formate lyase (Pfl), the enzyme that dissimilates pyruvate under anaerobic conditions, contains a glycyl radical that is essential for catalytic function. *In vitro*, this radical reacts very rapidly with oxygen, creating a peroxyl radical that ultimately cleaves the polypeptide backbone (Neidhardt, 1996; Reddy *et al.*, 1998). Synthesis of this enzyme is controlled by the Fnr protein (below) and ceases when the cell is exposed to significant levels of oxygen (Sawers and Suppmann, 1992). Interestingly, at the same time all extant Pfl is deactivated to a non-radical form through an interaction with alcohol dehydrogenase complex (Kessler *et al.*, 1992; Sawers and Watson, 1998). The deactivated enzyme can be reactivated when anaerobic conditions are restored, thereby sparing the cell the expense of synthesizing new enzyme. *E. coli* uses two other glycyl-radical enzymes – anaerobic ribonucleotide reductase (Sun *et al.*, 1996; Frey, 2001) and 2-ketoacid formate lyase (Hesslinger *et al.*, 1998) – when it grows anaerobically; like Pfl, both are extremely oxygen-sensitive and are repressed in aerobic cultures. The formate lyases are unnecessary in aerobic habitats: these reactions serve to preserve the dinucleotide redox balance when an external electron acceptor, like oxygen, is unavailable. In an aerobic habitat, dehydrogenation is a more desirable alternative. Anaerobic ribonucleotide reductase is displaced by an air-tolerant isozyme which, interestingly, has a stable tyrosyl radical at its active site (Sjoberg and Reichard, 1977). The factors that govern the oxygen sensitivity of protein radicals are not clear. In any case, it appears that by switching to aerobic catabolic strategies or oxygen-resistant isozymes, *E. coli* circumvents the growth problems that would otherwise ensue when oxygen damages its glycyl-radical enzymes. Bacteria that fail to make similar accomodations are constrained to grow in anaerobic habitats (below).

Aerobes retain some enzymes, however, that can be directly damaged by oxygen. For example, while the iron-sulphur clusters of dehydratases are most dramatically affected by superoxide and hydrogen peroxide, they are gradually degraded in aerobic extracts even when abundant SOD and catalase are provided (Kuo *et al.*, 1987; Gardner and Fridovich, 1991a,b; Liochev and Fridovich, 1992). This residual damage is presumably mediated by molecular oxygen. Like superoxide and H_2O_2, oxygen is a capable, although less reactive, univalent oxidant, and it is likely that it damages the clusters directly. (The aerobic ribonucleotide reductase is activated when molecular oxygen oxidizes its di-iron cluster (Stubbe and Riggs-Gelasco, 1998).) Surprisingly, experiments have not yet been reported that distinguish whether hyperoxia inactivates these enzymes *in vivo* because the excess oxygen accelerates superoxide production or because the oxygen itself directly oxidizes their clusters. If the latter is true, then this vulnerability must limit the extent to which SOD and peroxidase induction allow aerobes to resist hyperoxia.

10. CONCORDANCE BETWEEN INDUCIBLE DEFENCES AND TYPES OF OXIDATIVE DAMAGE

Another article in this volume reviews the mechanisms that are used by bacteria to sense and mitigate oxidative stress (Pomposiello and Demple, 2002). However, it is worth summarizing the major points in the context of the mechanisms by which oxidants damage the cell.

10.1. Responses to Molecular Oxygen

When *E. coli* moves from an anoxic to an aerobic habitat, molecular oxygen is sensed by Fnr, a transcription factor whose activity is lost upon the oxidation of its [4Fe-4S] cluster (Kiley and Beinert, 1999). The oxidized cluster degrades to a [2Fe-2S] form; by destabilizing the dimer interface, this change eliminates the DNA-binding activity of the protein. The cluster can be directly oxidized by oxygen *in vitro*, and it seems likely that superoxide does not mediate this reaction *in vivo*. Thus the reaction by which oxygen is sensed resembles the reactions by which it can damage biomolecules – a theme that is repeated in the systems that detect superoxide and hydrogen peroxide. Fnr is used to detect oxygen *per se*, rather than respiratory flux, which is the role of the Arc system, and most of the genes that it regulates interact directly with either oxygen or one of its derivatives (Gunsalus and Park, 1994). For example, Fnr ensures that a terminal quinol oxidase is the primary respiratory oxidase when oxygen is present (Cotter and Gunsalus, 1992). Fnr represses the synthesis of MnSOD and CuZnSOD under anaerobic conditions (Compan and Touati, 1993; Gort *et al.*, 1999), when superoxide is absent, and it shuts off the synthesis of fumarate reductase (Frd) in air (Jones and Gunsalus, 1987) when a better respiratory acceptor is available and when Frd would otherwise generate superoxide. As mentioned, the inactivation of Fnr by oxygen also blocks the synthesis of pyruvate:formate lyase in an environment in which it would be damaged.

10.2. Responses to Superoxide

Superoxide is sensed by the SoxR protein (Greenberg *et al.*, 1990; Tsaneva and Weiss, 1990; Ding and Demple, 1997; Gaudu *et al.*, 1997). SoxR has a [2Fe-2S] iron-sulphur cluster, and the fact that it is activated in SOD mutants indicates that superoxide directly oxidizes it (Liochev *et al.*, 1999). Like other [2Fe-2S] clusters, the oxidized cluster of SoxR is stable and does not decompose. However, the change in its redox status alters the protein conformation enough to convert it to a transcriptional activator. Several of the genes that are induced by Sox have obvious roles in avoiding or moderating the types of damage that superoxide is known to create. MnSOD is induced to diminish the amount of superoxide (Hassan and Fridovich, 1977). Increased levels of endonuclease IV facilitate the repair of non-priming termini at oxidatively generated DNA strand breaks (Chan and Weiss, 1987).

Interestingly, SoxRS activates synthesis of fumarase C (Liochev and Fridovich, 1992) and aconitase A (Gruer and Guest, 1994), which are normally minor isozymes. Unlike the housekeeping fumarase A, fumarase C does not have an iron-sulphur cluster and is therefore fully resistant to superoxide.

Aconitase A, like aconitase B, has an iron-sulphur cluster, but it binds an uncharacterized factor that protects it from superoxide *in vitro* and *in vivo* (S.M. Varghese and J.A. Imlay, unpublished data). Through the induction of these enzymes, *E. coli* circumvents the metabolic consequences that would otherwise follow from the inactivation of its housekeeping aconitase and fumarase. One might wonder why *E. coli* does not use these resistant enzymes, rather than their vulnerable isozymes, during normal growth. The most obvious hypothesis is that the resistant enzymes are poorer catalysts, but surprisingly, that appears not to be the case. The turnover number of fumarase C is similar to that of fumarase A (Flint and Allen, 1996) and the aconitases have similar values of k_{cat}/K_M (Jordan *et al.*, 1999). It appears that the lability of the aconitase B cluster is important in the control of iron metabolism, and cells may maintain this unstable enzyme for that reason. One can also speculate that the superoxide sensitivity of fumarase A is useful when anaerobic cells, containing abundant fumarate reductase, enter aerobic habitats. Under those conditions, the preformed Frd generates substantial superoxide that could debilitate the cell (Imlay, 1995). However, if the superoxide inactivates fumarase, fumarate will accumulate and competitively inhibit further superoxide formation at the active site of Frd. This scheme has not been tested *in vivo*.

Several of the genes that are induced as part of the SoxR response have not yet been neatly tied to the behaviour of superoxide. A ferredoxin/flavodoxin oxidoreductase is induced (Liochev *et al.*, 1994), as are two flavodoxins (Zheng *et al.*, 1999; Guadu and Weiss, 2000), but their roles are not known. The induction of glucose-6-phosphate dehydrogenase (Greenberg and Demple, 1989) increases the rate at which NADPH is formed, but it has not been shown why superoxide stress should increase the demand for it. Demple and colleagues recently used microarrays to identify other Sox-regulated genes; it is hoped that this effort will further illuminate the defensive strategy and indirectly provide clues to the impact of superoxide upon cell physiology (Pomposiello *et al.*, 2001).

The SoxR system is not substantially activated in conventional aerobic laboratory cultures, and it is possible that its 'set point' is calibrated to detect a level of superoxide stress that occurs only when *E. coli* is exposed to redox-cycling compounds. Consistent with this idea is the observation that Sox activation prompts the induction of AcrAB, an antibiotic exporter and *micF*, an antisense RNA that inhibits synthesis of the OmpF porin and thereby diminishes antibiotic flux into the cell (Hidalgo and Demple, 1996; Miller and Sulavik, 1996; White *et al.*, 1997). However, a recent study indicated that the Sox regulon is expressed when chemostatic cultures receive normal aeration (Tseng *et al.*, 2001).

10.3. Responses to Hydrogen Peroxide

The OxyR regulatory protein is activated when hydrogen peroxide oxidizes a key cysteine residue, triggering the formation of a disulphide bond and a large conformational shift (Zheng *et al.*, 1998; Choi *et al.*, 2001). The activated protein acts as a positive transcription factor of about two dozen genes. These include those that encode the two primary scavengers of H_2O_2, HPI catalase and alkyl hydroperoxide reductase (Zheng *et al.*, 2001). Glutathione reductase and glutaredoxin are induced, presumably as a compensatory response to the threat that hydrogen peroxide poses for intracellular sulphydryls. Interestingly, the synthesis of Fur protein is also increased (Zheng *et al.*, 1999). Fur acts as a repressor of the synthesis of iron import systems, and it may be induced to ensure that little iron is available inside the cell to react with H_2O_2. The role of another induced protein, Dps, is less clear. Dps is a structural homologue of ferritin (Grant *et al.*, 1998), and as a metal chelator it might also be expected to diminish the amount of iron available for Fenton chemistry. However, the activity of Dps that has been demonstrated *in vitro* is its ability to condense DNA into a crystal structure that excludes DNA-damaging agents (Almiron *et al.*, 1992). Both of these activities are consistent with its ability to protect DNA from oxidants (Martinez and Kolter, 1997). Further experiments are needed to determine which of these two activities is responsible.

Because iron-sulphur clusters are among the most important targets of oxygen species, it seems reasonable that cells have evolved ways to defend them when stress is high. Indeed, when H_2O_2-treated, paraquat-treated or SOD-deficient *E. coli* are moved into an anaerobic environment, the damaged dehydratases are reactivated within minutes (Kuo *et al.*, 1987; Gardner and Fridovich, 1992; Keyer and Imlay, 1997). The repair of clusters requires, at minimum, that an electron and a ferrous atom be provided to a $[3Fe-4S]^{1+}$ cluster, converting it in turn to $[3Fe-4S]^0$ and $[4Fe-4S]^{2+}$ forms. If the damaged cluster has degraded beyond the $[3Fe-4S]^{1+}$ state, a sulphur donor may also be needed. Enzymes encoded by the *isc* locus catalyse the *de novo* assembly of protein clusters (Zheng and Dean, 1994; Takahashi and Nakamura, 1999; Schwartz *et al.*, 2000), but studies of *isc* mutants have shown that these enzymes are not required for the repair of damaged clusters (O. Djaman and J.A. Imlay, unpublished data). Homologues of the *isc* genes are encoded by the *suf* operon, and these genes are in fact induced in an OxyR-independent fashion when cells are exposed to H_2O_2 (Zheng *et al.*, 2001). Experiments are underway to test whether they are responsible for the repair of damaged clusters.

11. WHEN DO CELLS EXPERIENCE OXIDATIVE STRESS?

While the preceding studies have elaborated mechanistic details of oxidative damage, they do not address a fundamental issue: when in the lifestyle of a bacterium are O_2^- and H_2O_2 abundant enough to substantially reduce its fitness? Several situations have been suggested. The oxidation of extracellular compounds, such as polyphenols, thiols and even carbohydrates, creates H_2O_2 that can then diffuse into cells. Reactive oxygen species are rapidly generated inside cells that have inadvertently imported redox-cycling compounds, including antibiotics that are excreted by plants or other bacteria in order to suppress the growth of competitors. Facultative and aerobic bacteria appear to have calibrated their OxyR and SoxRS control systems so that they are not activated by the doses of oxidants that are generated during normal aerobic metabolism, which suggests that the purpose of these systems may be to defend bacteria against these extracellular sources of stress.

Most consideration, however, has gone to whether *E. coli* is adversely affected by oxidants that it generates itself. One experimental approach to this question would be to measure growth rates as oxygen concentration is systematically varied across levels that are themselves adequate to saturate the cytochrome oxidases – essentially, to test whether *E. coli* is a subtle microaerophile. I am unaware of any attempt to do so that would have detected minor effects on growth rate. A second approach would be to overproduce scavenging enzymes to determine whether a diminution of H_2O_2 or O_2^- concentration would improve cell performance. We failed to observe any improvement in growth rates or yield when SOD was overproduced, although our experiments would have been adequate only to detect large effects (unpublished data).

An alternative strategy is to compare the 'normal' intracellular superoxide level with that which causes noticeable deficits in growth or mutation rate. In one study SOD synthesis was controlled by a heterologous promoter and systematically varied (Gort and Imlay, 1998). When SOD levels were lowered to less than half the usual value, the activities of superoxide-sensitive enzymes were affected. When they were less than one-fourth their usual value, the rate of growth was measurably diminished. The implication was that *E. coli* expresses just enough SOD and cluster-repair activity to tolerate the amount of superoxide that is formed during routine aerobiosis. This analysis agreed quantitatively with predictions that were based upon the known or estimated rate constants for superoxide production, scavenging, enzyme damage and cluster repair. The idea that *E. coli* does not experience undue stress during routine aerobiosis is consistent with the observation that the SoxRS regulon is not active under these conditions.

It may prove difficult to conduct a similar analysis for H_2O_2 toxicity. Mutants that lack catalase and Ahp accumulate micromolar intracellular H_2O_2

and grow poorly (Seaver and Imlay, 2001a). However, their poor growth could be largely corrected by the addition of extracellular catalase, indicating that H_2O_2 equilibrates between the cells themselves and their extracellular habitat. This is not surprising, since the permeability coefficient of H_2O_2, unlike that of superoxide, is high (Seaver and Imlay, 2001b). However, H_2O_2 is created in laboratory media by the autoxidation of medium components. Under room lights, for example, chromophores are photochemically activated to transfer electrons to oxygen (McCormick et al., 1976), creating oxidative stress that can inhibit growth. Glucose and sulphydryls in the growth media also oxidize, forming H_2O_2 at substantial rates. Comparison of the rate of H_2O_2 diffusion across membranes, with the rate of endogenous H_2O_2 production, reveals that most intracellular H_2O_2 arises from extracellular sources whenever the concentration of H_2O_2 in the medium exceeds 0.2 μM (Seaver and Imlay, 2001b). Typical laboratory media contain much more H_2O_2 than this, so that intracellular H_2O_2 stress is in some ways an artificial consequence of the laboratory medium. In nature, of course, H_2O_2 stress can arise in the same way, but there is no reason to believe that laboratory conditions quantitatively mimic the rate of extracellular H_2O_2 formation in native habitats. Thus one cannot infer from experimental systems whether the H_2O_2 defences of E. coli or any bacterium are adequate to deal with the H_2O_2 doses to which they are exposed in nature.

12. OBLIGATE ANAEROBIOSIS

Bacteria that are categorized as either microaerophiles or obligate anaerobes are, by definition, at least partially incapacitated by the oxidative stress that they experience in air-saturated laboratory media. In 1971, McCord and Fridovich noted that SOD and catalase levels tend to be low or absent from these organisms (McCord et al., 1971). That correlation prompted their suggestion that superoxide and H_2O_2 toxicity might be the root cause of obligate anaerobiosis. As the previous discussion has shown, it is only recently that workers have uncovered molecular bases of oxidative injuries in aerobes. Investigators are now turning to enquire whether the injuries that block the growth of anaerobes are of the same type.

12.1. Oxygen Diminishes the Availability of Reduced Growth Substrates

Bacteria that require low-potential substrates cannot grow in aerobic environments, irrespective of any damage that reactive oxygen species might incur within the cell. Sulphur and iron, the elements that are the foci of oxidative stress inside the cell, are also the most redox-sensitive substrates outside them. Molecular oxygen directly oxidizes reduced sulphur compounds and thereby starves cells that require them. The same is true of iron, which is notoriously unavailable in aerobic habitats. Anionic chelators bind ferric iron more strongly than ferrous iron. Therefore, when chelators bind ferrous iron, its redox potential drops by up to half a volt, and its oxidation by molecular oxygen accelerates (Koppenol, 1985). For example, while ferrous iron is only slowly oxidized in water, the half-time for its oxidation in 50 mM phosphate salts is <1 min (K.R. Messner and J.A. Imlay, unpublished data). As a result, in many aerobic environments, iron may be available only in its ferric form. Ferric iron is poorly soluble, and its adhesion to a plethora of incidental molecules further reduces its availability for transport systems. Aerobic bacteria solve this problem by exporting a variety of siderophores that displace adventitious ligands, bind ferric iron, and are co-transported with the iron into the cell (Neilands, 1993). In contrast, iron is typically reduced in anaerobic habitats. Organisms which have evolved to dwell in anaerobic habitats may lack ferric-harvesting transport systems and be poorly suited for growth when the environmental iron is oxidized by oxygen.

12.2. Is Aerobic Cell Death Artifactual?

When bacteria are classified for aerotolerance, they are typically incubated on plates – a procedure that does not distinguish whether oxygen merely inhibits their growth or irreversibly kills them. A more discriminating test is whether cells that are exposed to air in liquid medium can subsequently form colonies on anaerobic plates. Results vary. Many strains of *Bacteroides*, for example, retain viability for several days when they are exposed to air, whereas some Clostridial species die rapidly (Woods and Jones, 1986).

   Aerobic death seems likely to be caused by DNA lesions, since damaged enzymes could presumably be repaired or replaced. Yet this raises the possibility of an artifact: because oxidative DNA damage is mediated by H_2O_2, one must consider the possibility that substantial H_2O_2 is generated chemically when low-potential anaerobic medium is suddenly infused with oxygen. The pace may be especially large when redox indicators, such as resazurin, are present. These dyes are reducible by thiols and oxidizable by oxygen, so in effect they catalyse the production of H_2O_2. However, even in the absence of these dyes, the photooxidation of medium components and the salt-catalysed oxidation of sugars can generate amounts of H_2O_2 that exceed what these cells are likely to experience when oxygen infiltrates their real-world habitats.

   In the absence of oxygen, some fastidious anaerobes can tolerate substantial exposure to oxygen. Mukhopadhyay and Wolfe observed that the archaeon

Methanobacterium thermoautotrophicum efficiently recovered when an anaerobic culture was flushed with air for 6 h and then made anaerobic once again. The resumption of growth was most rapid when the culture was washed free of all reductants prior to oxygenation (B. Mukhopadhyay and R.S. Wolfe, unpublished data). The latter condition would presumably be more typical of natural environments, where substantial levels of oxygen and low-potential compounds are unlikely to be present simultaneously.

If the behaviour of *M. thermoautotrophicum* is typical of obligate anaerobes, it is more understandable how these microbes traverse aerobic habitats to find anaerobic niches and how they persist in environments that must be periodically influxed with aerobic water.

12.3. Inactivation of Enzymes by Oxidants

It appears that hyperbaric oxygen inhibits the growth of aerobic and facultative organisms because it accelerates the production of superoxide, which damages specific catabolic and biosynthetic enzymes. One theory of obligate anaerobiosis is that anaerobes cannot grow in air because they suffer the same kind of damage, albeit at normoxic rather than hyperoxic conditions (McCord *et al.*, 1971; Tally *et al.*, 1977; Gregory *et al.*, 1978). Their increased sensitivity to oxygen could simply result from their low SOD titres. Alternatively, it is possible that the low-potential chemistry conducted by many anaerobes is simply incompatible with the presence of molecular oxygen. The current data suggest that both ideas are substantially correct.

Evidence to date supports the precept that oxidative damage to central metabolism is the root cause of obligate anaerobiosis, and that central metabolism is vulnerable because of the low-potential biochemistry that enables these organisms to ferment effectively. Peripheral pathways may also employ oxygen-sensitive enzymes, but that is likely to be a secondary adaptation to the habitat in which the central energy-producing strategy places the organism. It is assumed that if peripheral enzymes were the sole sites of oxygen sensitivity, they would rapidly be expunged to enlarge the potential habitat of the organism.

Bacteroides strains have been one focus of studies of obligate anaerobiosis, because they are relatively oxygen-tolerant. When these bacteria are exposed to air, growth stops abruptly, but viability persists for days (Rocha *et al.*, 1996). These species have levels of SOD and catalase similar to those of facultative bacteria (Gregory *et al.*, 1978; Rocha and Smith, 1999), which makes their inability to grow more of a puzzle. The specific injuries that are caused by oxygen were investigated in *Bacteroides thetaiotaomicron* (Pan and Imlay, 2001). This Gram-negative bacterium catabolyses simple sugars through glycolysis to pyruvate, at which point metabolism bifurcates: some of the pyruvate is reduced to succinate in order to recycle the NADH to NAD$^+$, and the remaining pyruvate is oxidized to acetate, with additional ATP production. Exposure of *B. thetaiotaomicron* to air rapidly disabled catabolism because of the loss of activity of two central enzymes, fumarase and pyruvate:ferredoxin oxidoreductase (PFOR). The fumarase enzyme belongs to the [4Fe-4S] dehydratase family and, like the primary fumarase of *E. coli*, is sensitive to superoxide *in vitro*. While it seems reasonable that superoxide was the toxin *in vivo*, this was not established. Assuming this to be the case, however, this result poses the question of why fumarases in facultative bacteria retain activity when cells are aerated, while that in *B. thetaiotaomicron* does not. One possibility is that the rate of superoxide production might be greater inside air-exposed *B. thetaiotaomicron* than in air-exposed *E. coli*. *Bacteroides* contains high titres of fumarate reductase, which resembles the *E. coli* enzyme in that it generates abundant superoxide when it is exposed to oxygen.

PFOR catalyses the oxidative decarboxylation of pyruvate, with electrons being transferred to ferredoxin rather than NAD$^+$. Ferredoxin delivers the electrons to hydrogenase, evolving H_2. PFOR is used in preference to pyruvate dehydrogenase (Pdh) in many anaerobes, since otherwise more pyruvate must be expended to reconvert NADH to NAD$^+$. In *B. thetaiotaomicron*, the use of PFOR rather than Pdh allows more carbon flux to acetate, with improved yield of ATP. PFOR enzymes contain thiamine pyrophosphate (TPP), which is the site of decarboxylation and coenzyme A addition, and up to three iron-sulphur clusters in series, which direct the electrons from the active site to the protein surface, where ferredoxin docks. The enzyme from most anaerobes loses activity in aerobic buffers. Examination of the *B. thetaiotaomicron* PFOR revealed that it is inactivated *in vitro* by molecular oxygen, but not by superoxide (Pan and Imlay, 2001). That study did not establish the nature of the damage. However, the *Desulfovibrio africaans* enzyme provides an instructive exception to the rule that PFOR is oxygen-sensitive (Pieulle *et al.*, 1995). This enzyme retains activity in aerobic buffer, and it differs from other PFOR enzymes in that it contains an extra domain that is positioned to occlude the terminal iron-sulphur cluster (Chabriere *et al.*, 1999). Deletion of the extra domain causes sensitivity (Pieulle *et al.*, 1997). The implication is that oxygen damages the enzyme by oxidizing the solvent-exposed cluster. The PFOR enzyme in *B. thetaiotaomicron* was repaired *in vivo* when the cell is returned to anaerobic conditions, consistent with the ability of this organism to repair oxidatively damaged clusters (Pan and Imlay, 2001).

Thus, in *B. thetaiotaomicron*, the two events that disable central metabolism each involve the oxidation of an exposed iron-sulphur cluster. The fact that PFOR is inactivated by molecular oxygen, rather than by superoxide, is significant, because it limits the ability of the bacterium to protect itself from an oxidative environment by the synthesis of scavenging enzymes. From an evolutionary perspective, this is not surprising, since if superoxide alone were responsible for the oxygen intolerance of anaerobes, they would probably pay the small price of increased SOD synthesis in order to expand their potential habitat.

In few other anaerobes have the consequences of oxygen exposure been tracked *in vivo*. However, at this point, at least four families of enzymes can be distinguished whose members are highly oxygen-sensitive. The first is the large group of [Fe-S] dehydratases, such as aconitase and fumarase, that operate upon substrates that have activating carbonyl or carboxylate moieties proximal to the site of desaturation/saturation (Flint and Allen, 1996) (Fig. 5, right pathway). This family is present in aerobic and facultative organisms as well as anaerobes, and as shown in *E. coli*, the sensitivity of these enzymes seems to limit the oxygen resistance of aerobes.

PFOR belongs to a second family of ferredoxin-dependent dehydrogenases. The pyruvate, 2-ketoglutarate, indolepyruvate, and branched-chain 2-ketoacid-ferredoxin oxidoreductases are thiamine-dependent oxidative decarboxylases that shuttle electrons to ferredoxin through [4Fe-4S] clusters (Mai and Adams, 1994, 1996). These enzymes, when prepared from anaerobic eubacteria or archaea, are rapidly inactivated when exposed to oxygen *in vitro*. Given their similarity to PFOR, this is likely to involve the oxidation of the superficial iron-sulphur cluster. In contrast, the 2-oxoacid-ferredoxin oxidoreductases of aerobic archaea are oxygen-resistant (Zhang *et al.*, 1996). Their resistance correlates with the absence of two of the three [4Fe-4S] clusters that are found in the anaerobic enzymes. It is not clear why this spares them sensitivity to oxygen, or whether this protection is gained at the price of reduced catalytic efficiency.

Interestingly, a family of ferredoxin-dependent enzymes that oxidize glyceraldehyde-3-phosphate, formaldehyde, and glyceraldehyde has been identified in archaea. They contain tungsten and pterin cofactors as well as iron-sulphur clusters, and they are irreversibly damaged by oxygen *in vitro*.

A third family is comprised of flavin-dependent enzymes that catalyse the dehydration of refractory substrates. These enzymes operate upon lactyl-CoA, phenyllactyl-CoA, 2-hydroxyglutaryl-CoA, and 4-hydroxybutyryl-CoA, which are generated in the reductive branches of carbohydrate and peptide fermentations (Kuchta and Abeles, 1985; Cinkaya *et al.*, 1997; Dickert *et al.*, 2000; Locher *et al.*, 2001) (Fig. 5, left pathway). These substrates are difficult to dehydrate because they lack an electron-withdrawing carbonyl group adjacent to the C-H bond that must be broken. Aerobes do not need to conduct such dehydrations: they are free to operate upon more oxidized substrates and therefore configure their pathways so that only activated metabolites are dehydrated. The flavoenzyme dehydratases contain a stable flavosemiquinone and very low-potential iron-sulphur cluster. While details of the catalytic cycle are unclear, Buckel and co-workers have proposed that an electron is transferred from flavosemiquinone to substrate, generating a ketyl radical that labilizes the proton (Hans *et al.*, 1999). The dehydratase proper is activated when a low-potential electron is provided to it by an activating protein. The activating protein is a

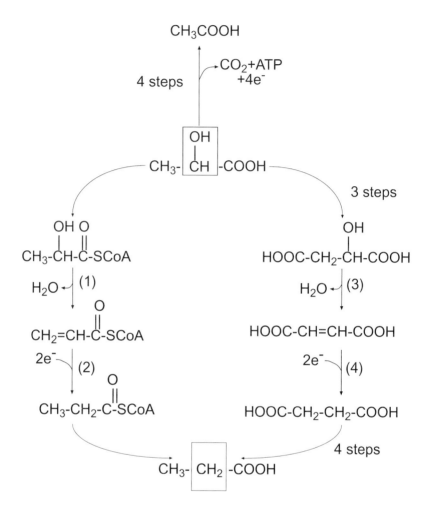

Figure 5   Alternative strategies to ferment lactate to propionate. (Top) The oxidation of lactate to acetate generates ATP. It also generates reducing equivalents that are dissipated by the reduction of lactate to propionate. (Left) The acrylyl-CoA reductive pathway is direct, but requires the dehydration of lactyl-CoA (reaction 1). The dehydratase employs a flavosemiquinone radical. Both the radical and the iron-sulphur clusters of the catalytic and activating enzymes are degraded upon exposure to oxygen. (Right) The succinate reductive pathway is more circuitous, but less oxygen-sensitive. The first three steps provide a carboxylate that activates the methylene carbon for deprotonation; the final four steps recycle the carboxylate. The dehydration (reaction 3) is catalysed by a [4Fe-4S]-containing fumarase. Reaction 4 is catalysed by fumarate reductase. When the latter enzyme is exposed to air, it generates superoxide, which can oxidatively inactivate the fumarase. *Bacteroides* species also utilize this pathway during glucose fermentation (see text), although most succinate is excreted before conversion to propionate.

homodimer with a [4Fe-4S] cluster that bridges the dimer interface, and it is this cluster that delivers the electron to the catalytic enzyme. In the presence of oxygen, this solvent-exposed cluster is rapidly degraded to an inactive [2Fe-2S]$^{2+}$ form (Hans *et al.*, 2000). This cluster conversion resembles that of air-exposed Fnr, the oxygen-sensing transcription factor of *E. coli*. By comparison, the [4Fe-4S]$^{2+}$ cluster of the dehydratase is relatively oxygen-stable, degrading more slowly to a [3Fe-4S]$^{+}$ form, similar to the clusters of the aconitase family. Interestingly, the activating protein exhibits structural and functional similarities to the oxygen-sensitive nitrogenase reductase protein, which also passes a low-potential electron from its exposed cluster to the metal centre of a recipient protein (Strop *et al.*, 2001).

The common element of all of these enzymes is that they employ solvent-exposed [4Fe-4S] clusters – in the simple dehydratases, to interact with metabolites, and in the ferredoxin oxidoreductases and flavin-dependent dehydratases, to interact with the electron-receiving clusters of protein substrates. Notably, it is the oxidized (2+) form of these clusters that is irreversibly damaged by oxidants, because the overoxidized [4Fe-4S]$^{3+}$ cluster is structurally unstable. Presumably, the reduced (1+) forms of these clusters can also be oxidized by oxygen, but that reaction leaves the enzyme in a catalytically competent (2+) form and will have the effect only of diverting a small fraction of the electron flux.

A final prominent class of oxygen-sensitive enzymes includes those which have active-site glycyl radicals (Sawers and Watson, 1998). Anaerobic *E. coli* synthesizes several members of this family: pyruvate-formate lyase, 2-ketoacid-formate lyase (for threonine catabolism), and anaerobic ribonucleotide reductase. Like PFOR, Pfl provides a pathway of pyruvate catabolism that avoids dinucleotide reduction and spares acetyl-CoA for ATP production. Because oxygen reacts avidly with radicals, carbohydrate-fermenting bacteria that rely upon Pfl excrete lactic acid and pyruvate when they are exposed to air (Abbe *et al.*, 1982; Svensater *et al.*, 1985; Yamada *et al.*, 1985; Takahashi *et al.*, 1991). Other glycyl-radical enzymes can be recognized in the genomes by sequence motifs, but as yet have no assigned functions. Interestingly, B12-dependent enzymes generate enzyme and substrate radicals during the catalytic cycle, but oxygen does not interfere, presumably because the radical lifetime is short and its solvent exposure is limited (Marsh, 2000).

12.4. Scavenging Systems in Obligate Anaerobes

Obligate anaerobes defend themselves against oxygen: recent work has established that not only relatively oxygen-tolerant anaerobes, such as *Bacteroides* species, but also *Clostridia* and anaerobic Archaea activate the synthesis of scavenger systems when they sense the presence of oxygen. A surprise has been the discovery that many obligate anaerobes contain superoxide reductase (SOR), rather than superoxide dismutase, as a scavenger of superoxide (Jenney *et al.*, 1999; Abreau *et al.*, 2000; Jovanovic *et al.*, 2000; Lombard *et al.*, 2000a,b; Lumppio *et al.*, 2001). The reductase works as a two-component system, in which low-potential electrons are delivered to the reductase by rubrerythrin. It has been proposed that fastidious anaerobes are better served by a reductase than a dismutase because the latter has the virtue of not generating molecular oxygen as an end product. In this respect, superoxide reductase resembles peroxidase systems, which reduce hydrogen peroxide fully to water rather than dismuting half to molecular oxygen, as catalase does. However, the process of superoxide reduction in place of dismutation trades the formation of oxygen for hydrogen peroxide. In most biochemical contexts the latter molecule is the more active oxidant. An alternative rationale for SOR might be that dismutases require two molecules of substrate to complete their catalytic cycles. When concentrations of the oxidant are low, dismuting enzymes may not scavenge them effectively.

12.5. Evolutionary Pressure to Escape Obligate Anaerobiosis

Thus, the original observation that obligate anaerobes contain little SOD or catalase now appears to be misleading: many use SOR and peroxidases instead. Their overall scavenging capacity may still be lower than that of aerobes, because the vulnerability of their enzymes to molecular oxygen establishes a limit on oxygen tolerance upon which high scavenger titres cannot improve.

Can anaerobes dispense with the reactions that are catalysed by oxygen-sensitive enzymes? Perhaps, but at a price. Anaerobic metabolism is most efficient when electrons can be moved at low potentials onto poor acceptors, processes that are typically catalysed by oxidant-sensitive cofactors. While lactic acid bacteria, for example, conduct fermentations in air, the absence of Pfl and PFOR prevents them from harvesting energy from pyruvate dissimilatory pathways, so that the net ATP yield is sharply diminished from that of anaerobes that contain these enzymes. This reduces the competitiveness of lactic acid bacteria in many anaerobic habitats and precludes their fermentation of some non-carbohydrate substrates.

Why are oxygen-sensitive enzymes not replaced by oxygen-resistant isozymes? It is very improbable that anaerobes lack selective pressure to evolve oxygen-resistant fermentation pathways: many of these organisms live at interfaces in which oxygen exposure is a common event, and the consequent suspension of metabolism disadvantages them. Further, the existence of oxidant-resistant isozymes of fumarase, aconitase, PFOR and even nitrogenase refutes the contention that these chemistries can be achieved only by enzyme structures that are inherently vulnerable to oxidants. Thus, the fact that oxidant-sensitive enzymes have not been displaced by resistant isozymes suggests that there is a down-side to the resistant enzymes that remains to be uncovered. While a reduced catalytic efficiency is one obvious guess, this has not been supported thus far.

The case of nitrogenase most clearly forces this issue. It stands alone among very oxygen-sensitive enzymes in that its activity would be just as useful in air as in anoxia. Yet most nitrogen-fixing bacteria have evolved elaborate, cumbersome and metabolically expensive strategies to achieve microoxic and anoxic conditions, so that the iron-sulphur clusters of this enzyme, and of the ferredoxin-reducing enzymes that provide electrons to it, are protected. This is true despite the recent discovery in *Streptomyces thermoautotrophicus* of an alternative nitrogenase that is evidently fully oxygen-resistant (Eibbe *et al.*, 1997). The reason that this enzyme is not in widespread use is currently a mystery – just as, in general, we have yet to fully illuminate the reason that anaerobes choose to remain anaerobes.

13. PROSPECTS

In the mid 1980s, the emphasis in research on oxidative stress turned from the debate over whether reactive oxygen species are toxic to an exploration of the mechanisms of their toxicity. While that work has progressed substantially in the past 15 years, particularly in model microbes such as *E. coli*, even in these familiar and experimentally accessible organisms major puzzles remain. For example, the field has not yet reached a conclusion as to whether oxidants damage membranes or how they compromise sulphur assimilation; the significance of protein carbonylation is unsettled, and the source of the free iron that generates hydroxyl radicals is unknown. The purpose of SOD in the periplasmic compartment is not understood at all. Furthermore, while workers have identified most of the enzymes that are induced during oxidative stress, in many cases their roles in combating stress are entirely unclear.

To a certain point, the reliance upon *E. coli* as a model organism can be justified by the truism that chemistry is chemistry – that is, that the adventitious reactions of oxygen species are likely to follow similar rules in all organisms. In many ways that is true – most of the effects of oxygen species upon mammalian cells, for example, were prefigured by observations made in *E. coli*. By now, however, work has progressed to the fine details of oxidative physiology, and it has become apparent that the lessons of *E. coli* cannot be projected upon other microbes without revision. As a first, important example, oxidative stress is both quantitatively and qualitatively different in obligate anaerobes. This is true of the defensive responses of these organisms, too. These differences will be critical, since the real applications of our knowledge of oxidative stress will often concern organisms that in these respects do not resemble model enterics.

As the biochemical details are further pinned down, questions have emerged at an accelerating rate about the ecology and evolution of oxidative stress. The experimental work done to date has been primarily conducted in pure laboratory cultures that are not designed to mimic natural habitats. This is a shortcoming, since community structures in natural environments may allow the function of oxidant signalling and scavenging systems that are ineffectual in dilute, planktonic suspensions. Furthermore, cell defences are presumably evolved to fit the slower and/or episodic patterns of growth that occur in nature. These considerations have prompted some workers to suggest that the unculturability of most bacteria may be due in part to their inability to tolerate oxidative stress under laboratory growth conditions. These issues will also become important when oxygen tension is manipulated for practical reasons – whether to influx oxygen into sediments to promote bioremediation or to maximize stress in antimicrobial medical treatments. Thus, as the focus of inquiry about oxidative stress shifted over time from chemistry to biochemistry, it is now moving to physiology and finally ecology.

ACKNOWLEDGEMENTS

I thank those colleagues who allowed me to cite their unpublished data. This work was supported by GM49640 and GM59030 from the National Institutes of Health.

REFERENCES

Abbe, K., Takahashi, S. and Yamada, T. (1982) Involvement of oxygen-sensitive pyruvate formate-lyase in mixed-acid fermentation by *Streptococcus mutans* under strictly anaerobic conditions. *J. Bacteriol.* **152,** 175–182.

Abreau, I.A., Saraiva, L.M., Carita, J., Huber, H., Stetter, K.O., Cabelli, D. and Teixeira, M. (2000) Oxygen detoxification in the strict anaerobic archaeon *Archaeoglobus fulgidus*: superoxide scavenging by neelaredoxin. *Mol. Microbiol.* **38,** 322–334.

Almiron, M., Link, A.J., Furlong, D. and Kolter, R. (1992) A novel DNA-binding protein with regulatory and protective roles in starved *Escherichia coli. Genes Dev.* **6,** 2646–2654.

Anderson, R.F. (1982) Flavin-oxygen complex formed on the reaction of superoxide ions with flavosemiquinone radicals. In: *Flavins and Flavoproteins* (V. Massey and C.H. Williams, eds). pp. 278–283. Elsevier North Holland, New York.

Aslund, F., Zheng, M., Beckwith, J. and Storz, G. (1999) Regulation of the OxyR transcriptional factor by hydrogen peroxide and the cellular thiol-disulphide status. *Proc. Natl Acad. Sci. USA* **96**, 6161–6165.

Beauchamp, C. and Fridovich, I. (1970) A mechanism for the production of ethylene from methional. The generation of the hydroxyl radical by xanthine oxidase. *J. Biol. Chem.* **245**, 5214–5222.

Benov, L. (1999) Why superoxide imposes an aromatic amino acid auxotrophy in *Escherichia coli. J. Biol. Chem.* **274**, 4202–4206.

Benov, L. and Fridovich, I. (1997) Superoxide imposes leakage of sulfite from *Escherichia coli. Arch. Biochem. Biophys.* **347**, 271–274.

Benov, L., Kredich, N.M. and Fridovich, I. (1996) The mechanism of the auxotrophy for sulfur-containing amino acids imposed upon *Escherichia coli* by superoxide. *J. Biol. Chem.* **271**, 21037–21040.

Bielski, B.H.J. and Richter, H.W. (1977) A study of the superoxide radical chemistry by stopped-flow radiolysis and radiation induced oxygen consumption. *J Am. Chem. Soc.* **99**, 3019–3023.

Bielski, B.H.J., Arudi, R.L. and Sutherland, M.W. (1983) A study of the reactivity of HO_2/O_2^- with unsaturated fatty acids. *J. Biol. Chem.* **258**, 4759–4761.

Bielski, B.H.J., Cabelli, D.E. and Arudi, R.L. (1985) Reactivity of HO_2/O_2^- radicals in aqueous solution. *J. Phys. Chem. Ref. Data* **14**, 1041–1062.

Blakely, W.F., Fuciarelli, A.F., Wegher, B.J. and Dizdaroglu, M. (1990) Hydrogen peroxide-induced base damage in deoxyribonucleic acid. *Radiat. Res.* **121**, 338–343.

Boehme, D.E., Vincent, K. and Brown, O.R. (1976) Oxygen and toxicity: inhibition of amino acid biosynthesis. *Nature* **262**, 418–420.

Boveris, A. and Cadenas, E. (1982) Production of superoxide radicals and hydrogen peroxide in mitochondria. In: *Superoxide Dismutase*, vol 2. (L.W. Oberley, ed.), pp. 15–30. CRC Press, Boca Raton.

Brawn, K. and Fridovich, I. (1981) DNA strand scission by enzymically generated oxygen radicals. *Arch. Biochem. Biophys.* **206**, 414–419.

Brown, O.R. and Seither, R.L. (1983) Oxygen and redox-active drugs: shared toxicity sites. *Fund. Appl. Toxicol.* **3**, 209–214.

Brown, O.R., Smyk-Randall, E., Draczynska-Lusiak, B. and Fee, J.A. (1995) Dihydroxy-acid dehydratase, a [4Fe-4S] cluster-containing enzyme in *Escherichia coli:* effects of intracellular superoxide dismutase on its inactivation by oxidant stress. *Arch. Biochem. Biophys.* **319**, 10–22.

Carlioz, A. and Touati, D. (1986) Isolation of superoxide dismutase mutants in *Escherichia coli:* is superoxide dismutase necessary for aerobic life? *EMBO J.* **5**, 623–630.

Carlson, B.W., Miller, L.L., Neta, P. and Grodkowski, J. (1984) Oxidation of NADH involving rate-limiting one-electron transfer. *J. Am. Chem. Soc.* **106**, 7233–7239.

Caselli, A., Marzocchini, R., Camici, G., Manao, G., Moneti, G., Pieraccini, G. and Ramponi, G. (1998) The inactivation mechanism of low molecular weight phospho-tyrosine-protein phosphatase by H_2O_2. *J. Biol. Chem.* **273**, 32554–32560.

Chabriere, E., Charon, M.H., Volbeda, A., Pieulle, L., Hatchikian, E.C. and Fontecilla-Camps, J.C. (1999) Crystal structures of the key anaerobic enzyme pyruvate:ferredoxin oxidoreductase, free and in complex with pyruvate. *Nat. Struct. Biol.* **6**, 182–190.

Chan, E. and Weiss, B. (1987) Endonuclease IV of *Escherichia coli* is induced by paraquat. *Proc. Natl Acad. Sci. USA* **84**, 3189–3193.

Chang, E.C., Crawford, B.F., Hong, Z., Bilinski, T. and Kosman, D.J. (1991) Genetic and biochemical characterization of Cu, Zn superoxide dismutase mutants in *Saccharomyces cerevisiae. J. Biol. Chem.* **266**, 4417–4424.

Chen, H., Tsuchida, S., Tamai, K. and Sato, K. (1993) Identification of cysteine residues involved in disulphide formation in the inactivation of glutathione transferase P-form by hydrogen peroxide. *Arch. Biochem. Biophys.* **300**, 137–141.

Choi, H., Kim, S., Mukhopadhyay, P., Cho, S., Woo, J., Storz, G. and Ryu, S. (2001) Structural basis of the redox switch in the OxyR transcription factor. *Cell* **105**, 103–113.

Cinkaya, I., Buckel, W., Medina, M., Gomez-Moreno, C. and Cammack, R. (1997) Electron-nuclear double resonance spectroscopy investigation of 4-hydroxybutyryl-CoA dehydratase from *Clostridium aminobutyricum*: comparison with other flavin radical enzymes. *J. Biol. Chem.* **378**, 843–849.

Compan, I. and Touati, D. (1993) Interaction of six global transcription regulators in expression of manganese superoxide dismutase in *Escherichia coli* K-12. *J. Bacteriol.* **175**, 1687–1696.

Cotter, P.A. and Gunsalus, R.P. (1992) Contribution of the *fnr* and *arcA* gene products in coordinate regulation of cytochrome o and d oxidase (*cyoABCDE* and *cydAB*) genes in *Escherichia coli*. *FEMS Microbiol. Lett.* **70**, 31–36.

Crow, J.P., Beckman, J.S. and McCord, J.M. (1995) Sensitivity of the essential zinc-thiolate moiety of yeast alcohol dehydrogenase to hypochlorite and peroxynitrite. *Biochemistry* **34**, 3544–3552.

Denu, J.M. and Tanner, K.G. (1998) Specific and reversible inactivation of protein tyrosine phosphatases by hydrogen peroxide: evidence for a sulfenic acid intermediate and implications for redox regulation. *Biochemistry* **37**, 5633–5642.

Dickert, S., Pierik, A.J., Linder, D. and Buckel, W. (2000) The involvement of coenzyme A esters in the dehydration of (R)-phenyllactate to (E)-cinnamate by *Clostridium sporogenes*. *Eur. J. Biochem.* **267**, 3874–3884.

Ding, H. and Demple, B. (1997) In vivo kinetics of a redox-regulated transcriptional switch. *Proc. Natl Acad. Sci. USA* **94**, 8445–8449.

Dukan, S. and Nystrom, T. (1999) Oxidative stress defense and deterioration of growth-arrested *Escherichia coli* cells. *J. Biol. Chem.* **274**, 26027–26032.

Dukan, S., Farewell, A., Ballesteros, M., Taddei, F., Radman, M. and Nystrom, T. (2000) Protein oxidation in response to increased transcriptional or translational errors. *Proc. Natl Acad. Sci USA* **97**, 5746–5749.

Eibbe, M., Gadkari, D. and Meyer, O. (1997) N_2 fixation by *Streptomyces thermoautotrophicus* involves a molybdenum-dinitrogenase and a manganese-superoxide oxidoreductase that couple N_2 reduction to the oxidation of superoxide produced from O_2 by a molybdenum-CO dehydrogenase. *J. Biol. Chem.* **272**, 26627–26633.

Ellis, H.R. and Poole, L.B. (1997) Roles for the two cysteine residues of AhpC in catalysis of peroxide reduction by alkyl hydroperoxide reductase from *Salmonella typhimurium*. *Biochemistry* **36**, 13349–13356.

Farr, S.B., D'Ari, R. and Touati, D. (1986) Oxygen-dependent mutagenesis in *Escherichia coli* lacking superoxide dismutase. *Proc. Natl Acad. Sci. USA* **83**, 8268–8272.

Farr, S.B., Touati, D. and Kogoma, T. (1988) Effects of oxygen stress on membrane functions in *Escherichia coli*: role of HPI catalase. *J. Bacteriol.* **170**, 1837–1842.

Fee, J.A. (1982) Is superoxide important in oxygen poisoning? *Trends Biochem. Sci.* **7**, 84–86.

Fitzsimons, D.W. (ed.). (1979) *Oxygen Free Radicals in Tissue Damage*, pp. 43–56. Ciba Foundation Series 65, Elsevier/North-Holland, Amsterdam.

Flint, D.H. and Allen, R.M. (1996) Iron-sulfur proteins with nonredox functions. *Chem. Rev.* **96**, 2315–2334.

Flint, D.H. and Emptage, M.H. (1990) Dihydroxyacid dehydratase: isolation, characterization as Fe-S proteins and sensitivity to inactivation by oxygen radicals. In: *Biosynthesis of Branched Chain Amino Acids* (D.C.Z. Barak and J.V. Schloss., eds), pp. 285–314. New York: VCH Publishers, New York.

Flint, D.H., Tuminello, J.F. and Emptage, M.H. (1993) The inactivation of Fe-S cluster containing hydro-lyases by superoxide. *J. Biol. Chem.* **268**, 22369–22376.

Frey, P.A. (2001) Radical mechanisms of enzymatic catalysis. *Ann. Rev. Biochem.* **70**, 121–148.

Fridovich, I. (1970) Quantitative aspects of the production of superoxide anion radical by milk xanthine oxidase. *J. Biol. Chem.* **245**, 4053–4057.

Gardner, P.R. and Fridovich, I. (1991a) Superoxide sensitivity of the *Escherichia coli* aconitase. *J. Biol. Chem.* **266**, 19328–19333.

Gardner, P.R. and Fridovich, I. (1991b) Superoxide sensitivity of the *Escherichia coli* 6-phosphogluconate dehydratase. *J. Biol. Chem.* **266**, 1478–1483.

Gardner, P.R. and Fridovich, I. (1992) Inactivation–reactivation of aconitase in *Escherichia coli*. A sensitive measure of superoxide radical. *J. Biol. Chem.* **267**, 8757–8763.

Gaudu, P. and Weiss, B. (2000) Flavodoxin mutants of *Escherichia coli* K-12. *J. Bacteriol.* **182**, 1788–1793.

Gaudu, P., Moon, N. and Weiss, B. (1997) Regulation of the soxRS oxidative stress regulon. Reversible oxidation of the Fe-S center of SoxR in vivo. *J. Biol. Chem.* **272**, 5082–5086.

Gonzalez-Flecha, B. and Demple, B. (1997) Homeostatic regulation of intracellular hydrogen peroxide concentration in aerobically growing *Escherichia coli*. *J. Bacteriol.* **179**, 382–388.

Gort, A.S. and Imlay, J.A. (1998) Balance between endogenous superoxide stress and antioxidant defenses. *J. Bacteriol.* **180**, 1402–1410.

Gort, A.S., Ferber, D.M. and Imlay, J.A. (1999) The regulation and role of the periplasmic copper, zinc superoxide dismutase of *Escherichia coli*. *Mol. Micro.* **32**, 179–191.

Grant, C.M., Quinn, K.A. and Dawes, I.W. (1999) Differential protein S-thiolation of glyceraldehyde-3-phosphate dehydrogenase isozymes influences sensitivity to oxidative stress. *Mol. Cell. Biol.* **19**, 2650–2656.

Grant, R.A., Filman, D.J., Finkel, S.E., Kolter, R. and Hogle, J.M. (1998) The crystal structure of Dps, a ferritin homolog that binds and protects DNA. *Nat. Struct. Biol.* **5**, 294–303.

Greenberg, J.T. and Demple, B. (1989) A global response induced in *Escherichia coli* by redox-cycling agents overlaps with that induced by peroxide stress. *J. Bacteriol.* **171**, 3933–3939.

Greenberg, J.T., Monach, P., Chou, J.H., Josephy, P.D. and Demple, B. (1990) Positive control of a global antioxidant defense regulon activated by superoxide-generating agents in *E. coli*. *Proc. Natl Acad. Sci. USA* **87**, 6181–6185.

Gregory, E.M., Moore, W.E. and Holdeman, L.V. (1978) Superoxide dismutase in anaerobes: survey. *Appl. Environ. Microbiol.* **35**, 988–991.

Gruer, M.J. and Guest, J.R. (1994) Two genetically-distinct and differentially-regulated aconitases (AcnA and AcnB) in *Escherichia coli*. *Microbiology* **140**, 2531–2541.

Gunsalus, R.J. and Park, S.J. (1994) Aerobic-anaerobic gene regulation in *Escherichia coli*: control by the ArcAB and Fnr regulons. *Res. Microbiol.* **145**, 437–450.

Gutteridge, J.M.C. and Halliwell, B. (1990) The measurement and mechanism of lipid peroxidation in biological systems. *TIBS* **15**, 129–135.

Halliwell, B. (1976) An attempt to demonstrate a reaction between superoxide and hydrogen peroxide. *FEBS Lett.* **72**, 8–10.

Halliwell, B. (1978) Superoxide-dependent formation of hydroxyl radicals in the presence of iron salts. Its role in degradation of hyaluronic acid by a superoxide-generating system. *FEBS Lett.* **96**, 238–242.

Hans, M., Sievers, J., Muller, U., Bill, E., Vorholt, J.A., Linder, D. and Buckel, W. (1999) 2-hydroxyglutaryl-CoA dehydratase from *Clostridium symbiosum*. *Eur. J. Biochem.* **265**, 404–414.

Hans, M., Buckel, W. and Bill, E. (2000) The iron-sulfur clusters in 2-hydroxygutaryl-CoA dehydratase from *Acidaminococcus fermentans*. Biochemical and spectroscopic investigations. *Eur. J. Biochem.* **267**, 7082–7093.

Hassan, H.M. and Fridovich, I. (1977) Regulation of the synthesis of superoxide dismutase in *Escherichia coli*. Induction by methyl viologen. *J. Biol. Chem.* **252**, 7667–7672.

Hausladen, A. and Fridovich, I. (1994) Superoxide and peroxynitrite inactivate aconitases, but nitric oxide does not. *J. Biol. Chem.* **269**, 29405–29408.

Hesslinger, C., Fairhurst, S.A. and Sawers, G. (1998) Novel keto acid formate-lyase and propionate kinase enzymes are components of an anaerobic pathway in *Escherichia coli* that degrades L-threonine to propionate. *Mol. Microbiol.* **27**, 477–492.

Hidalgo, E. and Demple, B. (1996) Adaptive responses to oxidative stress: the soxRS and oxyR regulons. In: *Regulation of Gene Expression in Escherichia coli* (E. Lin and A. Lynch, eds), pp. 435–452. R G Landes Company, Austin.

Hofmeister, A.E., Albracht, S.P. and Buckel, W. (1994) Iron-sulfur cluster-containing L-serine dehydratase from *Peptostreptococcus asaccharolyticus*: correlation of the cluster type with enzymatic activity. *FEBS Lett.* **351**, 416–418.

Hutchinson, F. (1985) Chemical changes induced in DNA by ionizing radiation. *Prog. Nucleic Acid Res.* **32**, 116–154.

Imlay, J.A. (1995) A metabolic enzyme that rapidly produces superoxide, fumarate reductase of *Escherichia coli*. *J. Biol. Chem.* **270**, 19767–19777.

Imlay, J.A. and Fridovich, I. (1991) Assay of metabolic superoxide production in *Escherichia coli*. *J. Biol. Chem.* **266**, 6957–6965.

Imlay, J.A. and Linn, S. (1986) Bimodal pattern of killing of DNA-repair-defective or anoxically grown *Escherichia coli* by hydrogen peroxide. *J. Bacteriol.* **166**, 519–527.

Imlay, J.A. and Linn, S. (1988a) DNA damage and oxygen radical toxicity. *Science* **240**, 1302–1309.

Imlay, J.A. and Linn, S. (1988b) Toxic DNA damage by hydrogen peroxide through the Fenton reaction *in vivo* and *in vitro*. *Science* **240**, 640–642.

Jenney, F.E., Jr., Verhagen, M.F., Cui, X. and Adams, M.W. (1999) Anaerobic microbes: oxygen detoxification without superoxide dismutase. *Science* **286**, 306–309.

John, G.S., Brot, N., Ruan, J., Erdjument-Bromage, H., Tempst, P., Weissbach, H. and Nathan, C. (2001) Peptide methionine sulfoxide reductase from *Escherichia coli* and *Mycobacterium tuberculosis* protects bacteria against oxidative damage from reactive nitrogen intermediates. *Proc. Natl Acad. Sci. USA* **98**, 9901–9906.

Jones, H.M. and Gunsalus, R.P. (1987) Regulation of *Escherichia coli* fumarate reductase (*frdABCD*) operon expression by respiratory electron acceptors and the *fnr* gene product. *J. Bacteriol.* **169**, 3340–3349.

Jordan, P.A., Tang, Y., Bradbury, A.J., Thomson, A.J. and Guest, J.R. (1999) Biochemical and spectroscopic characterization of *Escherichia coli* aconitases (AcnA and AcnB). *Biochem. J.* **344**, 739–746.

Jovanovic, T., Ascenso, C., Hazlett, K.R., Sikkink, R., Krebs, C., Litwiller, R., Benson, L.M., Moura, I., Moura, J.J., Radolf, J.D., Huynh, B.H., Naylor, S. and Rusnak, F. (2000) Neelaredoxin, an iron-binding protein from the syphilis spirochete, *Treponema pallidum*, is a superoxide reductase. *J. Biol. Chem.* **275**, 28439–28448.

Kessler, D., Herth, W. and Knappe, J. (1992) Ultrastructure and pyruvate formate-lyase radical quenching property of the multienzymic AdhE protein of *Escherichia coli*. *J. Biol. Chem.* **267**, 18073–18079.

Keyer, K. and Imlay, J.A. (1996) Superoxide accelerates DNA damage by elevating free-iron levels. *Proc. Natl Acad. Sci. USA* **93**, 13635–13640.

Keyer, K. and Imlay, J.A. (1997) Inactivation of dehydratase [4Fe-4S] clusters and disruption of iron homeostasis upon cell exposure to peroxynitrite. *J. Biol. Chem.* **272**, 27652–27659.

Keyer, K., Gort, A.S. and Imlay, J.A. (1995) Superoxide and the production of oxidative DNA damage. *J. Bacteriol.* **177**, 6782–6790.

Kiley, P.J. and Beinert, H. (1999) Oxygen sensing by the global regulator, FNR; the role of the iron-sulfur cluster. *FEMS Microbiol. Rev.* **22**, 341–352.

Kohen, R. and Chevion, M. (1988) Cytoplasmic membrane is the target organelle for transition metal mediated damage induced by paraquat in *Escherichia coli. Biochemistry* **27**, 2597–2603.

Koppenol, W.H. (1985) Energetics of interconversion reaction of oxyradicals. *Free Rad. Res. Commun.* **1**, 91–131.

Koppenol, W.H. (1990) Oxyradical reactions: from bond-dissociation energies to reduction potentials. *FEBS Letts.* **264**, 165–167.

Kuchta, R.D. and Abeles, R.H. (1985) Lactate reduction in *Clostridium propionicum.* Purification and properties of lactyl-CoA dehydratase. *J. Biol. Chem.* **260**, 13181–13189.

Kuo, C.F., Mashino, T. and Fridovich, I. (1987) α,β-dihydroxyisovalerate dehydratase: a superoxide-sensitive enzyme. *J. Biol. Chem.* **262**, 4724–4727.

Lauble, H., Kennedy, M.C., Beinert, H. and Stout, C.D. (1992) Crystal structures of aconitase with isocitrate and nitroisocitrate bound. *Biochemistry* **31**, 2735–2748.

Leven, S., Heimberger, A. and Eisenstark, A. (1990) Catalase HPI influences membrane permeability in *Escherichia coli* following near-UV stress. *Biochem. Biophys. Res. Commun.* **171**, 1224–1228.

Levin, D.E., Hollstein, M., Christman, M.F., Schwiers, E.A. and Ames, B.N. (1982) A new *Salmonella* tester strain (TA102) with A.T base pairs at the site of mutation detects oxidative mutagens. *Proc. Natl Acad. Sci. USA* **79**, 7445–7449.

Levine, R.L., Mosoni, L., Berlett, B.S. and Stadtman, E.R. (1996) Methionine residues as endogenous antioxidants in proteins. *Proc. Natl Acad. Sci. USA* **93**, 15036–15040.

Lind, C., Gerdes, R., Schuppe-Koistinen, I. and Cogreave, I.A. (1998) Studies on the mechanism of oxidative modification of human glyceraldehyde-3-phosphate dehydrogenase by glutathione: catalysis by glutaredoxin. *Biochem. Biophys. Res. Commun.* **247**, 481–486.

Liochev, S.I. and Fridovich, I. (1992) Fumarase C, the stable fumarase of *Escherichia coli*, is controlled by the soxRS regulon. *Proc. Natl Acad. Sci. USA* **89**, 5892–5896.

Liochev, S.I. and Fridovich, I. (1994) The role of superoxide in the production of hydroxyl radical: *in vitro* and *in vivo. Free Rad. Biol. Med.* **16**, 29–33.

Liochev, S.I., Hausladen, A., Beyer, W.F. and Fridovich, I. (1994) NADPH:ferredoxin oxidoreductase acts as a paraquat diaphorase and is a member of the soxRS regulon. *Proc. Natl Acad. Sci. USA* **91**, 1328–1331.

Liochev, S.I., Benov, L., Touati, D. and Fridovich, I. (1999) Induction of the soxRS regulon of *Escherichia coli* by superoxide. *J. Biol. Chem.* **274**, 9479–9481.

Locher, K., Hans, M., Yeh, A.P., Schmid, B., Buckel, W. and Rees, D.C. (2001) Crystal structure of the *Acidaminococcus fermentans* 2-hydroxyglutaryl-CoA dehydratase component A. *J. Mol. Biol.* **307**, 297–308.

Lombard, M., Fontecave, M., Touati, D. and Niviere, V. (2000a) Reaction of the desulfoferrodoxin from *Desulfoarculus baarsii* with superoxide anion. Evidence for a superoxide reductase activity. *J. Biol. Chem.* **275**, 115–121.

Lombard, M., Touati, D., Fontecave, M. and Niviere, V. (2000b) Superoxide reductase as a unique defense system against superoxide stress in the microaerophile *Treponema pallidum. J. Biol. Chem.* **275**, 27021–27026.

Lumppio, H.L., Shenvi, N.V., Summars, A.O., Voordouw, G. and Kurtz, Jr, D.M. (2001) Rubrerythrin and rubredoxin oxidoreductase in *Desulfovibrio vulgaris*: a novel oxidative stress protection system. *J. Bacteriol.* **183**, 101–108.

Luo, Y., Han, Z., Chin, S.M. and Linn, S. (1994) Three chemically distinct types of oxidants formed by iron-mediated Fenton reactions in the presence of DNA. *Proc. Natl Acad. Sci. USA* **91**, 12438–12442.

Mai, X. and Adams, M.W. (1994) Indolepyruvate ferredoxin oxidoreductase from the hyper-thermophilic archaeon *Pyrococcus furiosus*. A new enzyme involved in peptide fermentation. *J. Biol. Chem.* **269**, 16726–16732.

Mai, X. and Adams, M.W. (1996) Characterization of a fourth type of 2-keto acid-oxidizing enzyme from a hyperthermophilic archaeon: 2-ketoglutarate ferredoxin oxidoreductase from *Thermococcus litoralis*. *J. Bacteriol.* **178**, 5890–5896.

Marsh, E.N.G. (2000) Coenzyme-B12-dependent glutamate mutase. *Bioorg. Chem.* **28**, 176–189.

Martinez, A. and Kolter, R. (1997) Protection of DNA during oxidative stress by the non-specific DNA-binding protein Dps. *J. Bacteriol.* **179**, 5188–5194.

Massey, V. (1994) Activation of molecular oxygen by flavins and flavoproteins. *J. Biol. Chem.* **36**, 22459–22462.

Massey, V., Strickland, S., Mayhew, S.G., Howell, L.G., Engel, P.C., Matthews, R.G., Schuman, M. and Sullivan, P.A. (1969) The production of superoxide anion radicals in the reaction of reduced flavins and flavoproteins with molecular oxygen. *Biochem. Biophys. Res. Commun.* **36**, 891–897.

McClune, G.J. and Fee, J.A. (1976) Stopped flow spectrophotometric observation of super-oxide dismutation in aqueous solution. *FEBS Lett.* **67**, 294–298.

McCord, J.M. and Day, Jr., E.D. (1978) Superoxide-dependent production of hydroxyl rad-ical catalyzed by iron-EDTA complex. *FEBS Letts.* **86**, 139–142.

McCord, J. and Fridovich, I. (1969) Superoxide dismutase. An enzymic function for ery-throcuprein (hemocuprein). *J. Biol. Chem.* **244**, 6049–6055.

McCord, J.M., Keele Jr, B.B. and Fridovich, I. (1971) An enzyme-based theory of obligate anaerobiosis: the physiological function of superoxide dismutase. *Proc. Natl Acad. Sci. USA* **68**, 1024–1027.

McCormick, J.P., Fischer, J.R. and Pachlatko, J.P. (1976) Characterization of a cell-lethal prod-uct from the photooxidation of tryptophan: hydrogen peroxide. *Science* **191**, 468–469.

Meij, J.T., Suzuki, S., Panagia, V. and Dhalla, N.S. (1994) Oxidative stress modifies the activity of cardiac sarcolemmal phospholipase C. *Biochim. Biophys. Acta* **1199**, 6–12.

Messner, K.R. and Imlay, J.A. (1999) The identification of primary sites of superoxide and hydrogen peroxide formation in the aerobic respiratory chain and sulfite reductase com-plex of *Escherichia coli*. *J. Biol. Chem.* **274**, 10119–10128.

Miki, T., Yoshida, L.S. and Kakinuma, K. (1992) Reconstitution of superoxide-forming NADPH oxidase activity with cytochrome b558 purified from porcine neutrophils. *J. Biol. Chem.* **267**, 18695–18701.

Miller, P. and Sulavik, M. (1996) Overlaps and parallels in the regulation of intrinsic mul-tiple-antibiotic resistance in *Escherichia coli*. *Mol. Microbiol.* **21**, 441–448.

Minghetti, K.C. and Gennis, R.B. (1988) The two terminal oxidases of the aerobic respira-tory chain of *Escherichia coli* each yield water and not peroxide as a final product. *Biochem. Biophys. Res. Commun.* **155**, 243–248.

Morgan, A.R., Cone, R.L. and Elgert, T.M. (1976) The mechanism of DNA strand breakage by vitamin C and superoxide and the protective roles of catalase and superoxide dismu-tase. *Nucleic Acids Res.* **3**, 1139–1149.

Moskovitz, J., Rahman, M.A., Strassman, J., Yancey, S.O., Kushner, S.R., Brot, N. and Weissbach, H. (1995) *Escherichia coli* peptide methionine sulfoxide reductase gene: reg-ulation of expression and role in protecting against oxidative damage. *J. Bacteriol.* **177**, 502–507.

Muller, F. (1987) Flavin radicals: chemistry and biochemistry. *Free Rad. Biol. Med.* **3**, 215–230.

Naqui, A. and Chance, B. (1986) Reactive oxygen intermediates in biochemistry. *Ann. Rev. Biochem.* **55**, 137–166.

Natvig, D.O., Imlay, K., Touati, D. and Hallewell, R.A. (1987) Human copper-zinc superoxide dismutase complements superoxide dismutase-deficient *Escherichia coli* mutants. *J. Biol. Chem.* **262**, 14697–14701.

Neidhardt, F.C. (1996) Anaerobic dissimilation of pyruvate. In: *Escherichia coli and Salmonella. Cellular and Molecular Biology* (F.C. Neidhardt, ed.), pp. 199–205. ASM Press, Washington, D.C.

Neilands, J.B. (1993) Siderophores. *Arch. Biochem. Biophys.* **302**, 1–3.

Pan, N. and Imlay, J.A. (2001) How does oxygen inhibit central metabolism in the obligate anaerobe *Bacteroides thetaiotaomicron? Mol. Microbiol.* **39**, 1562–1571.

Pieulle, L., Guigliarelli, B., Asso, M., Dole, F., Bernadac, A. and Hatchikian, E.C. (1995) Isolation and characterization of the pyruvate-ferredoxin oxidoreductase from the sulfate-reducing bacterium *Desulfovibrio africanus. Biochim. Biophys. Acta* **1250**, 49–59.

Pieulle, L., Magro, V. and Hatchikian, E.C. (1997) Isolation and analysis of the gene encoding the pyruvate-ferredoxin oxidoreductase of *Desulfovibrio africanus*, production of the recombinant enzyme in *Escherichia coli* and effect of carboxy-terminal deletions on its stability. *J. Bacteriol.* **179**, 5684–5692.

Pomposiello, P.J. and Demple, B. (2002) Global adjustment of microbial physiology during free radical stress. *Adv. Microbiol. Physiol.* **46**, 319–341.

Pomposiello, P.J., Bennik, M.H. and Demple, B. (2001) Genome-wide transcriptional profiling of the *Escherichia coli* responses to superoxide stress and sodium salicylate. *J. Bacteriol.* **183**, 3890–3902.

Prinz, W.A., Aslund, F., Holmgren, A. and Beckwith, J. (1997) The role of the thioredoxin and glutaredoxin pathways in reducing protein disulphide bonds in the *Escherichia coli* cytoplasm. *J. Biol. Chem.* **272**, 15661–15667.

Rahman, M.A., Nelson, H., Weissbach, H. and Brot, N. (1992) Cloning, sequencing and expression of the *Escherichia coli* peptide methionine sulfoxide reductase gene. *J. Biol. Chem.* **267**, 15549–15551.

Rai, P., Cole, T.D., Wemmer, D.E. and Linn, S. (2001) Localization of Fe^{2+} at an RTGR sequence within a DNA duplex explains preferential cleavage by Fe^{2+} and H_2O_2. *J. Mol. Biol.* **312**, 1089–1101.

Reddy, S.G., Wong, K.K., Parast, C.V., Peisach, J., Magliozzo, R.S. and Kozarich, J.W. (1998) Dioxygen inactivation of pyruvate formate-lyase: EPR evidence for the formation of protein-based sulfinyl and peroxyl radicals. *Biochemistry* **37**, 558–563.

Rocha, E.R. and Smith, C.J. (1999) Role of the alkyl hydroperoxide reductase (*ahpCF*) gene in oxidative stress defense of the obligate anaerobe *Bacteroides fragilis. J. Bacteriol.* **181**, 5701–5710.

Rocha, E.R., Selby, T., Coleman, J.P. and Smith, C.J. (1996) Oxidative stress response in an anaerobe, *Bacteroides fragilis*: a role for catalase in protection against hydrogen peroxide. *J. Bacteriol.* **178**, 6895–6903.

Rowley, D.A. and Halliwell, B. (1982) Superoxide-dependent formation of hydroxyl radicals from NADH and NADPH in the presence of iron salts. *FEBS Lett.* **142**, 39–41.

Sawers, G. and Suppmann, B. (1992) Anaerobic induction of pyruvate formate-lyase gene expression is mediated by the ArcA and Fnr proteins. *J. Bacteriol.* **174**, 3474–3478.

Sawers, G. and Watson, G. (1998) A glycyl radical solution: oxygen-dependent interconversion of pyruvate formate-lyase. *Mol. Microbiol.* **29**, 945–954.

Sawyer, D.T. and Valentine, J.S. (1981) How super is superoxide? *Acc. Chem. Res.* **14**, 393–400.

Schuppe-Koistinen, I., Moldeus, P., Bergman, T. and Cotgreave, I.A. (1994) S-thiolation of human endothelial cell glyceraldehyde-3-phosphate dehydrogenase after hydrogen peroxide treatment. *Eur. J. Biochem.* **221**, 1033–1037.

Schwartz, C.J., Djaman, O., Imlay, J.A. and Kiley, P.J. (2000) The cysteine desulfurase, IscS, has a major role in *in vivo* Fe-S cluster formation in *Escherichia coli. Proc. Natl Acad. Sci. USA* **97**, 9009–9014.

Seaver, L.C. and Imlay, J.A. (2001a) Alkyl hydroperoxide reductase is the primary scavenger of endogenous hydrogen peroxide in *Escherichia coli. J. Bacteriol.* **183**, 7173–7181.

Seaver, L.C. and Imlay, J.A. (2001b) Hydrogen peroxide fluxes and compartmentalization inside growing *Escherichia coli. J. Bacteriol.* **183**, 7182–7189.

Sjoberg, B.M. and Reichard, P. (1977) Nature of the free radical in ribonucleotide reductase from *Escherichia coli. J. Biol. Chem.* **252**, 536–541.

Smirnova, G.V., Muzyka, N.G., Glukhovchenko, M.N. and Oktyabrsky, O.N. (1997) Effects of penetrating and non-penetrating oxidants on *Escherichia coli. Biochemistry (Mosc.)* **62**, 480–484.

Strop, P., Takahara, P.M., Chiu, H.-J., Haylen, C., Angove, C., Burgess, B.K. and Rees, D.C. (2001) Crystal structure of the all-ferrous [4Fe-4S]^0 form of the nitrogenase iron protein from *Azotobacter vinelandii. Biochemistry* **40**, 651–656.

Stubbe, J. and Riggs-Gelasco, P. (1998) Harnessing free radicals: formation and function of the tyrosyl radical in ribonucleotide reductase. *TIBS* **23**, 438–443.

Sun, X., Ollagnier, S., Schmidt, P.P., Atta, M., Mulliez, E., Lepape, L., Eliasson, R., Graslund, M., Fontecave, M., Reichard, P. and Sjoberg, B.M. (1996) The free radical of the anaerobic ribonucleotide reductase from *Escherichia coli* is at glycine 681. *J. Biol. Chem.* **271**, 6827–6831.

Svensater, G., Takahashi-Abbe, S., Abbe, K., Birkhed, D., Yamada, T. and Edwardsson, S. (1985) Anaerobic and aerobic metabolism of sorbitol in *Streptococcus sanguis* and *Streptococcus mitior. J. Dent. Res.* **64**, 1284–1289.

Takabe, T., Asami, S. and Akazawa, T. (1980) Glycolate formation catalyzed by spinach leaf transketolase utilizing the superoxide radical. *Biochemistry* **19**, 3985–3989.

Takahashi, Y. and Nakamura, M. (1999) Functional assignment of the ORF2-*iscS-iscU-iscA-hscB-hscA-fdx*-ORF3 gene cluster involved in the assembly of Fe-S clusters in *Escherichia coli. J. Biochem. (Tokyo)* **126**, 917–926.

Takahashi, N., Iwami, Y. and Yamada, T. (1991) Metabolism of intracellular polysaccharide in the cells of *Streptococcus mutans* under strictly anaerobic conditions. *Oral Microbiol. Immunol.* **6**, 299–304.

Tally, F.P., Goldin, B.R., Jacobus, N.V. and Gorbach, S.L. (1977) Superoxide dismutase in anaerobic bacteria of clinical significance. *Infect. Immun.* **16**, 20–25.

Touati, D., Jacques, M., Tardat, B., Bouchard, L. and Despied, S. (1995) Lethal oxidative damage and mutagenesis are generated by iron in Δ*fur* mutants of *Escherichia coli*: protective role of superoxide dismutase. *J. Bacteriol.* **177**, 2305–2314.

Tsaneva, I.R. and Weiss, B. (1990) *soxR*, a locus governing a superoxide response regulon in *Escherichia coli* K-12. *J. Bacteriol.* **172**, 4197–4205.

Tseng, C.-P., Yu, C.-C., Lin, H.-H., Chang, C.-Y. and Kuo, J.-T. (2001) Oxygen- and growth rate-dependent regulation of *Escherichia coli* fumarase (FumA, FumB and FumC) activity. *J. Bacteriol.* **183**, 461–467.

Vaidyanathan, V.V., Sastry, P.S. and Ramasarma, T. (1993) Regulation of the activity of glyceraldehyde-3-phosphate dehydrogenase by glutathione and H_2O_2. *Mol. Cell. Biochem.* **129**, 57–65.

van Loon, A.P.G.M., Pesold-Hurt, B. and Schatz, G. (1986) A yeast mutant lacking mitochondrial manganese-superoxide dismutase is hypersensitive to oxygen. *Proc. Natl Acad. Sci. USA* **83**, 3820–3824.

Walling, C. (1975) Fenton's reagent revisited. *Acc. Chem. Res.* **8**, 125–131.

White, D.G., Goldman, J.D., Demple, B. and Levy, S.B. (1997) Role of the *acrAB* locus in organic solvent tolerance mediated by expression of *marA*, *soxS* or *robA* in *Escherichia coli*. *J. Bacteriol.* **179**, 6122–6126.

Winterbourn, C.C. (1979) Comparison of superoxide with other reducing agents in the biological production of hydroxyl radicals. *Biochem. J.* **182**, 625–628.

Winterbourn, C.C. and Metodiewa, D. (1999) Reactivity of biologically important thiol compounds with superoxide and hydrogen peroxide. *Free Rad. Biol. Med.* **27**, 322–328.

Wood, P.M. (1987) The two redox potentials for oxygen reduction to superoxide. *TIBS* **12**, 250–251.

Woods, D.R. and Jones, D.T. (1986) Physiological responses of Bacteroides and Clostridium strains to environmental stress factors. *Adv. Microb. Physiol.* **28**, 1–64.

Yamada, T., Takahashi-Abbe, S. and Abbe, K. (1985) Effects of oxygen on pyruvate formate-lyase *in situ* and sugar metabolism of *Streptococcus mutans* and *Streptococcus sanguis*. *Infect. Immun.* **47**, 129–134.

Zhang, Q., Iwasaki, T., Wakagi, T. and Oshima, T. (1996) 2-oxoacid:ferredoxin oxidoreductase from the thermoacidophilic archaeon, *Sulfolobus* sp. strain 7. *J. Biochem.* **120**, 587–599.

Zheng, L. and Dean, D.R. (1994) Catalytic formation of a nitrogenase iron-sulfur cluster. *J. Biol. Chem.* **269**, 18723–18726.

Zheng, M., Aslund, F. and Storz, G. (1998) Activation of the OxyR transcription factor by reversible disulphide bond formation. *Science* **279**, 1718–1721.

Zheng, M., Doan, B., Schneider, T.D. and Storz, G. (1999) OxyR and SoxRS regulation of *fur*. *J. Bacteriol.* **181**, 4639–4643.

Zheng, M., Wang, X., Templeton, L.J., Smulski, D.R., LaRossa, R.A. and Storz, G. (2001) DNA microarray-mediated transcriptional profiling of the *Escherichia coli* response to hydrogen peroxide. *J. Bacteriol.* **183**, 4562–4570.

Drug Resistance in Yeasts – an Emerging Scenario

Rajendra Prasad, Sneh Lata Panwar and Smriti

Membrane Biology Laboratory, School of Life Sciences, Jawaharlal Nehru University, New Delhi-110067, India

ABSTRACT

In view of the increasing threat posed by fungal infections in immunocompromised patients and due to the non-availability of effective treatments, it has become imperative to find novel antifungals and vigorously search for new drug targets. Fungal pathogens acquire resistance to drugs (antifungals), a well-established phenomenon termed multidrug resistance (MDR), which hampers effective treatment strategies. The MDR phenomenon is spread throughout the evolutionary scale. Accordingly, a host of responsible genes have been identified in the genetically tractable budding yeast *Saccharomyces cerevisiae*, as well as in a pathogenic yeast *Candida albicans*. Studies so far suggest that, while antifungal resistance is the culmination of multiple factors, there may be a unifying mechanism of drug resistance in these pathogens. ABC (ATP binding cassette) and MFS (major facilitator superfamily) drug transporters belonging to two different superfamilies, are the most prominent contributors to MDR in yeasts. Considering the abundance of the drug transporters and their wider specificity, it is believed that these drug transporters may not exclusively export drugs in fungi. It has become apparent that the drug transporters of the ABC superfamily of *S. cerevisiae* and *C. albicans* are multifunctional proteins, which mediate important physiological functions. This review summarizes current research on the molecular mechanisms underlying drug resistance, the emerging regulatory circuits of MDR genes, and the physiological relevance of drug transporters.

ADVANCES IN MICROBIAL PHYSIOLOGY VOL 46
ISBN 0-12-027746-8

ABBREVIATIONS

See also Table 1, Gene nomenclature

CDR   *Candida* drug resistance
MFS   Major facilitator superfamily
NBD   Nucleotide-binding domain
NQO   4-nitroquinoline *N*-oxide
PDRE   Pleiotropic drug resistance element
TMS   Transmembrane stretch

1. INTRODUCTION

Opportunistic fungal infections, which are common in immunocompromised patients, are an important medical problem (Odds, 1988). While
infections due to *Candida* and *Aspergillus* species are the most predominant, rarely encountered opportunistic fungi have also emerged as
significant pathogens (Prasad, 1991; Ernst and Schmidt, 2000). For example, *Trichosporon beigelii*, *Fusarium* species, *Pseudollescheria boydii* and
moulds of the class Zygomycetes can cause invasive infections. In addition
to the opportunistic pathogenic fungi, a limited number of fungi like
Histoplasma capsulatum, *Paracoccidioides brasiliensis*, *Penicillium marn-
effei* and *Coccidioides immitis* exist with a true pathogenic potential for

healthy hosts and cause life-threatening infections (Marichal, 1999; Ernst and Schmidt, 2000).

In the current scenario, studies related to combating fungal infections have gained considerable importance. The treatment of fungal infections has been hampered by two commonly encountered problems: (1) the narrow spectrum and toxicity associated with the currently available antifungals; and (2) the rise in the number of fungi that develop resistance to the available antifungals. Of the two, the phenomenon of multidrug resistance (MDR) which is responsible for the emergence of the 'antifungal-resistant' species has been the principal cause of failure in the treatment of fungal infections. MDR, which is defined as the resistance of an organism to a spectrum of drugs that share neither a common target nor a common structure, is spread throughout the evolutionary scale (van Veen and Konings, 1997). Several genes involved in MDR have been identified in the genetically tractable budding yeast *Saccharomyces cerevisiae*, as well as in *Candida albicans* amongst several other pathogenic fungi. This review focuses on the current status of research in drug resistance in yeasts. A detailed discussion on drug extrusion pumps has been included in the review since they not only export drugs, but also appear to be versatile proteins, which perform multiple physiological functions.

2. MECHANISMS OF ANTIFUNGAL DRUG RESISTANCE IN YEASTS

Because of the limited availability of antifungals, treatment of fungal infections has not been very successful (Ghannoum and Rice, 1999) (Fig. 1). This is further compounded by the fact that the generation of newer antifungals has lagged behind when compared with the pace at which drug-resistant fungal species have emerged. The close similarity between fungal pathogens (lower eukaryotes) and mammalian cells (higher eukaryotes) poses problems for the development of antifungals. Notwithstanding this similarity, there are still unique components in the fungal cell that are being exploited as antifungal targets (Fostel and Lartey, 2000). The mannans, glucans and chitins – the components of the fungal cell wall – and a few of the enzymes of the ergosterol biosynthetic pathway are unique to the fungal cells (Farkas, 1979; St Georgiev, 2000; Munro *et al.*, 2001). Antifungal generation has, therefore, targeted enzymes involved in the synthesis of the above macromolecules of the cell wall (Hector, 1993; Buurman *et al.*, 1998; Timpel *et al.*, 1998; Borowski, 2000; Munro *et al.*, 2001). Among the enzymes of the ergosterol biosynthetic pathway, squalene epoxidase, P45014DM or CYP51 (*ERG11*), Δ^{14}-reductase (*ERG24*) and Δ^8-Δ^7-isomerase (*ERG2*) have been the targets of numerous antifungal agents (Table 1) (Fostel and Lartey, 2000). Recently, it has been found

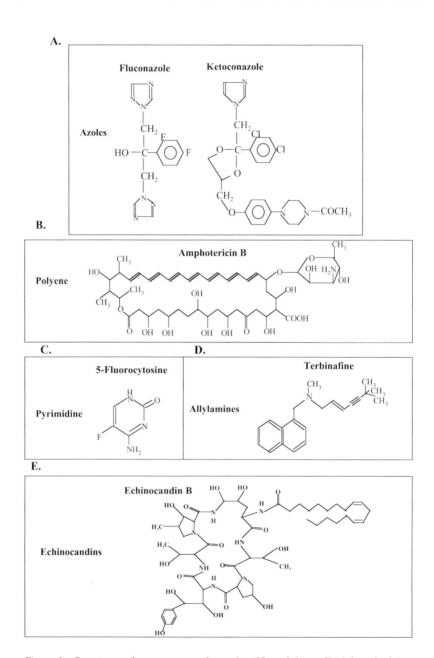

Figure 1   Structures of some commonly used antifungal drugs. Each box depicts representative structures of drugs of a particular type. See Figure 2 for the site of action of these antifungals in a yeast cell.

Table 1   Gene nomenclature.

Gene	Coding For
ABC	ATP-binding cassette
ADR1	*Aspergillus* drug resistance
AflMDR	*A. flavus* multidrug resistance
AfuMDR2	*A. fumigatus* multidrug resistance 2
ALDP	Adrenoleukodystrophy protein
AP-1	Activator protein 1
ATM1	ABC transporter of mitochondria 1
ATR1	Aminotriazole resistance 1
ATR	ATP-binding cassette transporter
bfr1$^+$	Brefeldin resistance 1
CaALK8	*C. albicans* alkane-inducible cytochrome P450 8
CaMDR1/BEN^r	*C. albicans* multidrug resistance 1/Benomyl resistance
CAP1	*Candida* activator protein
CAR1	Changed amiloride resistance 1
CaYCF1	*C. albicans* yeast cadmium factor 1
CaYOR1	*C. albicans* yeast oligomycin resistance 1
CdMDR1	*C. dubliniensis* multidrug resistance 1
CDRs	*Candida* drug resistance
CFTR	Cystic fibrosis transmembrane conductance regulator
CgMDR1	*C. glabrata* multidrug resistance 1
CneMDR1	*C. neoformans* multidrug resistance 2
CYHR	Cycloheximide resistance
ERGs	Ergosterol biosynthetic genes
FCR1	Fluconazole resistance 1
FLR1	Fluconazole resistance 1
FLU1	Fluconazole 1 (see Table 3)
GCNs	General control non-derepressible
GLR1	Glutathione reductase 1
GSH1	γ-Glutamylcysteine synthase 1
Hmt1	Heavy metal tolerance 1
HST6	Homologue *STE6*
HXTs	Hexose transporter
mam1	Secretion of mating factor M in *S. pombe*
MDR1	Multidrug resistance 1
MRP	Multidrug resistance-associated protein
P45014DM or	Cytochrome P450 dependent 14 α-demethylase
CYP51	Cytochrome P450 subfamily 51
P450alk	Alkane-inducible cytochrome P450
PDH1	Pleiomorphic drug resistance homologue
PDRs	Pleiotropic drug resistance
PMR1	*Penicillium* multidrug resistance 1
PXA	Peroxisomal ABC-transporter
RL1	RNase L inhibitor 1
SEC14	One of the mutants defective in protein secretion
SGE1	Suppressor of *gal1* mutations (see Table 3)
SNQ2	Sensitivity to 4-nitroquinoline *N*-oxide

Table 1   Cont.

Gene	Coding For
STE6	Sterile 6 (gene involved in mating – a factor transport)
TPS2	Trehalose 6-phosphate synthase 2
TRR1	Thioredoxin reductase 1
TRX2	Thioredoxin 2
YAP1	Yeast activator protein 1
YCF1	Yeast cadmium factor 1
YEF3	Yeast elongation factor 3
YOR1	Yeast oligomycin resistance 1
YRR1	Yeast reveromycin resistance 1
YRS1	Yeast reveromycin sensitivity

that *C. albicans* harbours enzymes involved in the glyoxylate cycle, which help its survival inside the macrophages during infection. Since mammalian cells lack glyoxylate cycle, the enzymes of this pathway probably represent potential antifungal targets (Lorenz and Fink, 2001).

In spite of the fact that there are many targets (mentioned above), with the exception of a few, most of the currently used antifungal drugs are directed against the enzymes of the ergosterol biosynthetic pathway. Fig. 2 depicts the targets of the commonly used antifungals and the mechanism by which resistance to these antifungals develops. Although the molecular bases of antifungal resistance in fungi are not very clear, evidence accumulated so far suggests that MDR is a multifactorial phenomenon comprising a combination of several mechanisms (White *et al.*, 1997, 1998; Ghannoum and Rice, 1999). A few of the well-known molecular mechanisms of antifungal resistance in *C. albicans*, which are discussed below, are shown in Fig. 3.

2.1. Target Alteration or Overexpression

Amongst the known antifungals, azole derivatives like fluconazole, ketoconazole and itraconazole have been the most widely used triazoles for combating fungal infections. Azoles specifically inhibit P45014DM enzyme of the ergosterol biosynthetic pathway (Wilkinson *et al.*, 1972, 1974; Vanden Bossche *et al.*, 1989). This inhibition results in the accumulation of 14-methylated sterols that cause disruption of membrane structure and function. Azole derivatives interact with the haem of P45014DM where the unhindered nitrogen atom of the azole ring (N3 in imidazole or N4 in azole derivatives) binds to the haem iron at its sixth co-ordinate position. The blocking of this position which is normally occupied by activated oxygen prevents initiation of the hydroxylation

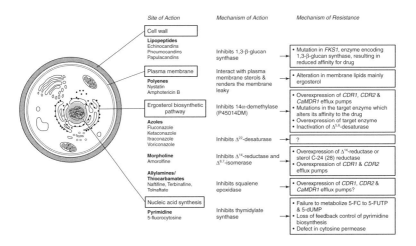

Figure 2   A cartoon of the yeast cell depicting the sites of action of the currently used antifungal drugs and the mechanisms by which *C. albicans* develops resistance to these antifungals. Refer to Vanden Bossche *et al.*, 1994; White *et al.*, 1998; Ghannoum and Rice, 1999; Marichal *et al*, 1999.

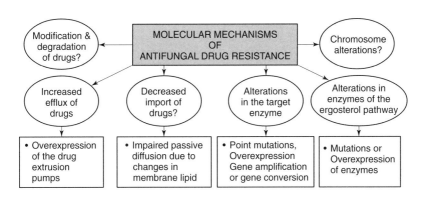

Figure 3   A schematic representation of the known molecular factors that contribute to antifungal drug resistance in *C. albicans*. For details on the mechanisms see the text and refer to White *et al.*, 1998; Ghannoum and Rice, 1999; Marichal *et al.*, 1999.

reaction. The structure, lipophilicity and stereochemical orientation of the N-1 side chain of the azoles also determine the affinity and the selectivity of azoles for their target (Sanglard *et al.*, 1998a; Vanden Bossche and Koymans, 1998). The mechanism by which azoles are imported into the cell still remains to be ascertained, although it is suggested that the hydrophobicity of these drugs could facilitate their entry. Once the azole drug enters the cell, its interaction with P45014DM can be modified in two ways: target alteration and overexpression. Both are discussed in the following sections.

2.1.1. Alterations in P45014DM

Several point mutations in the *ERG11* gene coding for P45014DM enzyme, which reduce its affinity for azoles have been identified (Lamb *et al.*, 1997; Loffler *et al.*, 1997; White, 1997a; Sanglard *et al.*, 1998a; Favre *et al.*, 1999; Kelly *et al.*, 1999a; Marichal *et al.*, 1999). Amongst the point mutations which are identified, the mutation Y132H represents a novel change in protein leading to fluconazole resistance (Kelly *et al.*, 1999b). This mutation does not permit normal binding of fluconazole to protein as is revealed from spectral studies with P45014DM. The most characteristic feature of cytochrome P450s is their absorbance at 450 nm in the reduced CO difference spectrum. Upon interaction with various compounds, cytochrome P450 exhibits substrate-induced changes which are of two types. A type I spectrum is obtained when the cytochrome P450 interacts with its substrate, while a type II spectrum is obtained when cytochrome P450 interacts with compounds that interfere with their haem group. Under normal circumstances, binding of fluconazole to the native protein results in a type II spectrum due to its co-ordination with haem as a sixth ligand, but an interaction with the Erg11p harbouring the Y132H mutation results in a type I spectrum. However, there is no effect of this mutation on ergosterol biosynthesis. This suggests that Y132H substitution occurs without significant perturbation of the haem environment and thus represents a novel change in protein leading to fluconazole resistance (Kelly *et al.*, 1999b). A compilation to show the frequency and positions of the so far identified amino acid substitutions shows that the mutations D116E, K128T, E266D and G464S occur with highest frequency, of which only the G464S mutation is exclusively present in azole-resistant isolates. The other three mutations are also found in azole-sensitive isolates (Marichal *et al.*, 1999). The I-helix stretch of Erg11p, which is highly conserved in the cytochrome P450 family, does not show any spontaneous mutations. The exact placement of all the identified mutations in a 3D model of the protein shows that these mutations are not randomly distributed but are rather clustered in three hot spot regions between amino acid residues 105–165, 266–287 and 405–488 (Marichal *et al.*, 1999).

Molecular modelling of P45014DM has provided an additional tool to understand its interactions with its natural substrate, lanosterol and inhibitors, the azoles (Lewis *et al.*, 1999). Two models of P45014DM, based on its homology to P450cam of *Pseudomonas putida CYP101* and P450BM-3 of *Bacillus megaterium* have been built to illustrate these interactions (Boscott and Grant 1994; Lewis *et al.*, 1999). Using the above two models, predicting substrate–enzyme interactions has a limitation since the models are deduced from the bacterial P450s which unlike the yeast P45014DM are soluble enzymes and do not interact with azole antifungals. In this regard, it is pertinent to mention that a crystal structure of P45014DM of *Mycobacterium tuberculosis*, which interacts with azole antifungals, is now available (Podust *et al.*, 2001). The crystal structure of this prokaryotic P45014DM could improve our understanding of the interactions between azoles and P45014DM of yeasts.

Another type of genetic alteration that involves the *ERG11* gene is the phenomenon of gene conversion and mitotic recombination. In the clinical isolates of *C. albicans*, where a R467K mutation was identified, it was observed that in addition to this mutation there were other allelic differences present in the sensitive isolates that were subsequently removed in the resistant isolates (White 1997b).

2.1.2. Overexpression of P45014DM

Resistance to fluconazole in many clinical isolates is often associated with the overexpression of *ERG11* (Sanglard *et al.*, 1995; Albertson *et al.*, 1996; White, 1997a). Although gene amplification is one of the common mechanisms of drug resistance in mammalian cells, overexpression of P45014DM in *C. albicans* is not linked to gene amplification (Stark and Wahl, 1984). The only example where an increased level of P45014DM is associated with the amplification of its (*ERG11*) gene, is observed in a clinical isolate of *C. glabrata* (Vanden Bossche *et al.*, 1992, 1994; Marichal *et al.*, 1997). Because of the simultaneous existence of mutations or overexpression of the efflux pumps in the same isolates, it has been rather difficult to ascertain the contribution of overexpression of P45014DM in azole resistance.

2.2. Alterations in Other Enzymes of the Ergosterol Biosynthetic Pathway

In addition to P45014DM, alterations in other enzymes of the ergosterol biosynthetic pathway also lead to drug resistance. One such study indicates that a defective sterol $\Delta^{5,6}$-desaturase (*ERG3*) leads to the development of fluconazole resistance in clinical isolates (Kelly *et al.*, 1997b). This defect in these isolates results in the accumulation of 14α-methylfecosterol, which compensates for ergosterol depletion. As a consequence of ergosterol depletion, these isolates are also cross-resistant to amphotericin B (Kelly *et al.*, 1997b).

The cytochrome P450 spectral studies performed in a system reconstituted with purified *ERG5* (Δ^{22}-desaturase or CYP61) of *C. glabrata* revealed interaction between azole antifungals and the haem protein. These results suggest that *ERG5* could also be a target for azoles and may contribute to antifungal resistance (Lamb *et al.*, 1999).

The azole-induced upregulation of *ERG11* along with other genes of the ergosterol biosynthetic pathway is suggestive of the existence of a common mechanism of upregulation of the *ERG* genes in *C. albicans* (Henry *et al.*, 2000). Recent transcript profiling in *C. albicans*, using microarrays, demonstrates the regulation of almost 300 genes in response to a 24-h exposure to 10 μM itraconazole. While a global regulation of *ERG* genes was evident from this transcript profiling, several genes of diverse functions as well as of unknown functions were either up- or down-regulated by the drug (De Backer *et al.*, 2001). This means that azoles could contribute to multiple, yet unknown, phenotypes that still remain to be identified. The dissection of the mechanisms mediating these phenotypes would provide new insights into the phenomenon of MDR.

2.3. Chromosome Alterations

With the possibility of undergoing karyotype variability, fungi have the advantage of being flexible with their genome, which enables them to adapt to environmental variations (Fierro and Martin, 1999). An alteration in the copy number of chromosomes as a mechanism of regulating gene expression in *C. albicans* has been observed (Janbon *et al.*, 1998), wherein nondisjunction of two specific chromosomes has been proposed as the cause of *in vitro* development of fluconazole resistance. It appears that chromosomal nondisjunction, which awaits further proof, points to a new mechanism of drug resistance in the pathogenic yeasts (Perepnikhatka *et al.*, 1999).

2.4. Modification or Degradation of Drugs

The detoxification of drugs by enzymes like the cytochrome P450s, which is a common method of rendering the drugs non-toxic in prokaryotes and higher eukaryotes, has not been observed in yeasts (Kappeli, 1986; Omura, 1999). Failure to detect any metabolite of azole antifungals in *Candida* cells has further reaffirmed the above notion (Hitchcock, 1993; White *et al.*, 1998). *Candida* species possess a unique class of cytochrome which comprises a family of proteins called the alkane-inducible cytochrome P450s (CYP52 gene family or P450alk) (Seghezzi *et al.*, 1992; Ohkuma *et al.*, 1995). Unlike the P45014DM cytochrome of yeast, the P450alk genes are involved in the hydroxylation of fatty acids and alkanes, thereby catalysing a step that precedes their further metabolism (Kappeli, 1986; Seghezzi *et al.*, 1992; Ohkuma *et al.*, 1995). Recently, the overexpression of *CaALK8*, one of the members of the P450alk gene family in a hypersensitive *C. albicans* host (disrupted in *CDR1* and *CDR2*, the two drug extrusion pump encoding genes) (Sanglard *et al.*, 1997), has been shown to confer resistance to at least three drugs including azoles. Interestingly, the hydroxylation of lauric acid mediated by *CaALK8* is competed out with these drugs implying a direct interaction of drugs with this P450alk protein (Panwar *et al.*, 2001). This suggests that *CaALK8* mediated hydroxylation of drugs could be a possible mechanism by which drugs are rendered non-toxic in *C. albicans*. However, no metabolic product of azoles has so far been detected in *CaALK8*-expressing cells. It also remains to be established how a P450alk-mediated drug resistance mechanism fits in with the established contributors of MDR. Δ^{22}-desaturase of *S. cerevisiae*, which appears to be a low activity hydroxylase, has already been shown to metabolize xenobiotics (Kelly *et al.*, 1993, 1997a). Taken together, the modification of drugs could well be a mechanism which might contribute to the overall drug resistance of *C. albicans*.

2.5. Drug Import

It is presumed that the hydrophobic nature of drugs permits easy import of these compounds by passive diffusion. However, the contribution of drug import to MDR is not well-established, since technically it has not been possible to separate efflux of drugs from their import. Nonetheless, there are a few studies, particularly with mammalian cells, where passive diffusion of drugs through lipid bilayer has been shown to be an important determinant of MDR (Ferte, 2000). The fluctuations in membrane fluidity are thus expected to affect passive diffusion and sensitivity to drugs. This was observed in a study wherein *erg* mutants of *S. cerevisiae* defective in different steps of ergosterol biosynthesis were used. These mutants were shown to possess high membrane fluidity and were hypersensitive to several drugs. The enhanced fluidity was linked to enhanced diffusion of drugs (Kaur and Bachhawat, 1999; Kohli and Prasad, unpublished observations). There are factors other than membrane fluidity which can also influence passive diffusion of drugs across the membrane bilayer and thus can affect drug susceptibilities. As an example, the deletion of *PDR16* and *PDR17*, homologues of *SEC14*, which are involved in regulating lipid synthesis in *S. cerevisiae*, results in changes in lipid composition without affecting the membrane fluidity (Van Den Hazel *et al.*, 1999). As a consequence of the changes in lipid composition, the uptake (passive diffusion) of rhodamine 6-G into de-energized cells was found to be increased in a hypersensitive $\Delta pdr16\Delta pdr17$ double mutant as compared with the wild-type strain (Van Den Hazel *et al.*, 1999). In this context, the role of import of drugs needs to be analysed more carefully. It is expected that with better experimental design and more studies, the contribution of import of drugs in MDR can be established.

2.6. Drug Efflux

Among several mechanisms of drug resistance, the overexpression of genes encoding drug extrusion pumps belonging to the ATP-binding cassette (ABC) and the major facilitator superfamily (MFS) of proteins is most common (Kolaczkowski and Goffeau, 1997; van Veen and Konings, 1997; Higgins, 2001; Jack *et al.*, 2001). An overexpression of P-glycoprotein (P-gp)/human MDR1 remains the most documented and well-characterized example of a drug extrusion pump belonging to the ABC superfamily that results in the failure of chemotherapy in tumour cells (Ambudkar *et al.*, 1999; Ueda *et al.*, 1999). The presence of proteins homologous to P-gp in all organisms ranging

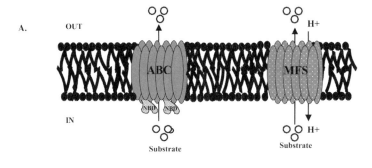

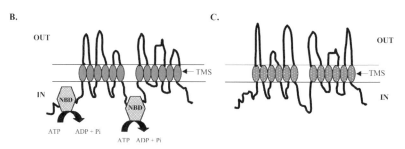

Figure 4 (A) A cartoon of an ABC and a MFS transporter of yeast. (B) and (C) Hydropathy-based predictions of topology of a typical ABC and MFS drug transporter, respectively. The ABC and the MFS transporters depicted here have (NBD-TMS$_6$)$_2$ and 12 TMS (transmembrane stretches), respectively. For the other types of topological arrangements refer to Tables 1 and 2. The NBDs (nucleotide binding domains) of the ABC transporters are responsible for the hydrolysis of ATP, which facilitates drug extrusion while the MFS transporters utilize a proton gradient to expel drugs.

from prokaryotes to eukaryotes portrays extrusion of drugs as a general theme of the mechanism of MDR (Michaelis and Berkower, 1995; Smart and Fleming, 1996; Tommasini *et al.*, 1997; van Veen and Konings, 1997; Urban *et al.*, 1999; Del Sorbo *et al.*, 2000; Theodoulou, 2000). MDR in *S. cerevisiae* is designated as PDR network, which comprises genes encoding for the drug extrusion pumps and their regulators. Homologues of the PDR genes, which contribute to antifungal resistance, are also present in the pathogenic yeasts (Prasad *et al.*, 1995; Sanglard *et al.*, 1997; Walsh *et al.*, 1997).

Amongst the two superfamilies of drug transporters, the ABC superfamily of transporters is characterized by the presence of nucleotide-binding domains (NBDs) that by binding and hydrolysing ATP facilitate the efflux of drugs across the membrane. On the other hand, proteins of the MFS lack the NBDs and utilize a proton motive force for drug efflux. One common feature of the two classes of proteins is the presence of multiple transmembrane stretches (TMS) that traverse the membranes (Fig. 4). The entire genome sequencing of *S. cerevisiae* revealed the existence of at least 30 ABC and 28 putative MFS transporters (Michaelis and Berkower, 1995; Decottignies and Goffeau, 1997; Goffeau *et al.*, 1997). While the ongoing genome-sequencing project of *C. albicans* will unravel the exact number of ABC and MFS transporter encoding genes, a few ABC drug transporters or *CDRs* (*Candida* drug resistance genes) for example, *CDR1* and *CDR2*, and MFS for example *CaMDR1*, have already been implicated in azole resistance. In addition, 14 new, partial gene sequences displaying homology to the NBD of human MDR1 have also been identified in *C. albicans*. These sequences do not show any significant homology to the known *CDRs*, but nonetheless might have an effect on the drug susceptibilities of *C. albicans* (Walsh *et al.*, 1997). In view of the importance of these pumps in multidrug resistance, the following sections detail various aspects of these transporter proteins.

2.6.1. ABC Drug Transporters of Yeasts

The ABC protein has a modular structure, wherein a typical 'full-length' protein of this superfamily comprises two intra-cytoplasmic domains consisting of the NBDs and two hydrophobic regions comprising six TMS. The sequence of domain organization varies amongst the proteins of the superfamily. The NBDs contain the Walker A and Walker B motifs (Walker *et al.*, 1982) and an ABC signature or C motif consensus sequence, (L, V)-SGG- $(X)_3$-R-hydrophobic residue-X-hydrophobic residue-A, which precedes the Walker B motif in these transporters (Michaelis and Berkower, 1995). Sequence comparison of the NBDs shows that these stretches of the ABC proteins are highly conserved throughout the evolutionary scale (Michaelis and Berkower, 1995; Decottignies and Goffeau, 1997).

Table 2   ABC transporters of fungi.

Organism[a]	Subfamily[b]	Gene name	Chr.no.	Size[c]	Location	Topology[d]	Function	Reference
Saccharomyces cerevisiae	PDR	PDR5	XV	1511	PM	$(NBD-TMS_6)_2$	Drug efflux pump	Balzi et al., 1994
		PDR12	XVI	1511	PM	$(NBD-TMS_6)_2$	Resistance to water-soluble, monocarboxylic acids with chain lengths of from C-1 to C-7	Bissinger and Kuchler, 1994; Piper et al., 1998
		PDR15	IV	1529	PM	$(NBD-TMS_6)_2$	Drug efflux pump, stress inducible	Wolfger et al., 1997
		SNQ2	IV	1501	PM	$(NBD-TMS_6)_2$	Drug efflux pump	Servos et al., 1993
	ALDP	PXA1	XVI	758	Peroxisomes	TMS_6-NBD	Required for β-oxidation of fatty acids	Shani et al, 1996
		PXA2	XI	853	Peroxisomes	TMS_6-NBD	Required for β-oxidation of fatty acids	Shani et al., 1996
	MRP/CFTR	YCF1	IV	1515	Vacuoles	$(TMS_6-NBD)_2$	Required for glutathione conjugated cadmium resistance	Szczypca et al., 1994
		YOR1	VII	1477	PM	$(TMS_6-NBD)_2$	Oligomycin resistance	Katzmann et al., 1995
		YRS1	VII	1477	?	$(TMS_6-NBD)_2$	Tolerance to organic anions	Cui et al., 1996
	MDR	ATM1	XIII	690	Mitochondria	TMS_6-NBD	Essential mitochondrial protein	Leighton and Schatz, 1995
		STE6	XI	1290	PM	$(TMS_6-NBD)_2$	a-factor export	McGrath and Varshavski, 1989
	YEF3	YEF3	XII	1044	?	$(NBD)_2$	Interaction with aminoacyl-tRNA	Sandbaken et al.,1990
		GCN20	VI	752	?	$(NBD)_2$	Interaction with tRNA and Gcn2p	Vasquez de Aldana et al., 1995
Schizosaccharomyces pombe	PDR	Abc1	II	1427	?	$(TMS_6-NBD)_2$	Unknown	Christensen et al., 1997a
	MDR	bfr1+	III	1530	?	$(NBD-TMS_6)_2$	Brefeldin A transport	Nagao et al., 1995
		pmd1+	III	1362	?	$(TMS_6-NBD)_2$	Drug efflux pump	Nishi et al., 1992
		Mam1	II	1336	?	$(TMS_6-NBD)_2$	M-factor transport	Christensen et al., 1997b
Candida albicans	PDR	CaCDR1	3	1501	PM	$(NBD-TMS_6)_2$	Drug efflux pump, phospholipid translocator	Prasad et al., 1995; Sanglard et al., 1995, Dogra et al., 1999

Table 2   Cont.

Organism[a]	Subfamily[b]	Gene name	Chr.no.	Size[c]	Location	Topology[d]	Function	Reference
		CaCDR2	3	1499	PM	$(NBD\text{-}TMS_6)_2$	Drug efflux pump, phospholipid translocator	Sanglard et al., 1997; Dogra et al., 1999
		CaCDR3	4	1501	PM	$(NBD\text{-}TMS_6)_2$	Phospholipid translocator	Balan et al., 1997; Smriti et al., 2001
		CaCDR4	1	1490	?	$(NBD\text{-}TMS_6)_2$	Phospholipid translocator?	Franz et al., 1998; Sanglard et al., 1999c
		CaCDR5	6	—	?	?	Drug efflux pump?	
	MDR	HST6	3	1323	?	$(TMS_6\text{-}NBD)_2$	Transport of a-factor, drugs?	Raymond et al., 1998
	MRP/CFTR	CaYOR1	?	ND	PM	?	Drug efflux pump?	Ogawa et al., 1998
		CaYCF1	7	1606	Vac?	?	Drug efflux pump?	Theiss et al., 1999
Candida glabrata	PDR	CgCDR1	?	1499	PM	$(NBD\text{-}TMS_6)_2$	Drug efflux pump	Sanglard et al., 1999a, 1999b
		CgCDR2	?	—	?	$(NBD\text{-}TMS_6)_2$	Drug efflux pump	Sanglard et al., 1999a, 1999b
		PDH1	?	1542	?	$(NBD\text{-}TMS_6)_2$	Drug efflux pump	Miyazaki et al., 1998
Candida dubliniensis	PDR	CdCDR1	?	?	?	$(NBD\text{-}TMS_6)_2$	Drug efflux pump?	Moran et al., 1998
		CdCDR2	?	?	?	$(NBD\text{-}TMS_6)_2$	Drug efflux pump?	Moran et al., 1998
Candida krusei	PDR	ABC1	?	Pcds[e]	?	?	Drug efflux pump	Katiyar and Edlind, 2001
		ABC2	?	Pcds[e]	?	?	Drug efflux pump	Katiyar and Edlind, 2001
Aspergillus flavus	MDR	Aflmdr1	?	1307	?	$(TMS_6\text{-}NBD)_2$	Unknown	Tobin et al., 1997
Aspergillus fumigatus	MDR	Afumdr1	?	1349	?	$(TMS_6\text{-}NBD)_2$	Resistance to cilofungin	Tobin et al., 1997
		Afumdr2	?	791	?	$TMS_6\text{-}NBD$	Not involved in MDR	Tobin et al., 1997
	-	Adr1	?	?	?	?	Upregulated by itraconazole	Slaven et al., 1999

Table 2　Cont.

Organism[a]	Subfamily[b]	Gene name	Chr.no.	Size[c]	Location	Topology[d]	Function	Reference
Aspergillus nidulans	PDR	*AtrA*	?	1466	?	$(NBD-TMS_6)_2$	Unknown	Del Sorbo *et al.*, 1997
		AtrB	?	1426	?	$(NBD-TMS_6)_2$	Drug efflux pump	Del Sorbo *et al.*, 1997
	MDR	*AtrC*	?	1284	?	$(TMS_6-NBD)_2$	Upregulated by various toxicants	Andrade *et al.*, 2000a
		AtrC2	?	1293	?	$(TMS_6-NBD)_2$	Upregulated by cycloheximide	Angermayr *et al.*, 1999
		AtrD	?	1348	?	$(TMS_6-NBD)_2$	MDR	Andrade *et al.*, 2000b
	–	*AbcA- D*	?	?	?	?	Unknown	Do Nascimiento *et al.*, 1999
Penicillium digitatum	PDR	*PMR1*	?	1619	?	$(NBD-TMS_6)_2$	Azole resistance	Nakaune *et al.*, 1998
Magnaporthe grisea	PDR	*ABC1*	?	1619	?	$(NBD-TMS_6)_2$	Essential for pathogenicity	Urban *et al.*, 1999
Cryptococcus neoformans	PDR	*CneMDR1*	?	1408	?	$(TMS_6-NBD)_2$	Drug efflux pump?	Thornewell *et al.*, 1997

[a] For *S. cerevisiae*, out of the 30 putative ABC proteins, only those with identified functions are listed.
[b] Names of subfamilies based on sequence similarity with human ALDP, MRP/CFTR, MDR and yeast (PDR, YEF3) ABC transporters.
[c] Number of amino acid residues.
[d] NBD, nucleotide-binding domain; TMS, transmembrane segment.
[e] Pcds, partial coding sequence available in the data base.

On the basis of sequence homologies, the 30 ABC proteins of *S. cerevisiae* are divided into six clusters, the PDR, MDR, MRP/CFTR, RL1, YEF3 and the ALDP subfamilies (Table 2) (Michaelis and Berkower, 1995; Decottignies and Goffeau, 1997; Taglicht and Michaelis, 1998). The proteins of the PDR subfamily that are involved in conferring drug resistance are well-characterized ABC transporters in *S. cerevisiae*. Proteins belonging to the MRP/CFTR subfamily are involved in conferring resistance to drugs or inorganic ions (Cole and Deeley, 1998). In addition, the MRP/CFTR subfamily of proteins has gained considerable attention as some proteins of this subfamily are of clinical significance (Michaelis and Berkower, 1995; Decottignies and Goffeau, 1997). The proteins of the other two subfamilies RL1 and YEF3 probably are incapable of acting as membrane transporters due to the lack of TMS (Sandbaken *et al.*, 1990; Michaelis and Berkower, 1995; Vasquez de Aldana *et al.*, 1995). ALDP members, on the other hand, similar to their mammalian counterparts, are half-transporters proposed to be involved in the oxidation of very long-chain fatty acids (Shani *et al.*, 1995, 1996).

Among all the ABC proteins identified so far in *S. cerevisiae*, five belonging to the PDR subfamily are involved in the extrusion of drugs (Table 2). *PDR5*, the first drug transporter to be identified in *S. cerevisiae* by several

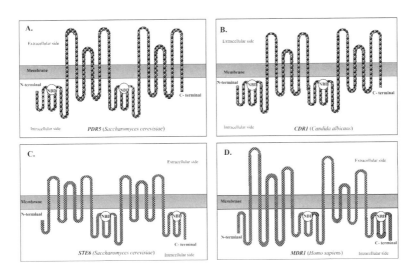

Figure 5   Predicted topological arrangement of *PDR5* (A), *STE6* (C) of *S. cerevisiae*, *CDR1* of *C. albicans* (B) and the human MDR1 (D). Pdr5p and Cdr1p consist of two duplicated halves, each comprising one hydrophilic domain followed by a hydrophobic domain. The two hydrophilic domains are cytoplasmic and each contains one nucleotide-binding domain (NBD). The two hydrophobic domains are considered to be spanning the membrane at least 12 times (TMS). Note the sequence of domain inversion in Pdr5p and Cdr1p $(NBD\text{-}TMS_6)_2$ with respect to Ste6p and Mdr1p $(TMS_6\text{-}NBD)_2$.

research groups gained considerable attention due to its sequence similarity to the human MDR1 (Balzi *et al.*, 1994; Bissinger and Kuchler 1994; Hirata *et al.*, 1994; Kralli *et al.*, 1995). Interestingly, in spite of high sequence similarity, Pdr5p (protein encoded by *PDR5*) displays a $(NBD\text{-}TMS_6)_2$ topology that mirrors the $(TMS_6\text{-}NBD)_2$ topology of human MDR1 (Fig. 5). The significance of this inversion of domains between the two close homologues of drug transporters is not known. The substrate profile of Pdr5p ranges from antifungals, clinical antimycotics, herbicides, agricultural fungicides to anticancer drugs and human steroid hormones (Kolaczkowski *et al.*, 1998). *SNQ2* of the PDR subfamily and a close homologue of Pdr5p was isolated as a gene that specifically conferred resistance to the mutagen 4-nitroquinoline *N*-oxide (NQO) (Servos *et al.*, 1993). In spite of being close homologues, both Pdr5p and Snq2p have an overlapping, as well as a specific substrate profile (Decottignies *et al.*, 1994, 1995).

PDR12, another homologue of *PDR5* and *SNQ2* is characterized by its specific substrate profile (Piper *et al.*, 1998). Though Pdr12p shares a >37% identity with Pdr5p and Snq2p, it neither confers resistance to NQO, a substrate specific for Snq2p, nor to cycloheximide, a substrate specific to Pdr5p. Pdr12p confers resistance to weak organic acids like sorbate, benzoate, acetate and propionate usually used as food preservatives. The involvement of an energy-dependent extrusion pump capable of expelling weak acids out of the cell provides a basis for food spoilage by some yeast species, in spite of the usage of weak organic acids as food preservatives (Piper *et al.*, 1998).

YOR1 and *YCF1* of the MRP/CFTR subfamily are homologous to human CFTR and MRP, which were isolated by their ability to confer resistance to oligomycin and cadmium, respectively (Szczypka *et al.*, 1994; Katzmann *et al.*, 1995). Ycf1p is localized in the vacuolar membranes and acts as a glutathione conjugate transporter, a key step in the detoxification pathway mediated by this gene. *YRS1* is another gene that resembles *YOR1* and complements the phenotype of the mutant sensitive to the anionic drug reveromycin A. *YRS1* is also capable of providing tolerance to a wide range of the organic anions that contain carboxyl group (Cui *et al.*, 1996).

In contrast to *S. cerevisiae*, the MDR pumps of *S. pombe*, the fission yeast are poorly characterized. Two drug extrusion pumps, $bfr1^+$ and $pmd1^+$ responsible for drug resistance in *S. pombe* have been partially characterized. While $bfr1^+$ confers resistance to compounds like brefeldin A, cerulenin, and cytochalasin B, $pmd1^+$ is responsible for conferring resistance to leptomycin B and other drugs (Nishi *et al.*, 1992, 1993; Nagao *et al.*, 1995). $hmt1^+$, a homologue of Pdr5p, is another protein that confers heavy metal tolerance to this yeast by sequestrating metal–phytochelatin complexes to the vacuole (Ortiz *et al.*, 1992) (Table 2).

The first ABC efflux pump characterized in any known pathogenic yeast, designated as *CDR1*, was isolated as a gene implicated in conferring resistance to cycloheximide in a *PDR5* disruptant hypersensitive strain of *S. cerevisiae* (Prasad *et al.*, 1995). Further studies extended the substrate spectrum of *CDR1* to various other drugs including the triazoles such as fluconazole, ketoconazole and itraconazole (Sanglard *et al.*, 1995, 1997). *CDR1* codes for a protein of 1501 amino acid residues (169.9 kDa), with a topology similar to that of Pdr5p and Snq2p (Prasad *et al.*, 1995). On the other hand, its topology mirrors that of *STE6* – the *S. cerevisiae* a-mating pheromone transporter (McGrath and Varshavski, 1989), as well as of the human MDR1 and CFTR (Prasad *et al.*, 1996). Despite a high structural and functional similarity between Cdr1p and Pdr5p, some distinct functional features tend to distinguish them. For example, both genes share overlapping specificities for cycloheximide and chloramphenicol, but *CDR1* affects sensitivity to oligomycin, while neither amplification nor disruption of *PDR5* alters susceptibilities to this mitochondrial inhibitor (Prasad *et al.*, 1995). It is worth mentioning here that some of the close homologues of *CDR1* in *C. albicans* are also functionally distinct. As an example, Cdr2p that exhibits 84% identity with Cdr1p has a distinct drug resistance profile as compared with *CDR1* (Sanglard *et al.*, 1997). The overexpression or deletion of *CDR3* and *CDR4*, the homologues of Cdr1p and Cdr2p, interestingly do not affect drug susceptibilities of yeast cells (Balan *et al.*, 1997; Franz *et al.*, 1998). The hydropathy plots of Cdr1p and Cdr3p, show that both the proteins have similar topological arrangements where the hydrophilic domain containing the NBDs precedes the hydrophobic TMS (Table 2). The only apparent difference between the two proteins appears to be in the C-terminal region, where Cdr3p has an extended loop connecting TM11 and TM12. In addition, there is a distinct stretch of 21 amino acids in the C-terminal portion of Cdr3p which apparently is absent in Cdr1p (Fig. 6). Keeping in view the importance of these regions in drug binding, the subtle differences in the structures of these proteins could be responsible for affecting their substrate specificity, thereby enabling some of them (Cdr1p and Cdr2p) but not others (Cdr3p and Cdr4p), to bind and transport drugs (Loo and Clarke, 1994, 1999a; Zhang *et al.*, 1995; Krishnamurthy *et al.*, 1998a).

Among non-*albicans* species, *C. glabrata* has emerged as an important nosocomial pathogen, wherein homologues of *CDRs, CgCDR1* and *CgCDR2* have been isolated (Table 2) (Sanglard *et al.*, 1999a, b). *CgCDR1* is also found to be up-regulated in the azole-resistant *C. glabrata* isolates (Sanglard *et al.*, 2001). Another putative ABC transporter encoded by *PDH1* gene is implicated in azole resistance in *C. glabrata* (Miyazaki *et al.*, 1998). Similarly, the existence of *CdCDR1* and *CdCDR2* in *C. dubliniensis* – a species closely related to *C. albicans* – contributes to the reduced accumulation of fluconazole in its azole resistant isolates (Moran *et al.*, 1998). A potential role of two putative ABC transporters *ABC1* and *ABC2* in drug resistance has also been suggested for *C. krusei* (Katiyar and Edlind, 2001).

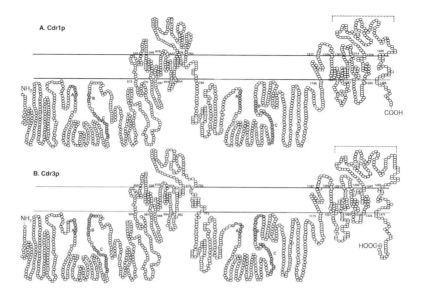

Figure 6   A hypothetical two-dimensional model of Cdr1p (A) and Cdr3p (B). The model is based on the hydrophobicity profiles of amino acid sequences and functional domains. Small circles represent amino acid residues. The shaded circles represent the Walker A, B and C motif – another consensus sequence within the nucleotide binding domains (NBD). The numbers indicate the beginning and the end of the transmembrane stretches (TMS). The extra stretch of 21 amino acids (as observed from a multiple sequence alignment) at the C-terminal in Cdr3p which is absent in Cdr1p is boxed. The loops between TM11 and TM12 (marked with a dashed line) are larger in Cdr3p (104 amino acids) as compared with Cdr1p (89 amino acids).

2.6.2. MFS Drug Transporters of Yeasts

The MFS proteins are proton motive force (PMF)-dependent secondary transporters that ferry substrates across the membrane (Paulsen *et al.*, 1996). These transporters are classified into five distinct clusters or families involved in (1) drug transport; (2) sugar uptake; (3) uptake of Krebs cycle intermediates; (4) phosphate ester/phosphate antiport; and (5) oligosaccharide uptake (Marger and Saier, 1993; Paulsen *et al.*, 1996; Pao *et al.*, 1998; Saier *et al.*, 1998). On the basis of hydropathy and phylogenetic analyses, MFS proteins, which comprise several hundred members, can be divided into two distinct groups containing either 12 or 14 TMS (Paulsen *et al.*, 1996). The structure–function relationships of MFS in yeast have not been studied in detail due to the diversity in their nucleotide and amino acid sequences. Few studies though postulate that the N-terminal halves of different MF subfamilies share greater similarities than their C-terminal halves. This suggests that C-terminal regions are probably involved in substrate recognition and N-terminal regions are involved in proton translocation (Saier and Reizer, 1991; Paulsen *et al.*, 1996).

In *S. cerevisiae*, out of the 28 putative MFS proteins identified, only a few such as *FLR1*, *ATR1* and *SGE1* have been implicated to play a role in drug resistance (Table 3) (Kanazawa *et al.*, 1988; Gompel-Klein and Brendel, 1990; Ehrenhofer-Murray *et al.*, 1998). Interestingly, there are other MFS proteins, such as *HXT9* and *HXT11* of the hexose transporter family, which have been implicated in contributing to the MDR scenario in *S. cerevisiae*. Both *HXT9* and *HXT11*, whose overexpression leads to sensitivity to drugs, belong to the monosaccharide transporters of the MF superfamily. Conversely, their deletion leads to resistance to a number of drugs like cycloheximide, sulfomethuron methyl and NQO (Nourani *et al.*, 1997). This effect is opposite to the generally observed mode of drug resistance exhibited by overexpression of the ABC or MFS drug extrusion pumps. Furthermore, the expression of the two-hexose transporters is under the control of *PDR1* and *PDR3*, the master regulatory proteins regulating the expression of the ABC drug extrusion pumps. The co-regulation of these hexose transporters, along with the drug extrusion pumps of the ABC superfamily, points towards intricacies in the phenomenon of MDR in yeast (discussed below).

The two MFS genes that play important roles in antifungal resistance in *C. albicans* are *CaMDR1* (previously known as BEN^r) and *FLU1*. *CaMDR1* was initially identified as a gene that conferred resistance to benomyl and methotrexate, the tubulin-binding agent and a tetrahydrofolate reductase inhibitor, respectively (Fling *et al.*, 1991). *FLU1* on the other hand, was initially isolated as a clone that conferred resistance to fluconazole, although recently mycophenolic acid has been shown to be its specific substrate (Calabrese *et al.*, 2000). *CaMDR1* is highly homologous to *FLR1*, while *FLU1* displays a high similarity to a *S. cerevisiae* ORF YLL028wp, which does not encode a transporter (Calabrese *et al.*, 2000). *CaMDR1* expression in *S. cerevisiae* confers resistance to several unrelated drugs and its overexpression is linked to fluconazole resistance in *C. albicans* (Gupta *et al.*, 1998; Marr *et al.*, 1998). Deletion of *CaMDR1* in *C. albicans* host CAI4 does not render it susceptible to drugs (Morschhauser *et al.*, 1999). The deletion of *FLU1* in CAI4 also does not affect its drug susceptibilities. However, its deletion in a hypersensitive host background of *C. albicans* that is deleted in *CDR1*, *CDR2* and *CaMDR1* genes, affects the drug susceptibilities of the host (Calabrese *et al.*, 2000). The expression of *CaMDR1* in *C. albicans* cells is enhanced by benomyl, methotrexate and several other unrelated drugs, and it is constitutively overexpressed in some of the azole-resistant clinical isolates (Gupta *et al.*, 1998). Unlike *CaMDR1*, the expression of *FLU1* does not vary between the clinical *C. albicans*-resistant and -sensitive isolates (Calabrese *et al.*, 2000). The identification of seven polymorphic mutant alleles of *CaMDR1* (*CaMDR1-1* to *1-7*), displaying distinct drug-resistance profiles, reveals

Table 3 MFS drug transporters of yeasts.

Organism	Gene	Amino acids	Topology[a]	Substrates	References
Saccharomyces cerevisiae	*FLR1*	548	12	Fluconazole, cycloheximide and NQO	Alarco et al., 1997
	SGE1	543	14	Anionic drugs like crystal violet, ethidium bromide and methylmethane sulfonate	Ehrenhofer-Murray et al., 1998
	ATR1	547	14	Aminotriazole and NQO	Kanazawa et al., 1988; Gompel-Klein and Brendel, 1990
Schizosaccharomyces pombe	*CAR1*	526	12	Amiloride	Jia et al., 1993
Candida albicans	*CaMDR1/BEN[r]*	564	12	Benomyl, methotrexate, sulfomethuron methyl, NQO and fluconazole	Fling et al., 1991; Gupta et al., 1998
	FLU1	610	12	Mycophenolic acid	Calabrese et al., 2000
Candida glabrata	*CgMDR1*	–	12	Fluconazole, benomyl, cycloheximide and NQO	Sanglard et al., 1999b
Candida dubliniensis	*CdMDR1*	557	12	Fluconazole, amorolfine, brefeldin A, cerulenin, cycloheximide, fluphenazine and NQO	Moran et al., 1998
Candida maltosa	*Cyh[r]*	552	12	Cycloheximide	Sasnauskas et al., 1992

[a] Indicates the number of transmembrane stretches in these drug transporters.

several in-frame point mutations leading to changes in amino acid residues. Most noteworthy of these changes is an insertion/replacement of an aspartate residue in a serine-asparagine-aspartate-rich domain (Gupta *et al.*, 1998). However, the relevance of such alleles of *CaMDR1* in azole resistance in *C. albicans* remains to be ascertained.

Homologues of *CaMDR1* have also been identified in *C. dubliniensis* (*CdMDR1*) and *C. glabrata* (*CgMDR1*), where their increased expression is responsible for fluconazole resistance (Moran *et al.*, 1998; Sanglard *et al.*, 1999b). Interestingly, the Ser-Asn-Asp-rich domain present in CaMdr1p and its alleles is not present in CdMdr1p (Moran *et al.*, 1998), Cyhrp (Sasnauskas *et al.*, 1992) and Flr1p (Alarco *et al.*, 1997; Broco *et al.*, 1999; Oskouian and Saba, 1999), which are close homologues of CaMdr1p (Gupta *et al.*, 1998). This implies that the Ser-Asn-Asp domain per se may not be responsible for drug resistance, though its exact role in drug resistance needs further investigation.

3. REGULATION OF MULTIDRUG RESISTANCE GENES

In order to develop strategies for dissecting the mechanisms of drug resistance, it is important to understand the molecular mechanisms regulating the expression of the drug extrusion pumps. As a consequence, a number of transcription factors belonging to either the PDR or YAP regulatory network have been characterized in *S. cerevisiae* and *C. albicans* (Fig. 7). However, the precise mechanism(s) by which the signal(s) generated by the divergent family of drugs, which would activate these transcription factors leading to the up-regulation of the battery of genes involved in MDR, is still unknown.

3.1. Regulation in *S. cerevisiae*

3.1.1. The PDR Network

The PDR network in *S. cerevisiae* comprises the Cys_6-Zn $(II)_2$ cluster family of transcription regulators namely, *PDR1*, *PDR3*, *PDR7* and *PDR9* (Balzi and Goffeau, 1991, 1994, 1995; Dexter *et al.*, 1994). Of the above-mentioned regulators, the involvement of at least *PDR1* and *PDR3* in regulating the expression of *PDR5*, *SNQ2* and *YOR1* is well-established (Meyers *et al.*, 1992; Balzi *et al.*, 1994; Dexter *et al.*, 1994; Katzmann *et al.*, 1994, 1995; Decottignies *et al.*, 1995; Mahe *et al.*, 1996a). *PDR1* and *PDR3* regulate the expression of these drug extrusion pumps encoding genes by binding to *cis*-acting elements, called the PDREs (pleiotropic drug resistance elements) in their promoter

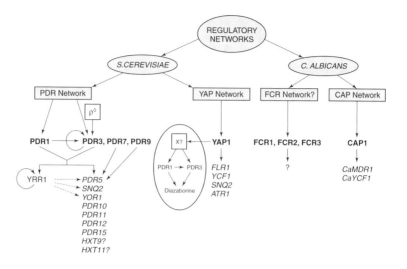

Figure 7   The regulatory circuits of MDR in yeasts. The transcription factors of the two regulatory circuits identified so far are marked in bold. *YAP1* is also known as *PAR1, SNQ3* or *PDR4*. *PDR1* and *PDR3* together regulate a number of ABC and MFS drug transporters and also regulate the transcription factor *YRR1*. The link between *YAP1* and PDR1/3 is represented within the open oval. The arrows on *YRR1* and *PDR3* indicate autoregulation of these transcription factors. The loss of mitochrondrial genome, indicated by ρ^0, also leads to the activation of *PDR3*. Refer to the text for further details.

regions (DeRisi *et al.*, 2000). PDREs are also present in the *PDR3* promoter implying its autoregulation (Delahodde *et al.*, 1995). Single amino acid substitutions in the genes coding for these regulators make them hyperactive, which in turn results in the overexpression of the downstream target genes that are involved in MDR (Meyers *et al.*, 1992; Balzi and Goffeau, 1994; Carvajal *et al.*, 1997). A number of different hyperactive alleles of Pdr1p have been identified (Carvajal *et al.*, 1997), wherein each allelic form of Pdr1p has been found to elicit a specific alteration in the levels of resistance mediated by its effector genes, for example *PDR5, SNQ2* and *YOR1*. A combination of the allelic form *pdr1-3* and *PDR5* promoter results in the highest level of transcriptional activation with a concomitant increase in the levels of the *PDR5* protein (Carvajal *et al.*, 1997). A genetic screen to identify the negative regulators of *PDR5* expression showed that loss of the mitochondrial genome (ρ^0) resulted in the up-regulation of *PDR3*. This study suggests the possibility of existence of another regulatory pathway linking the expression of the drug extrusion pumps with mitochondrial functions (Hallstorm and Moye-Rowley, 2000).

Apart from being regulated by *PDR1* and *PDR3*, the efflux pump-encoding genes *SNQ2* and *YOR1* are also regulated by another regulator *YRR1* (Hertle *et al.*, 1991), which was earlier isolated as a gene conferring high level of resistance to a cell cycle inhibitor, reveromycin A (Cui *et al.*, 1998). Interestingly, the expression of *YRR1* in turn is controlled by Pdr1p (and probably Pdr3p) implying cross-talk between the regulatory circuits that control the expression of the drug extrusion pumps in this yeast (Hertle *et al.*, 1991; Zhang *et al.*, 2001). The targets of *PDR1* and *PDR3* also include the two hexose MFS transporters *HXT9* and *HXT11* that are involved in the phenomenon of PDR in *S. cerevisiae* (discussed above). This study for the first time demonstrated the existence of genetic interactions between genes coding for two classes of transporters (ABC and MFS), to control the multidrug resistance process (Nourani *et al.*, 1997).

3.1.2. The YAP Network

The YAP regulatory proteins belong to the bZIP protein family and are related to the mammalian AP-1 transcription factor (Moye-Rowley *et al.*, 1989). The YAP regulatory network is involved in mediating resistance to a variety of toxicants, including cycloheximide, NQO, sulphomethuron methyl and cadmium, and in providing tolerance to oxidative stress (Moye-Rowley *et al.*, 1989). Eight alleles of the *YAP* family of regulators (*YAP1–8*) exist in *S. cerevisiae* of which three are implicated in regulating the expression of genes coding for drug extrusion pumps (Fernandes *et al.*, 1997). *YAP1* (also called *PAR1*, *SNQ3* or *PDR4*) binds to a conserved DNA sequence termed the YRE (*YAP1* response element) within the promoter of its targets to activate transcription (Moye-Rowley *et al.*, 1989). Yap1p functions by shuttling between the nucleus and cytoplasm in response to oxidative stress (Yan *et al.*, 1998). Overexpression of *YAP1* or *YAP2* leads to an increased resistance to a variety of drugs and metals. Interestingly, Δ*yap1* strains have the most pronounced phenotypes (sensitivity to hydrogen peroxide and low temperature), indicating that the other allelic forms cannot efficiently substitute for the *YAP1* functions (Fernandes *et al.*, 1997). Yap1p, controls the expression of *YCF1* and *FLR1* through the YREs (Alarco *et al.*, 1997; Jungwirth *et al.*, 2000; Nguyen *et al.*, 2000). While Ycf1p belongs to the ABC superfamily of proteins, Flr1p is a member of the MF superfamily implying that Yap1p, similar to PDR1/3, is also capable of regulating drug extrusion pumps of both the superfamilies. The other targets of Yap1p include the MFS protein *ATR1* (Coleman *et al.*, 1997), *TPS2* (Gounalaki and Thireos, 1994), *GSH1* (Wu and Moye-Rowley, 1994), *TRX2* (Kuge and Jones, 1997), *TRR1* (Morgan *et al.*, 1997) and *GLR1* (Grant *et al.*, 1996).

Diazaborine exhibits antibacterial activity as well as inhibiting the growth of yeast cells, by some unknown mechanism (Wendler *et al.*, 1997). It was shown that the overexpression of Yap1p in *S. cerevisiae* leads to diazaborine resistance only in a wild-type, but not in a $\Delta pdr1\Delta pdr3$ background (Wendler *et al.*, 1997). In addition to diazaborine resistance, *PDR3* is also required for Yap1p-mediated resistance to NQO through Snq2p. Interestingly, no obvious YRE in the *SNQ2* promoter is identified which suggests the possibility of the involvement of other unknown regulatory elements contributing to the Yap1p-mediated regulation (Bauer *et al.*, 1999). Nonetheless, the link between *YAP1* and *PDR1*, *PDR3* demonstrates the interplay between the two families of regulators and emphasizes the complexity in the regulatory circuits that control the expression of the drug extrusion pumps in yeasts.

3.2. Regulation in *C. albicans*

Unlike *S. cerevisiae*, the regulation of MDR in *C. albicans* or in other pathogenic yeasts is poorly understood. It would be evident from the following that the studies conducted so far already reveal the complexities of regulation of drug resistance in this pathogenic yeast.

3.2.1. The FCR Network

Functional complementation studies in a $\Delta pdr1\Delta pdr3$ strain of *S. cerevisiae* using a *C. albicans* genomic library led to the identification of three genes, designated as *FCRs* (Talibi and Raymond, 1999). In *S. cerevisiae*, the phenotype observed with *FCR* is mediated by an overexpression of *PDR5*, implying that *FCR1* behaves as a positive regulator of *PDR5* in this host (Talibi and Raymond, 1999). On the contrary, *FCR1* deletion in *C. albicans* leads to resistance to a number of drugs, which implies that *FCR1* probably acts as a negative regulator in a homologous environment. The target genes of this regulatory protein, as well as the question as to whether Fcr1p acts as a positive or a negative regulator in *C. albicans*, awaits further investigation (Fig. 7).

3.2.2. The CAP Network

Another regulatory gene, *CAP1* of *C. albicans* was isolated which showed homology to the YAP family of transcription regulators of *S. cerevisiae* (Alarco *et al.*, 1997). The hypersensitivity of the *S. cerevisiae* $\Delta yap1$ strain to cadmium or hydrogen peroxide is restored by *CAP1* expression. The Cap1p-mediated phenotype in *S. cerevisiae* is through the *FLR1* gene product, which is the closest homologue of *CaMDR1* of *C. albicans* (Alarco *et al.*, 1997). *CaMDR1* (Fling *et al.*, 1991), *CaYCF1* (Alarco and Raymond, 1999), *CaGLR1* (Grant *et al.*, 1996), and *CaTRR1* (Morgan *et al.*, 1997) are amongst the putative targets of *CAP1* in *C. albicans*. Further studies on *CAP1* show that it is also involved in multidrug resistance and in the oxidative stress response in *C. albicans*. The deletion of *CAP1* leads to resistance to fluconazole and other toxicants which implies that similar to *FCR1*, it also acts as a negative regulator of its target genes in *C. albicans* (Alarco and Raymond, 1999). That Cap1p could be a negative regulator in *C. albicans* was further confirmed by deleting the *CAP1* gene in a strain FR2 (Albertson *et al.*, 1996), which was selected *in vitro* for fluconazole resistance and which constitutively expressed *CaMDR1* (Alarco and Raymond, 1999). The deletion of *CAP1* did not suppress the level of *CaMDR1* expression: rather, it led to an elevated expression of this gene. This observation suggests that *CAP1* controls the suppression rather than the activation of *CaMDR1* in FR2. In such a situation, the overexpression of *CaMDR1* in FR2 could have been the result of mutations in the promoter of this gene or in a *trans*-regulatory factor, which could control the expression of this gene (Alarco and Raymond, 1999). The possibility that mutation(s) in a *trans*-regulatory factor could control the expression of *CaMDR1* in a set of clinical isolates was also suggested in a recent study (Wirsching *et al.*, 2000).

4. DRUG EFFLUX PUMPS ARE MULTIFUNCTIONAL PROTEINS

The rapidly growing families of transporters, particularly belonging to the ABC superfamily (traffic ATPases), are involved in performing diverse functions. The functional diversity of these proteins is reflected in their ever-emerging additional roles in absorption, excretion, signal transduction, bacterial pathogenesis and most importantly in drug and antibiotic resistance (Chakraburtty, 2001; Detmers *et al.*, 2001; Goldman and Kranz, 2001; Mendez and Salas, 2001; Schneider, 2001) (Fig. 8). Understandably, considerable attention is now being given to the physiological relevance of ABC transporters in general, and the drug transporters in particular. While some of the ABC transporters are specific for a substrate, many of them are multispecific. Wider substrate specificity, wherein a single transporter is capable of recognizing a variety of unrelated xenobiotics, is the most intriguing aspect of these drug transporters. As discussed above, there are already several energy-dependent drug transporters that have been identified in *S. cerevisiae*, *Candida* and other pathogenic fungi (Table 2). Considering the abundance of the drug transporters and their wider specificity, it is believed that these drug transporters may not exclusively export drugs. Recent evidence confirms that the drug transporters particularly of the ABC superfamily of *S. cerevisiae* and *C. albicans* are multifunctional proteins, which mediate important physiological functions.

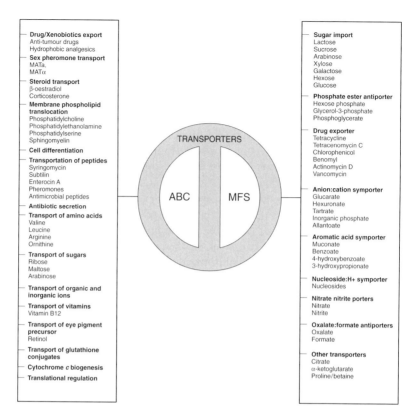

Figure 8 A schematic representation of the myriad functions performed by the ABC (ATP-binding cassette) and MFS (major facilitator superfamily) proteins. For details on the functions, the reader is referred to Pao *et al.*, 1998; Saier, 2000; Borges-Walmsley and Walmsley, 2001; Chakrabuitty, 2001; Detmers *et al.*, 2001; Goldman and Kranz, 2001; Marger and Saier, 1993; Mendez and Salas, 2001; Schneider, 2001.

4. 1. ABC Drug Extrusion Pumps

4.1.1. As Drug Transporter

Many laboratories report that the azole-resistant clinical isolates of *C. albicans*, owing to their ATP-dependent rapid efflux, accumulate lower amounts of drugs inside the cell (Albertson *et al.*, 1996; Clark *et al.*, 1996; Cannon *et al.*, 1998). Genes encoding drug extrusion pumps like *PDR5* of *S. cerevisiae* and *CDR1* and *CDR2* of *C. albicans* mediate the ATP-dependent rapid efflux of drugs (Albertson *et al.*, 1996; Clark *et al.*, 1996; Kolaczkowski *et al.*, 1996; Cannon *et al.*, 1998). An enhanced accumulation of fluconazole in the homozygous disruptants, for example Δ*cdr1* and Δ*cdr1*Δ*cdr2* of *C. albicans*, as compared with their wild-type host confirmed a causal relationship between efflux mediated by these two proteins and fluconazole accumulation (Sanglard *et al.*, 1997).

In order to understand the molecular mechanism of drug transport, it is important to identify domain(s) or amino acid residue(s) involved in substrate interaction and transport. Domain dissection of mammalian P-gp has revealed the importance of several amino acid residues and the putative TMS that contribute to its substrate specificity (Kajiji *et al.*, 1993; Loo and Clarke 1994, 1996, 1999a, 1999b, 1999c, 2000; Zhang *et al.*, 1995; Hanna *et al.*, 1996; Ueda *et al.*, 1997). Reports of such studies with the yeast transporters are only beginning to appear. In one such study, the deletion of 79 amino acids from the carboxy terminal of Cdr1p (which encompasses the TM12 of this transporter) does not result in the total loss of its ability to efflux cytotoxic agents (Krishnamurthy *et al.*, 1998a). While the expression of the truncated Cdr1p in *S. cerevisiae* resulted in impaired sensitivity to drugs like cycloheximide, anisomycin, sulfomethuron methyl and nystatin, its ability to confer resistance to drugs like o-phenanthroline, NQO, cerulenin, azoles, oligomycin, erythromycin, chloramphenicol and benomyl remained unaltered. Thus, it appears that the binding sites for certain drugs in Cdr1p, like in the human Mdr1p (Zhang *et al.*, 1995), could be localized within the TM 12. Interestingly, the truncated Cdr1p displayed no impairment in the NTPase activities and its ability to efflux Rhodamine 123, which is an anionic fluorescent substrate of Cdr1p. The truncation also did not affect the localization of Cdr1p in the plasma membrane. Additionally, when the truncated Cdr1p was overexpressed in a baculovirus–insect cell system, it elicited a drug-stimulated ATPase activity that was otherwise not detectable when Cdr1p was expressed in *S. cerevisiae* (Krishnamurthy *et al.*, 1998a). It has been observed that drugs stimulate the ATPase activity of human Mdr1p, which in turn is closely linked to its drug transport (resistance) activity (Ambudkar *et al.*, 1992; Ambudkar, 1998). However, such a link with yeast MDR proteins has not been established as yet (Decottignies *et al.*, 1994; Krishnamurthy *et al.*, 1998a).

Several amino acids that are important for providing substrate specificity to Pdr5p of *S. cerevisiae* have also been identified (Egner *et al.*, 1998, 2000). The mutations in several Pdr5p residues in the most highly conserved NBDs, in TM10 and in the predicted extracellular loop 6, which lies between the TM 11 and TM 12 led to drastic changes in substrate specificity. It appears from these mutational analyses that the overall structure of the folded ABC transporter might be an important determinant for their broad substrate specificity (Egner *et al.*, 1998, 2000).

Purification of the drug extrusion pumps is essential for understanding their important features including the kinetics of drug transport. Efforts directed to this effect have led to the partial purification of Pdr5p and Snq2p of *S. cerevisiae* (Decottignies *et al.*, 1994, 1995). NTPase activities performed with partially purified Pdr5p show that it has broad nucleotide specificity and is capable of hydrolysing CTP, GTP, UTP, ATP, and ADP but not AMP (Decottignies *et al.*, 1994). This broad specificity towards nucleotides is also shown by Cdr1p (Krishnamurthy *et al.*, 1998a), human Mdr1p (Hamada and Tsururo, 1988), and CFTR (Anderson *et al.*, 1991) and by bacterial permeases like the histidine permease (Bishop *et al.*, 1989). By using partially purified Pdr5p and Snq2p proteins, it was also observed that unlike oligomycin-sensitive NTPase activity of Pdr5p, Snq2p activity is insensitive to this inhibitor (Decottignies *et al.*, 1995). These studies with purified proteins further highlighted the differences between the two ABC transporters. Thus, despite being close homologues, Pdr5p and Snq2p transporters can be distinguished by their substrate specificity, as well as by the differences in their mechanism of functioning.

4.1.2. As Human Steroid Transporter

The ability of *PDR5* and *SNQ2* of *S. cerevisiae* and *CDR1* of *C. albicans* to efflux steroid hormones, like β-estradiol and corticosterone suggests that human steroid hormones could also be the substrates of these drug transporters (Kolaczkowski *et al.*, 1996; Mahe *et al.*, 1996b; Krishnamurthy *et al.*, 1998b). It must be noted that these hormones also upregulate the expression of *CDR1* in *C. albicans* cells (Krishnamurthy *et al.*, 1998c). Progesterone, which induces the overexpression of *CDR1* is not transported by Cdr1p (Krishnamurthy *et al.*, 1998b, 1998c). Interestingly, similar to Cdr1p, progesterone can only bind to Mdr1p, but is not transported by it (Ueda *et al.*, 1992). Notably, some of the drugs, such as cycloheximide, chloramphenicol, fluconazole and o-phenanthroline, to which *CDR1* confers resistance, reduce the efflux of β-estradiol and corticosterone probably by competition, thus implying commonality in binding site(s) between steroids and drugs (Krishnamurthy *et al.*, 1998b).

Since multidrug transporters of yeast can selectively mediate energy-dependent transport of human steroid hormones with high affinity and specificity, it is very likely that these hormones serve as physiological substrates of these proteins. Ergosterol – the analogue of mammalian cholesterol – is the closest molecule to steroids in yeast cells. It is thus quite possible that these multidrug transporters, like human Mdr1p, could be involved in the total sterol homeostasis in the yeast cells (Kontoyiannis, 2000). In this regard, it is also pertinent to mention that progesterone, corticosteroid and oestrogen binding proteins in *C. albicans* have already been identified (Feldman *et al.*, 1982; Stover *et al.*, 1983; Das and Datta, 1985; Skowronski and Feldman, 1989; Zhao *et al.*, 1995; Buckman and Miller, 1998). The interaction of ketoconazole with corticosteroid-binding protein has also been observed (Stover *et al.*, 1983). Additionally, the upregulation of *CDR1* transcription by steroid hormones and recent identification of steroid-responsive region in its promoter strongly point to a possible link between the steroid response cascade and multidrug resistance in *C. albicans* (Krishnamurthy *et al.*, 1998c).

4.1.3. As Phospholipid Translocator

Membrane phospholipids are asymmetrically distributed across the plasma membrane of all cell types in various organisms (Bevers *et al.*, 1996, 1999). Most of the phosphatidylethanolamine and phosphatidylserine are located on the inner leaflet of the lipid bilayer, while phosphatidylcholine and other lipids are predominantly localized on the outer leaflet of bilayer (Diaz and Schroit, 1996; Bevers *et al.*, 1999). The asymmetrical distribution of membrane lipids is very specific and its loss has been linked to various physiological consequences (Bevers *et al.*, 1996, 1999). The lipid asymmetry is maintained by membrane-bound phospholipid translocators divided into three classes: (1) bidirectional energy-dependent scramblase; (2) energy-dependent translocator that moves lipids towards the cytoplasmic surface of the plasma membrane (flippase); or (3) away (floppase) from the cytoplasmic surface of the membrane (Bevers *et al.*, 1999). Recent reports suggest that some of the ABC drug extrusion pumps down the evolutionary scale can mediate phospholipid translocation. Thus, human Mdr1p, Mdr2p are shown to translocate (flip/flop) membrane phospholipids between the two monolayers of the lipid bilayer of the plasma membrane (Ruetz and Gros 1994; Van Helvoort *et al.*, 1996). Pdr5p and Yor1p of *S. cerevisiae* and Cdr1p and Cdr2p of *C. albicans* can also mediate phospholipid translocation and therefore this property of ABC protein represents another conserved physiological function of these drug transporters (Decottignies *et al.*, 1998; Dogra *et al.*, 1999) (Table 4). However, this function does not appear to be a general feature of the proteins involved in drug extrusion since Snq2p, the close homologue of Pdr5p, is unable to mediate phospholipid translocation (Decottignies *et al.*, 1998). In a recent study, it is demonstrated that the gain-of-function mutants of *PDR1* and *PDR3*, down-regulate the flip (out-to-in) movement, while they up-regulate the flop (in-to-out) movement of phospholipids in response to nutrient starvation. These results, therefore, imply that regulation of the NBD–phospholipid translocation occurring through a phospholipid translocator, is a physiological response to environmental stress (Hanson and Nicholas, 2001).

The flopping or translocation of phosphatidylethanolamine by Cdr1p is an energy-dependent process (Dogra *et al.*, 1999). The decrease in the availability of this lipid in the outer half of the plasma membrane of a Δ*cdr1* strain further confirmed the involvement of Cdr1p in phosphatidylethanolamine

Table 4   ABC transporters of yeasts that function as phospholipid translocators.

Organism	Drug transporter[a]	Substrate	References
S. cerevisiae	PDR5 (floppase)	Phosphatidylethanolamine	Decottignies et al., 1998
	YOR1 (floppase)?[b]	Phosphatidylethanolamine?	Decottignies et al., 1998
C. albicans	CDR1 (floppase)	Phosphatidylethanolamine, Phosphatidylcholine, Phosphatidylserine	Dogra et al., 1999; Smriti et al., 2001
	CDR2 (floppase)	Phosphatidylethanolamine, Phosphatidylcholine, Phosphatidylserine	Smriti et al., 2001
	CDR3 (flippase)	Phosphatidylethanolamine, Phosphatidylcholine, Phosphatidylserine	Smriti et al., 2001
	CDR4 (floppase)?[b]	Phosphatidylethanolamine?	Sanglard et al., 1999c

[a] Floppase indicates an in-to-out movement of phospholipids.
Flippase indicates an out-to-in movement of phospholipids.
[b] Their involvement in phospholipid translocation needs further investigation.

translocation in *C. albicans*. Interestingly, Cdr1p and Cdr2p whose overexpression leads to drug resistance elicit outwardly directed phospholipid transbilayer exchange (floppases), while Cdr3p which does not confer drug resistance, is involved in inwardly directed translocation of phospholipids (flippase) (Smriti *et al.*, 2001). In addition to the differences in the directionality of phospholipid translocation, the floppase activities of Cdr1p and Cdr2p and flippase activity of Cdr3p are further distinguishable. Most importantly, drugs like fluconazole, cycloheximide and miconazole affect transbilayer movement of phospholipids mediated by Cdr1p, Cdr2p but do not affect the Cdr3p-mediated transbilayer exchange. These findings suggest that Cdr1p and Cdr2p presumably have common binding sites for drugs and phospholipids, while flippase activity of Cdr3p could be independent of drug binding (Smriti *et al.*, 2001). The difference in the directionality of phospholipid transfer between Cdrps might be linked to their ability to efflux cytotoxic drugs. It is very likely that phospholipid translocation could represent a normal physiological role of ABC proteins irrespective of their participation in drug extrusion. *CDR4*, a homologue of *CDR1*, *CDR2* and *CDR3* of *C. albicans* has no known role so far in conferring drug resistance, but it also could be a putative phospholipid translocator (Sanglard *et al.*, 1999c). Homologues of *CDR1* found in other non-*albicans* species have not been checked for their ability to translocate phospholipids, but considering their structural and functional homology, it is very likely that some of them will also be functioning as phospholipid translocators in those organisms.

4.2. MFS Drug Extrusion Pumps

Unlike ABC proteins, MFS proteins were identified as nutrient transporters involved in the accumulation of solutes like sugars, metabolites, amino acids, vitamins and both inorganic and organic ions (Fig. 8) (Marger and Saier, 1993; Pao *et al.*, 1998; Saier, 2000). That they can also be involved in expelling drugs is a rather recently recognized role (Goffeau *et al.*, 1997; Borges-Walmsley and Walmsley, 2001; Wolfger *et al.*, 2001). However, unlike the ABC drug transporters, the drug transporters of the MF superfamily have not been identified with other known functions. In one report *CaMDR1* has been related to virulence of *C. albicans*. The mouse models show that the Δ*camdr1* mutants colonize, but are deficient in causing mortality (Becker *et al.*, 1995). The exact mechanism by which the disruption of *CaMDR1* leads to reduced virulence in *C. albicans* is not known. The possibility that the Δ*camdr1* strain is unable to secrete a virulence factor, which could be important for interaction with the host factors, cannot be ruled out (Becker *et al.*, 1995). These results therefore suggest a role of *CaMDR1* in a function other than the transport of drugs and offer a new vista for the development of therapeutic measures. The completion of genome sequencing of yeasts followed by functional genomics would certainly highlight the unexplored physiological role of such proteins.

5. CONCLUDING REMARKS

It is apparent from the foregoing that drug resistance in yeasts is a rapidly growing area wherein new contributors are being identified at an enormous pace. The scene is further complicated by the realization that drug resistance is attributed to the interplay between multiple mechanisms. Among several mechanisms of antifungal resistance, drug extrusion pumps have acquired a prominent place. The broad specificity of drug transporters, though intriguing, is clearly suggestive of their wider physiological relevance in mediating important cellular functions. In these regards, the knowledge of drug transporter proteins, their drug-binding specificities and domain dissection represent important aspects, which await further investigation. The ability of *Candida* to grow as biofilms and the recent realization that the sessile cells therein are notoriously more resistant to azoles, warrant a closer look at the mechanisms of antifungal drug resistance operating in these circumstances. Whether the known drug resistance mechanisms or appearance of yet unique mechanisms in sessile cells of biofilms contribute to hyper-resistance, remains to be ascertained. Identification of those signals perceived by the various MDR genes in response to drugs that lead to their overproduction, is another upcoming area, which would provide further insight into the regulatory mechanisms of these genes. The identification of transcription regulators and the realization of cross talk between the regulators and their targets already suggest complexity in the regulatory networks in yeasts. Thus, a better understanding of molecular mechanisms of drug resistance will certainly pave the way to improved prevention strategies, diagnostic tools and treatment programmes, so that combating drug resistance in pathogenic fungi becomes more vigorous and focused than was hitherto possible.

ACKNOWLEDGEMENTS

The work reported from the authors' laboratory was supported in part by grants to one of us (R.P.) from the Department of Biotechnology (DBT-BT/PRO798/HRD20/8/98; BT/PR1110/MED/09/186/98), Department of Science and Technology (SP/SO/D57/97), Council of Scientific and Industrial Research (CSIR) (60(0028)/98-EMR-II), India. S.L.P. and Smriti acknowledge the fellowships awarded by DBT and CSIR, respectively.

REFERENCES

Alarco, A.M. and Raymond, M. (1999) The bZip transcription factor Cap1p is involved in multidrug resistance and oxidative stress response in *Candida albicans*. *J. Bacteriol.* **181**, 700–708.

Alarco, A.M., Balan, I., Talibi, D., Mainville, N. and Raymond, M. (1997) AP1-mediated multidrug resistance in *Saccharomyces cerevisiae* requires *FLR1* encoding a transporter of the major facilitator superfamily. *J. Biol. Chem.* **272**, 19304–19313.

Albertson, G.D., Niimi, M., Cannon, R.D. and Jenkinson, H.F. (1996) Multiple efflux mechanisms are involved in *Candida albicans* fluconazole resistance. *Antimicrob. Agents Chemother.* **40**, 2835–2841.

Ambudkar, S.V. (1998) Drug-stimulatable ATPase activity in crude membranes of human *MDR1*-transfected mammalian cells. *Methods Enzymol.* **292**, 504–514.

Ambudkar, S.V., Lelong, I.H., Zhang, J., Cardarelli, C.O., Gottesman, M.M. and Pastan, I. (1992) Partial purification and reconstitution of the human multidrug-resistance pump: Characterisation of the drug-stimulatable ATP hydrolysis. *Proc. Natl Acad. Sci. USA* **89**, 8472–8476.

Ambudkar, S.V., Dey, S., Hrycyna, C.A., Ramachandran, M., Pastan, I. and Gottesman, M.M. (1999) Biochemical, cellular, and pharmacological aspects of the multidrug transporter. *Ann. Rev. Pharmacol. Toxicol.* **39**, 361–398.

Anderson, M.P., Berger, H.A., Rich, D.P., Gregory, R.J., Smith, A.E. and Welsh, M.J. (1991) Nucleoside triphosphates are required to open the CFTR chloride channel. *Cell* **67**, 775–784.

Andrade, A.C., Del Sorbo, G., Van Nistelrooy, J.G.M. and De Waard, M.A. (2000a) The ABC transporter AtrB from *Aspergillus nidulans* mediates resistance to all major classes of fungicides and some natural toxic compounds. *Microbiology* **146**, 1987–1997.

Andrade, A.C., Van Nistelrooy, J.G.M., Peery, J.G.M., Skatrud, P.L. and De Waard, M.A. (2000b) The role of ABC transporters from *Aspergillus nidulans* in protection against cytotoxic agents and in antibiotic production. *Mol. Gen. Genet.* **263**, 966–977.

Angermayr, K., Parson, W., Stoffler, G. and Haas, H. (1999) Expression of *atrC*-encoding a novel member of the ATP binding cassette transporter family in *Aspergillus nidulans* is sensitive to cycloheximide. *Biochim. Biophys. Acta* **1453**, 304–310.

Balan, I., Alarco, A.M. and Raymond, M. (1997) The *Candida albicans CDR3* gene codes for an opaque-phase ABC transporter. *J. Bacteriol.* **179**, 7210–7218.

Balzi, E. and Goffeau, A. (1991) Multiple or pleiotropic drug resistance in yeast. *Biochim. Biophys. Acta* **1073**, 241–252.

Balzi, E. and Goffeau, A. (1994) Genetics and biochemistry of yeast multidrug resistance. *Biochim. Biophys. Acta* **1187**, 152–162.

Balzi, E. and Goffeau, A. (1995) Yeast multidrug resistance: The PDR network. *J. Bioenerget. Biomemb.* **27**, 71–76.

Balzi, E., Wang, M., Leterme, S., Van Dyck, L. and Goffeau, A. (1994) *PDR5*, a novel yeast multidrug resistance conferring transporter controlled by the transcription regulator *PDR1*. *J. Biol. Chem.* **269**, 2206–2214.

Bauer, B.E., Wolfger, H. and Kuchler, K. (1999) Inventory and function of yeast ABC proteins: about sex, stress, and pleiotropic drug and heavy metal resistance. *Biochim. Biophys. Acta* **1461**, 217–236.

Becker, J.M., Henry, L.K., Jiang, W. and Koltin, Y. (1995) Reduced virulence of *Candida albicans* mutants affected in multidrug resistance. *Infect. Immun.* **63**, 4515–4518.

Bevers, E.M., Confurius, P. and Zwaal, R.F.A. (1996) Regulatory mechanisms in maintenance and modulation of transmembrane lipid asymmetry: pathophysiological implications. *Lupus* **5**, 480–487.

Bevers, E.M., Confurius, P., Dekkers, D.W.C. and Zwaal, R.F.A. (1999) Lipid translocation across the plasma membrane of mammalian cells. *Biochim. Biophys. Acta.* **1439**, 317–330.

Bishop, L., Agbayani, R. Jr., Ambudkar, S.V., Maloney, P.C. and Ames, G.F. (1989) Reconstitution of a bacterial periplasmic permease in proteoliposomes and demonstration of ATP hydrolysis concomitant with transport. *Proc. Natl Acad. Sci. USA* **86**, 6953–6957.

Bissinger, P.H. and Kuchler, K. (1994) Molecular cloning and expression of *Saccharomyces cerevisiae STS1* gene product, a yeast ABC transporter conferring mycotoxin resistance. *J. Biol. Chem.* **269**, 4180–4186.

Borges-Walmsley, M.I. and Walmsley, A.R. (2001) The structure and function of drug pumps. *Trends Microbiol.* **9**, 71–79.

Borowski, E. (2000) Novel approaches in the rational design of antifungal agents of low toxicity. *Farmaco* **55**, 206–208.

Boscott, P.E. and Grant, G.H. (1994) Modeling cytochrome P450 14α-demethylase (*Candida albicans*) from P450cam. *J. Mol. Graph.* **12**, 185–192.

Broco, N., Tenreiro, S., Viegas, C.A. and Sa-Correja, I. (1999) FLR1 gene (ORFYBR008c) is required for benomyl and methotrexate resistance in *Saccharomyces cerevisiae* and its benomyl induced expression is dependent on Pdr3 transcriptional regulator. *Yeast* **15**, 1595–1608.

Buckman, J. and Miller, S.M. (1998) Binding and reactivity of *Candida albicans* estrogen binding protein with steroid and other substrates. *Biochemistry* **37**, 14326–14336.

Buurman, E.T., Westwater, C., Hube, B., Brown, A.J.P., Odds, F.C. and Gow, N.A.R. (1998) Molecular analysis of CaMnt1p, a mannosyl transferase important for adhesion and virulence of *Candida albicans*. *Proc. Natl Acad. Sci. USA* **95**, 7670–7675.

Calabrese, D., Bille, J. and Sanglard, D. (2000) A novel multidrug efflux transporter gene of the major facilitator superfamily from *Candida albicans (FLU1)* conferring resistance to fluconazole. *Microbiology* **146**, 2743–2754.

Cannon, R.D., Fischer, F.J., Niimi, K., Niimi, M. and Arisawa, M. (1998) Drug pumping mechanisms in *Candida albicans*. *Jap. J. Med. Mycol.* **39**, 73–78.

Carvajal, E., Van Den Hazel, H.B., Cybularz-Kolaczkowska, A., Balzi, E. and Goffeau, A. (1997) Molecular and phenotypic characterisation of yeast *PDR1* mutants that show hyperactive transcription of various ABC multidrug transporter genes. *Mol. Gen. Genet.* **256**, 406–415.

Chakraburtty, K. (2001) Translational regulation by ABC systems. *Res. Microbiol.* **152**, 391–399.

Christensen, P.U., Davis, K., Nielsen, O. and Davey, J. (1997a) Abc1: a new ABC transporter from the fission yeast *Schizosaccharomyces pombe*. *FEMS Microbiol. Lett.* **147**, 97–102.

Christensen, P.U., Davey, J. and Nielsen, O. (1997b) The *Schizosaccharomyces pombe* mam1 gene encodes an ABC transporter mediating secretion of M-factor. *Mol. Gen. Genet.* **255**, 226–236.

Clark, F.S., Parkinson, T., Hitchcock, C.A. and Gow, N.A.R. (1996) Correlation between Rhodamine 123 accumulation and azole sensitivity in *Candida* species: possible role for drug efflux in drug resistance. *Antimicrob. Agents Chemother.* **40**, 419–425.

Cole, S.P.C. and Deeley, R.G. (1998) Mutidrug resistance mediated by the ATP-binding cassette transporter protein MRP. *BioEssays* **20**, 931–940.

Coleman, S.T., Tseng, E. and Moye-Rowley, W.S. (1997) *Saccharomyces cerevisiae* basic-leucine zipper protein regulatory networks converge at the *ATR1* structural gene. *J. Biol. Chem.* **272**, 23224–23230.

Cui, Z., Hirata, D., Tsuchiya, E., Osada, H. and Miyakawa, T. (1996) The multidrug resistance-associated protein (MRP) subfamily (*Yrs1/Yor1*) of *Saccharomyces cerevisiae* is important for the tolerance to a broad range of organic anions. *J. Biol. Chem.* **271**, 14712–14716.

Cui, Z., Shiraki, T., Hirata, D. and Miyakawa, T. (1998) Yeast gene *YRR1*, which is required for resistance to 4-nitroquinoline N-oxide, mediates transcriptional activation of the multidrug resistance transporter gene *SNQ2*. *Mol. Microbiol.* **29**, 1307–1315.

Das, M. and Datta, A. (1985) Steroid binding protein(s) in yeasts. *Biochem. Int.* **11**, 171–176.

De Backer, M.D., Ilyina, T., Ma, X.-J., Vandoninck, S., Luyten, W.H.M.L. and Vanden Bossche, H. (2001) Genomic profiling of the response of *Candida albicans* to itraconazole treatment using a DNA microarray. *Antimicrob. Agents Chemother.* **45**, 1660–1670.

Decottignies, A. and Goffeau, A. (1997) Complete inventory of the yeast ABC proteins. *Nat. Genet.* **15**, 137–145.

Decottignies, A., Kolaczkowski, M., Balzi, E. and Goffeau, A. (1994) Solubilisation and characterisation of the overexpressed PDR5 multidrug resistance nucleotide triphosphatase of yeast. *J. Biol. Chem.* **269**, 12797–12803.

Decottignies, A., Lambert, L., Catty, P., Degand, H., Eppings, E.A., Moye-Rowley, W.S., Balzi, E. and Goffeau, A. (1995) Identification and characterisation of SNQ2, a new multidrug ATP binding cassette transporter of the yeast plasma membrane. *J. Biol. Chem.* **270**, 18150–18157.

Decottignies, A., Grant, A.M., Nichols, J.W., De Wet, H., McIntosh, D.B. and Goffeau, A. (1998) ATPase and multidrug transport activities of the overexpressed yeast ABC protein Yor1p. *J. Biol. Chem.* **273**, 12612–12622.

Del Sorbo, G., Andrade, A.C., Van Nisterlooy, J.G., Van Kan, J.A., Balzi, E. and De Waard, M.A. (1997) Multidrug resistance in *Aspergillus nidulans* involves novel ATP-binding cassette transporters. *Mol. Gen. Genet.* **254**, 417–426.

Del Sorbo, G., Schoonbeek, H-J. and De Waard, M.A. (2000) Fungal transporters involved in efflux of natural toxic compounds and fungicides. *Fungal Genet. Biol.* **30**, 1–15.

Delahodde, A., Delaveau, T. and Jacq, C. (1995) Positive autoregulation of the yeast transcription factor Pdr3p, which is involved in control of drug resistance. *Mol. Cell. Biol.* **15**, 4043–4051.

DeRisi, J., van den Hazel, B., Marc, P., Balzi, E., Brown, P., Jacq, C. and Goffeau, A. (2000) Genome microarray analysis of transcriptional activation in multidrug resistance yeast mutants. *FEBS Lett.* **470**, 156–160.

Detmers, F.J.M., Lanfermeijer, F.C. and Poolman, B. (2001) Peptides and ATP binding cassette peptide transporters. *Res. Microbiol.* **152**, 245–258.

Dexter, D., Rowley, S.M., Wu, A. and Golin, J. (1994) Mutations in the yeast *PDR3*, *PDR4*, *PDR7* pleiotropic (multiple) drug resistance loci affect the transcript level of an ATP binding cassette transporter encoding gene, *PDR5*. *Genetics* **136**, 505–515.

Diaz, C. and Schroit, A.J. (1996) Role of translocases in the generation of phosphatidylserine asymmetry. *J. Memb. Biol.* **151**, 1–9.

Do Nascimiento, A.M., Terenzi, M.F., Goldman, M.M.H. and Glodman, G.H. (1999) A novel ATP-binding cassette transporter involved in multidrug resistance in the filamentous fungus *Aspergillus nidulans*. *Fungal Genet. Newslett.* **46**, 44.

Dogra, S., Krishnamurthy, S., Gupta, V., Dixit, B.L., Gupta, C.M., Sanglard, D. and Prasad, R. (1999) Asymmetric distribution of phosphatidylethanolamine in *C. albicans*: possible mediation by *CDR1*, a multidrug transporter belonging to ATP binding cassette (ABC) superfamily. *Yeast* **15**, 111–121.

Egner, R., Rosenthal, F.E., Kralli, A., Sanglard, D. and Kuchler, K. (1998) Genetic separation of FK506 susceptibility and drug transport in the yeast Pdr5 ATP-binding cassette multidrug resistance transporter. *Mol. Biol. Cell* **9**, 523–543.

Egner, R., Bauer, B.E. and Kuchler, K. (2000) The transmembrane domain 10 of the yeast

Pdr5p ABC antifungal efflux pump determines both substrate specificity and inhibitor susceptibility. *Mol. Microbiol.* **35**, 1255–1263.

Ehrenhofer-Murray, A.E., Keller, M.U., Seitz, K. and Sengstag, C. (1998) The Sge1 protein of *Saccharomyces cerevisiae* is a membrane associated multidrug transporter. *Yeast* **14**, 49–65.

Ernst, J.F. and Schmidt, A. (eds). (2000) *Dimorphism in Human Pathogenic and Apathogenic Yeasts*, Karger, Switzerland.

Farkas, V. (1979) Biosynthesis of cell walls of fungi. *Microbiol. Rev.* **43**, 117–144.

Favre, B., Didmon, M. and Ryder, N.S. (1999) Multiple amino acid substitutions in lanosterol 14α-demethylase contribute to azole resistance in *Candida albicans*. *Microbiology* **145**, 2715–2725.

Feldman, D., Li, Y., Burshell, A., Stathis, P. and Loose, D.S. (1982) An estrogen-binding protein and endogenous ligand in *Saccharomyces cerevisiae*: possible hormone receptor system. *Science* **218**, 297–298.

Fernandes, L., Rodrigues-Pousada, C. and Struhl, K. (1997) Yap, a novel family of eight bZIP proteins in *Saccharomyces cerevisiae* with distinct biological functions. *Mol. Cell. Biol.* **17**, 6982–6993.

Ferte, J. (2000) Analysis of the tangled relationships between P-glycoprotein-mediated mutlidrug resistance and the lipid phase of the cell membrane. *Eur. J. Biochem.* **267**, 277–294.

Fierro, F. and Martin, J.F. (1999) Molecular mechanisms of chromosomal rearrangement in fungi. *Crit. Rev. Microbiol.* **25**, 1–17.

Fling, M.E., Kopf, J., Tamarkin, A., Gorman, J.A., Smith, H.A. and Koltin, Y. (1991) Analysis of a *Candida albicans* gene that encodes a novel mechanism for resistance to benomyl and methotrexate. *Mol. Gen. Genet.* **227**, 318–329.

Fostel, J.M. and Lartey, P.A. (2000) Emerging novel antifungal agents. *Drug Discov. Today* **52**, 25–32.

Franz, R., Michel, S. and Morschhauser, J. (1998) A fourth gene from the *Candida albicans* CDR family of ABC transporters. *Gene* **220**, 91–98.

Ghannoum, M. and Rice, L.B. (1999) Antifungal agents: Mode of action, mechanism of resistance, and correlation of these mechanisms with bacterial resistance. *Clin. Microbiol. Rev.* **12**, 501–512.

Goffeau, A., Park, J., Paulsen, I.T., Jonniaux, J.-L., Dinh, T., Mordant, P. and Saier, M.H. Jr. (1997) Multidrug-resistant transport proteins in yeast: complete inventory and phylogenetic characterisation of yeast open reading frames within the major facilitator superfamily. *Yeast* **13**, 43–54.

Goldman, B.S. and Kranz, R.G. (2001) ABC transporters associated with cytochrome *c* biogenesis. *Res. Microbiol.* **152**, 323–329.

Gompel-Klein, P. and Brendel, M. (1990) Allelism of *SNQ1* and *ATR1*, genes of the yeast *Saccharomyces cerevisiae* required for sensitivity to 4-nitroquinoline-N-oxide and aminotriazole. *Curr. Genet.* **18**, 93–96.

Gounalaki, N. and Thireos, G. (1994) Yap1p, a yeast transcriptional activator that mediates multidrug resistance, regulates the metabolic stress response. *EMBO J.* **13**, 4036–4041.

Grant, C.M., Collinson, L.P., Roe, J.-H. and Dawes, I.W. (1996) Yeast glutathione reductase is required for protection against oxidative stress and is a target gene for yAP-1 transcriptional regulation. *Mol. Microbiol.* **21**, 171–179.

Gupta, V., Kohli, A.K., Krishnamurthy, S., Puri, N., Aalamgeer, S.A., Panwar, S.L. and Prasad, R. (1998) Identification of mutant alleles of *CaMDR1*, a major facilitator of *Candida albicans* which confers multidrug resistance and its *in vitro* transcriptional activation. *Curr. Genet.* **34**, 192–199.

Hallstorm, T.C. and Moye-Rowley, W.S. (2000) Multiple signals from dysfunctional mitochondria activate the pleiotropic drug resistance pathway in *Saccharomyces cerevisiae*. *J. Biol. Chem.* **275**, 37347–37356.

Hamada, H. and Tsururo, T. (1988) Purification of the 170- to 180-kilodalton membrane glycoprotein associated with multidrug resistance. 170- to 180-kilodalton membrane glycoprotein is an ATPase. *J. Biol. Chem.* **263**, 1454–1458.

Hanna, M., Brault, M., Kwan, T., Kast, C. and Gros, P. (1996) Mutagenesis of transmembrane domain 11 of P-glycoprotein by alanine scanning. *Biochemistry* **35**, 3625–3635.

Hanson, P.K. and Nicholas, J.W. (2001) Energy-dependent flip of fluorescent-labeled phospholipids is regulated by nutrient starvation and transcription factors, PDR1 and PDR3. *J. Biol. Chem.* **276**, 9861–9867.

Hector, R.F. (1993) Compounds active against cell walls of medically important fungi. *Clin. Microbiol. Rev.* **6**, 1–21.

Henry, K.W., Nickels, J.T. and Edlind, T.D. (2000) Upregulation of *ERG* genes in *Candida* species by azoles and other sterol biosynthesis inhibitors. *Antimicrob. Agents Chemother.* **44**, 2693–2700.

Hertle, K., Haase, E. and Brendel, M. (1991) The *SNQ3* gene of *Saccharomyces cerevisiae* confers hyper-resistance to several functionally unrelated chemicals. *Curr. Genet.* **19**, 429–433.

Higgins, C.F. (2001) ABC transporters: physiology, structure and mechanism – an overview. *Res. Microbiol.* **152**, 205–210.

Hirata, D., Yano, K., Miyahara, K. and Miyakawa, T. (1994) *Saccharomyces cerevisiae YDR1*, which encodes a member of the ATP-binding cassette (ABC) superfamily, is required for multidrug resistance. *Curr. Genet.* **26**, 285–294.

Hitchcock, C.A. (1993) Resistance of *Candida albicans* to azole antifungal agents. *Biochem. Soc. Trans.* **21**, 1039–1047.

Jack, D.J., Yang, N.M. and Saier, M.H. Jr. (2001) The drug/metabolite transporter superfamily. *Eur. J. Biochem.* **268**, 3620–3639.

Janbon, G., Sherman, F. and Rustchenko, E. (1998) Monosomy of a specific chromosome determines L-sorbose utilization: a novel regulatory mechanism in *Candida albicans*. *Proc. Natl Acad. Sci. USA* **95**, 5150–5155.

Jia, Z.P., McCullough, N., Wong, L. and Young, P.G. (1993) The amiloride resistance gene, *car1* of *Schizosaccharomyces pombe*. *Mol. Gen. Genet.* **241**, 298–304.

Jungwirth, H., Wendler, F., Platzer, B., Bergler, H. and Hogenauer, G. (2000) Diazaborine resistance in yeast involves the efflux pumps Ycf1p and Flr1p and is enhanced by a gain-of-function allele of gene *YAP1*. *Eur. J. Biochem.* **267**, 4809–4816.

Kajiji, S., Talbot, F., Grizzuti, K., Van Dyke-Phillips, V., Agresti, M., Safa, A.R. and Gros, P. (1993) Functional analysis of P-glycoprotein mutants identifies predicted transmembrane domain 11 as a putative drug binding site. *Biochemistry* **32**, 4185–4194.

Kanazawa, S., Driscoll, M. and Struhl, K. (1988) *ATR1*, a *Saccharomyces cerevisiae* gene encoding a transmembrane protein required for aminotriazole resistance. *Mol. Cell. Biol.* **8**, 664–673.

Kappeli, O. (1986) Cytochromes P-450 of yeasts. *Microbiol. Rev.* **50**, 244–258.

Katiyar, S.K. and Edlind, T.D. (2001) Identification and expression of multidrug resistance-related ABC transporter gene in *Candida krusei*. *Med. Mycol.* **39**, 109–116.

Katzmann, D.J., Burnett, P.E., Golin, J., Mahe, Y. and Moye-Rowley, W.S. (1994) Transcriptional control of the yeast *PDR5* gene by the *PDR3* gene product. *Mol. Cell. Biol.* **14**, 4653–4661.

Katzmann, D.J., Hallstrom, T.C., Voet, M., Wysock, W., Golin, J., Volckert, G. and Moye-Rowley, W.S. (1995) Expression of an ATP-binding cassette transporter-encoding gene (*YOR1*) is required for oligomycin resistance in *Saccharomyces cerevisiae*. *Mol. Cell. Biol.* **15**, 6875–6883.

Kaur, R. and Bachhawat, A.K. (1999) The yeast multidrug resistance pump, Pdr5p, confers reduced drug resistance in erg mutants of *Saccharomyces cerevisiae*. *Microbiology* **145**, 809–818.

Kelly, S.L., Lamb, D.C., Baldwin, B.C. and Kelly, D.E. (1993) Benzo(a)pyrene hydroxylase activity in yeast is mediated by P450 other than sterol 14α-demethylase. *FEBS Lett.* **197**, 428–432.

Kelly, S.L., Lamb, D.C. and Kelly, D.E. (1997a) Sterol 22-desaturase, cytochrome P45061, possesses activity in xenobiotic metabolism. *FEBS Lett.* **412**, 233–235.

Kelly, S.L., Lamb, D.C., Kelly, D.E., Manning, N.J., Loeffler, J., Hebart, H., Schumacher, U. and Einsele, H. (1997b) Resistance to fluconazole and cross-resistance to amphotericin B in *Candida albicans* from AIDS patients caused by defective sterol $\Delta^{5,6}$-desaturation. *FEBS Lett.* **400**, 80–82.

Kelly, S.L., Lamb, D.C., Juergen, L., Einsele, H. and Kelly, D.E. (1999a) The G464S amino acid substitution in *Candida albicans* sterol 14α-demethylase causes fluconazole resistance in the clinic through reduced affinity. *Biochem. Biophys. Res. Commun.* **262**, 174–179.

Kelly, S.L., Lamb, D.C. and Kelly, D.E. (1999b) Y132H substitution in *Candida albicans* sterol 14α-demethylase confers fluconazole resistance by preventing binding to haem. *FEMS Microbiol. Lett.* **180**, 171–175.

Kolaczkowski, M. and Goffeau, A. (1997) Active efflux by multidrug transporters as one of the strategies to evade chemotherapy and novel practical implications of yeast pleiotropic drug resistance. *Pharmacol. Ther.* **76**, 219–242.

Kolaczkowski, M., van der Rest, M.E., Cybularz-Kolaczkowska, A., Soumillion, J.-P., Konings, W.N. and Goffeau, A. (1996) Anticancer drugs, ionophoric peptides, and steroids as substrates of the yeast multidrug transporter Pdr5p. *J. Biol. Chem.* **271**, 31543–31548.

Kolaczkowski, M., Kolaczkowska, A., Luczynski, J., Witek, S. and Goffeau, A. (1998) *In vivo* characterization of the drug resistance profile of the major ABC transporters and other components of the yeast pleiotropic drug resistance network. *Microb. Drug Resist.* **4**, 143–158.

Kontoyiannis, D.P. (2000) Efflux mediated resistance to fluconazole could be modulated by sterol homeostasis in *Saccharomyces cerevisiae*. *J. Antimicrob. Chemother.* **46**, 199–203.

Kralli, A., Bohen, S.P. and Yamamoto, K.R. (1995) LEM1, an ATP-binding-cassette transporter, selectively modulates the biological potency of steroid hormones. *Proc. Natl Acad. Sci. USA* **92**, 4701–4705.

Krishnamurthy, S., Chatterjee, U., Gupta, V., Prasad, R., Das, P., Snehlata, P., Hasnain, S.E. and Prasad, R. (1998a) Deletion of transmembrane domain 12 of *CDR1*, a multidrug transporter from *Candida albicans*, leads to altered drug specificity: expression of a yeast multidrug transporter in Baculovirus expression system. *Yeast* **14**, 535–550.

Krishnamurthy, S., Gupta, V., Snehlata, P. and Prasad, R. (1998b) Characterisation of human steroid hormone transport mediated by Cdr1p, multidrug transporter of *Candida albicans*, belonging to the ATP binding cassette super family. *FEMS Microbiol. Lett.* **158**, 69–74.

Krishnamurthy, S., Gupta, V., Prasad, R., Panwar, S.L. and Prasad, R. (1998c) Expression of *CDR1*, a multidrug resistance gene of *Candida albicans*: In vitro transcriptional activation by heat shock, drugs and human steroid hormones. *FEMS Microbiol. Lett.* **160**, 191–197.

Kuge, S. and Jones, N. (1994) YAP1 dependent activation of *TRX2* is essential for the response of *Saccharomyces cerevisiae* to oxidative stress by hyperoxides. *EMBO J.* **13**, 655–664.

Lamb, D.C., Kelly, D.E., Schunck, W.-H., Shyadehi, A.Z., Akhtar, M., Lowe, D.J., Baldwin, B.C. and Kelly, S.L. (1997) The mutation T315A in *Candida albicans* sterol 14α-demethylase causes reduced enzyme activity and fluconazole resistance through reduced affinity. *J. Biol. Chem.* **272**, 5682–5688.

Lamb, D.C., Maspahy, S., Kelly, D.E., Manning, N.J., Geber, A., Bennett, J.E. and Kelly,

S.L. (1999) Purification, reconstitution, and inhibition of cytochrome P-450 sterol Δ^{22}-desaturase from the pathogenic fungus *Candida glabrata*. *Antimicrob. Agents Chemother.* **43**, 1725–1728.

Leighton, J. and Schatz, G. (1995) An ABC transporter in the mitochondrial inner membrane is required for normal growth of yeast. *EMBO J.* **14**, 188–195.

Lewis, D.F.V., Wiseman, A. and Tarbit, M.H. (1999) Molecular modelling of lanosterol 14α-demethylase (CYP51) from *Saccharomyces cerevisiae* via homology with CYP102, a unique bacterial cytochrome P450 isoform: quantitative structure–activity relationships (QSARs) within two related series of antifungal azole derivatives. *J. Enzyme Inhibit.* **14**, 175–192.

Loffler, J., Kelly, S.L., Hebart, H., Schumacher, U., Lass-Florl, C. and Einsele, H. (1997) Molecular analysis of *cyp51* from fluconazole-resistant *Candida albicans* strains. *FEMS Microbiol. Lett.* **151**, 263–268.

Loo, T.W. and Clarke, D.M. (1994) Mutations to amino acids located in predicted transmembrane segment 6 (TM6) modulate the activity and substrate specificity of human P-glycoprotein. *Biochemistry* **33**, 14049–14057.

Loo, T.W. and Clarke, D.M. (1996) Inhibition of oxidative cross-linking between engineered cysteine residues at positions 332 in predicted transmembrane segments (TM) 6 and 975 in predicted TM 12 of human P-glycoprotein by drug substrates. *J. Biol. Chem.* **271**, 27482–27487.

Loo, T.W. and Clarke, D.M. (1999a) Identification of residues in the drug binding domain of human P-glycoprotein. *J. Biol. Chem.* **274**, 35388–35392.

Loo, T.W. and Clarke, D.M. (1999b) Molecular dissection of the human multidrug resistance P-glycoprotein. *Biochem. Cell Biol.* **77**, 11–23.

Loo, T.W. and Clarke, D.M. (1999c) The transmembrane domains of the human multidrug resistance P-glycoprotein are sufficient to mediate drug binding and trafficking to the cell surface. *J. Biol. Chem.* **274**, 24759–24765.

Loo, T.W. and Clarke, D.M. (2000) The packing of the transmembrane segments of human multidrug resistance P-glycoprotein is revealed by disulfide cross-linking analysis. *J. Biol. Chem.* **275**, 5253–5256.

Lorenz, M.C. and Fink, G.R. (2001) The glyoxylate cycle is required for fungal virulence. *Nature* **412**, 83–86.

Mahe, Y., Parle-McDermott, A., Nourani, A., Delahodde, A., Lamprecht, A. and Kuchler, K. (1996a) The ATP-binding cassette multidrug transporter Snq2 of *Saccharomyces cerevisiae*: a novel target for the transcription factors Pdr1 and Pdr3. *Mol. Microbiol.* **20**, 109–117.

Mahe, Y., Lemoine, Y. and Kuchler, K. (1996b) The ATP binding cassette transporters Pdr5 and Snq2 of *Saccharomyces cerevisiae* can mediate transport of steroids *in vivo. J. Biol. Chem.* **271**, 26167–26172.

Marger, M.D. and Saier, M.H. Jr. (1993) A major superfamily of transmembrane facilitators that catalyse uniport, symport and antiport. *Trends Biochem. Sci.* **18**, 13–20.

Marichal, P. (1999) Mechanisms of resistance to azole antifungal compounds. *Curr. Opin. Anti-infect. Invest. Drugs* **1**, 318–333.

Marichal, P., Vanden Bossche, H., Odds, F.C., Nobels, G., Warnock, D.W., Timmerman, V., Van Broeckhoven, C., Fay, S. and Mose-Larsen, P. (1997) Molecular biological characterization of an azole-resistant *Candida glabrata* isolate. *Antimicrob. Agents Chemother.* **41**, 2229–2237.

Marichal, P., Koymans, L., Willemsens, S., Bellens, D., Verhasselt, P., Luyten, W., Borgers, M., Ramaekers, F.C.S., Odds, F.C. and Vanden Bossche, H. (1999) Contribution of mutations in the cytochrome P450 14α-demethylase (Erg11p, Cyp51p) to azole resistance in *Candida albicans*. *Microbiology* **145**, 2701–2713.

Marr, K.A., Lyons, C.N., Bowden, R.A. and White, T.C. (1998) Rapid, transient fluconazole resistance in *Candida albicans* is associated with increased mRNA levels of CDR. *Antimicrob. Agents Chemother.* **42**, 2584–2589.

McGrath, J.P. and Varshavski, A. (1989) The yeast *STE6* gene encodes a homologue of the mammalian multidrug resistance P-glycoprotein. *Nature* **340**, 400–404.

Mendez, C. and Salas, J.A. (2001) The role of ABC transporters in antibiotic-producing organisms: drug secretion and resistance mechanisms. *Res. Microbiol.* **152**, 341–350.

Meyers, S., Schauer, W., Balzi, E., Wagner, M., Goffeau, A. and Golin, J. (1992) Interaction of the yeast pleitropic drug resistance genes *PDR1* and *PDR5*. *Curr. Genet.* **21**, 431–436.

Michaelis, S. and Berkower, C. (1995) Sequence comparison of yeast ATP-binding cassette proteins. *Cold Spring Harbor Symp. Quant. Biol.* **LX**, 291–307.

Miyazaki, H., Miyazaki, Y., Geber, A., Parkinson, T., Hitchcock, C.A., Falconer, D.J., Ward, D.J., Marsden, K. and Bennett, J.E. (1998) Fluconazole resistance associated with drug efflux and increased transcription of a drug transporter gene, *PDH1*, in *Candida glabrata*. *Antimicrob. Agents Chemother.* **42**, 1695–1701.

Moran, G.P., Sanglard, D., Donnelly, S.M., Shanley, D.B., Sullivan, D.J. and Coleman, D.C. (1998) Identification and expression of multidrug transporters responsible for flucona-zole resistance in *Candida dubliniensis*. *Antimicrob. Agents Chemother.* **42**, 1819–1830.

Morgan, B.A., Banks, G.R., Toone, W.M., Raitt, D., Kuge, S. and Johnston, L.H. (1997) The Skn7 response regulator controls gene expression in the oxidative stress response of the budding yeast *Saccharomyces cerevisiae*. *EMBO J.* **16**, 1035–1044.

Morschhauser, J., Michel, S. and Staib, P. (1999) Sequential gene disruption in *Candida albicans* by FLP-mediated site-specific recombination. *Mol. Microbiol.* **32**, 547–556.

Moye-Rowley, W.S., Harshmann, K.D. and Parker, C.S. (1989) Yeast YAP1 encodes a novel form of the Jun family of transcriptional activator protein. *Genes Dev.* **3**, 283–292.

Munro, C.A., Winter, K., Buchan, A., Henry, K., Becker, J.M., Brown, A.J.P., Bulawa, C.E. and Gow, N.A.R. (2001) Chs1 of *Candida albicans* is an essential chitin synthase required for synthesis of the septum and for cell integrity. *Mol. Microbiol.* **39**, 1414–1426.

Nagao, K., Taguchi, Y., Arioka, M., Kadokura, H., Takatsuki, A., Yoda, K. and Yamasaki, M. (1995) *bfr1+*, a novel gene of *Schizosaccharomyces pombe* which confers brefeldin A resistance, is structurally related to the ATP-binding cassette superfamily. *J. Bacteriol.* **177**, 1536–1543.

Nakaune, R., Adachi, K., Nawata, O., Tomiyama, M., Akutsu, K. and Hibi, T. (1998) A novel ATP-binding cassette transporter involved in multidrug resistance in the phy-topathogenic fungus *Penicillium digitatum*. *Appl. Environ. Microbiol.* **64**, 3983–3988.

Nguyen, D.-T., Alarco, A.-M. and Raymond, M. (2000) Multiple Yap1p-binding sites medi-ate induction of the yeast major facilitator *FLR1* gene in response to drugs, oxidants, and alkylating agents. *J. Biol. Chem.* **276**, 1138–1145.

Nishi, K., Yoshida, M., Nishimura, M., Nishikawa, M., Nishiyama, M., Horinouchi, S. and Beppu, T. (1992) A leptomycin B resistance gene of *Schizosaccharomyces pombe* encodes a protein similar to the mammalian P-glycoproteins. *Mol. Microbiol.* **6**, 761–769.

Nishi, K., Yoshida, M., Horinouchi, S. and Beppu, T. (1993) Mating of the fission yeast independently of *pmd1+* gene product, a structural homologue of budding yeast STE6 and mammalian P-glycoproteins. *Arch. Microbiol.* **160**, 162–165.

Nourani, A., Wesolowski-Louvel, M., Delaveau, T., Jacq, C. and Delahodde, A. (1997) Multiple-drug-resistance phenomenon in the yeast *Saccharomyces cerevisiae*: involve-ment of two hexose transporters. *Mol. Cell. Biol.* **17**, 5453–5460.

Odds, F.C. (ed.) (1988) Candida *and Candidosis: A Review and Bibliography*, Ballière Tindall, London.

Ogawa, A., Hashida-Okado, T., Endo, M., Tsuruo, T., Takesako, K. and Kato, I. (1998) Role of ABC transporters in aureobasidin A resistance. *Antimicrob. Agents Chemother.* **42**, 755–761.

Ohkuma, M., Muraoka, S.-I., Tanimoto, T., Fujii, M., Ohta, A. and Takagi, M. (1995) *CYP52* (cytochrome P450alk) multigene family in *Candida maltosa*: Identification and characterisation of eight members. *DNA Cell Biol.* **14**, 163–173.

Omura, T. (1999) Forty years of cytochrome P450. *Biochem. Biophys. Res. Commun.* **266**, 690–698.

Ortiz, D., Kreppel, L., Speiser, D.M., Scheel, G., McDonald, G. and Ow, D.W. (1992) Heavy metal tolerance in the fission yeast requires an ATP-binding cassette-type vacuolar membrane transporter. *EMBO J.* **11**, 3491–3499.

Oskouian, B. and Saba, J.D. (1999) *YAP1* confers resistance to the fatty acid synthase inhibitor cerulenin through the transporter Flr1p in *Saccharomyces cerevisiae*. *Mol. Gen. Genet.* **261**, 346–353.

Panwar, S.L., Krishnamurthy, S., Gupta, V., Alarco, A.-M., Raymond, M., Sanglard, D. and Prasad, R. (2000) *CaALK8*, an alkane assimilating cytochrome P450 confers multiple drug resistance when expressed in a hypersensitive strain of *Candida albicans*. *Yeast* **18**, 1117–1129.

Pao, S.S., Paulsen, I.T. and Saier, M.H. Jr. (1998) Major facilitator superfamily. *Microbiol. Mol. Biol. Rev.* **62**, 1–34.

Paulsen, I.T., Brown, M.H. and Skurray, R.A. (1996) Proton-dependent multidrug efflux systems. *Microbiol. Rev.* **60**, 575–608.

Perepnikhatka, V., Fischer, F.J., Niimi, M., Baker, R.A., Cannon, R.D., Wang, Y.-K., Sherman, F. and Rustchenko, E. (1999) Specific chromosome alterations in fluconazole-resistant mutants of *Candida albicans*. *J. Bacteriol.* **181**, 4041–4049.

Piper, P., Mahe, Y., Thompson, S., Pandjaitan, R., Holyoak, C., Egner, R., Muhlbauer, M., Coote, P. and Kuchler, K. (1998) The Pdr12 ABC transporter is required for the development of weak organic acid resistance in yeast. *EMBO J.* **17**, 4257–4265.

Podust, L.M., Poulos, T.L. and Waterman, M.R. (2001) Crystal structure of cytochrome P450 14α-sterol demethylase (CYP51) from *Mycobacterium tuberculosis* in complex with azole inhibitors. *Proc. Natl Acad. Sci. USA* **98**, 3068–3073.

Prasad, R. (ed.) (1991) *Candida Albicans: Cellular and Molecular Biology*, Springer-Verlag, Berlin.

Prasad, R., Worgifosse, P.D., Goffeau, A. and Balzi, E. (1995) Molecular cloning and characterisation of a novel gene of *C. albicans*, *CDR1*, conferring multiple resistance to drugs and antifungals. *Curr. Genet.* **27**, 320–329.

Prasad, R., Krishnamurthy, S., Prasad, R., Gupta, V. and Lata, S. (1996) Multidrug resistance: an emerging threat. *Curr. Sci.* **71**, 205–213.

Raymond, M., Dignard, D., Alarco, A.M., Mainville, N., Magee, B.B. and Thomas, D.Y. (1998) A Ste6p/P-glycoprotein homologue from the asexual yeast *Candida albicans* transports the a-factor mating pheromone in *Saccharomyces cerevisiae*. *Mol. Microbiol.* **27**, 587–598.

Ruetz, S. and Gros, P. (1994) Phosphatidylcholine translocase: a physiological role for the *mdr2* gene. *Cell* **77**, 1071–1081.

Saier, M.H. Jr (2000) A functional-phylogenetic classification system for transmembrane solute transporters. *Microbiol. Mol. Biol. Rev.* **64**, 354–411.

Saier, M.H. Jr. and Reizer, J. (1991) Families and superfamilies of transport proteins common to prokaryotes and eukaryotes. *Curr. Opin. Struct. Biol.* **1**, 362–368.

Saier, M.H. Jr, Paulsen, I.T., Sliwinski, M.K., Pao, S.S., Skurray, R.A. and Nikaido, H. (1998) Evolutionary origins of multidrug and drug-specific efflux pumps. *FASEB J.* **12**, 265–274.

Sandbaken, M.G., Lupisella, J.A., Di Domenico, B. and Chakraburthy, K. (1990) Protein synthesis in yeast: Structural and functional analysis of the gene encoding elongation factor 3. *J. Biol. Chem.* **265**, 15838–15844.

Sanglard, D., Kuchler, K., Ischer, F., Pagani, J.-L., Monod, M. and Bille, J. (1995) Mechanisms of resistance to azole antifungal agents in *Candida albicans* isolates from AIDS patients involve specific multidrug transporters. *Antimicrob. Agents Chemother.* **39**, 2378–2386.

Sanglard, D., Ischer, F., Monod, M. and Bille, J. (1997) Cloning of *Candida albicans* genes conferring resistance to azole antifungal agents: Characterisation of *CDR2*, a new multidrug ABC transporter gene. *Microbiology* **143**, 405–416.

Sanglard, D., Ischer, F. and Koymans, L. (1998a) Amino acid substitutions in the cytochrome P-450 lanosterol 14-α-demethylase (CYP51A1) from azole-resistant *Candida albicans* clinical isolates contribute to resistance to azole antifungals. *Antimicrob. Agents Chemother.* **42**, 241–255.

Sanglard, D., Ischer, F., Calabrese, D., de Michelle, M. and Bille, J. (1998b). Multiple resistance mechanism to azole antifungals in yeast clinical isolates. *DRUG Resist Updates* **1**, 255–265.

Sanglard, D., Ischer, F. and Bille, J. (1999a) The role of the ATP binding cassette (ABC)-transporter genes *CgCDR1* and *CgCDR2* in the resistance of *Candida glabrata* to azole antifungals. In: *ASM Conference on Candida and Candidiasis*. Charleston, South Carolina, USA, 1–4 March, 55.

Sanglard, D., Ischer, F., Calabrese, D., Majcherczyk, P.A. and Bille, J. (1999b) The ATP binding cassette transporter gene *CgCDR1* from *Candida glabrata* is involved in the resistance of clinical isolates to azole antifungal agents. *Antimicrob. Agents Chemother.* **43**, 2753–2765.

Sanglard, D., Ischer, F., Monod, M., Dogra, S., Prasad, R. and Bille, J. (1999c) Analysis of the ATP-binding cassette (ABC)-transporter gene *CDR4* from *Candida albicans*. In: *ASM Conference on Candida and Candidiasis*. Charleston, South Carolina, USA, 1–4 March, 56.

Sanglard, D., Ischer, F. and Bille, J. (2001) Role of ATP-binding-cassette transporter genes in high-frequency acquisition of resistance to azole antifungals in *Candida glabrata*. *Antimicrob. Agents Chemother.* 45, 1174–1183.

Sasnauskas, K., Jomantiene, R., Lebediene, E., Lebedys, J., Januska, A. and Janulaitis, A. (1992) Cloning and analysis of a *Candida maltosa* gene which confers resistance to cycloheximide. *Gene* **116**, 105–108.

Schneider, E. (2001) ABC transporters catalyzing carbohydrate uptake. *Res. Microbiol.* **152**, 303–310.

Seghezzi, W., Meili, C., Ruffiner, R., Kuenzi, R., Sanglard, D. and Fiechter, A. (1992) Identification and characterisation of additional members of the cytochrome P450 multigene family *CYP52* of *Candida tropicalis*. *DNA Cell Biol.* **11**, 767–780.

Servos, J., Haase, E. and Brendel, M. (1993) Gene *SNQ2* of *Saccharomyces cerevisiae*, which confers resistance to 4-nitroquinoline-N-oxide and other chemicals, encodes a 169 kDa protein homologous to ATP-dependent permeases. *Mol. Gen. Genet.* **236**, 214–218.

Shani, N., Watkins, P.A. and Valle, D. (1995) *PXA1*, a possible *Saccharomyces cerevisiae* ortholog of the human adrenoleukodystrophy gene. *Proc. Natl Acad. Sci. USA* **92**, 6012–6016.

Shani, N., Sapag, A. and Valle, D. (1996) Characterisation and analysis of conserved motifs in a peroxisomal ATP-binding cassette transporter. *J. Biol. Chem.* **271**, 8725–8730.

Skowronski, R. and Feldman, D. (1989) Characterisation of an estrogen-binding protein in the yeast *Candida albicans*. *Endocrinology* **124**, 1965–1972.

Slaven, J.W., Anderson, M.J., Sanglard, D., Dixon, G.K., Bille, J., Roberts, I.A. and Denning, D.W. (1999) Induced expression of a novel *Aspergillus fumigatus* putative drug efflux gene in response to itraconazole. *Fungal Genet. Newslett.* **46**, 64.

Smart, C.C. and Fleming, A.J. (1996) Hormonal and environmental regulation of a plant PDR5-like ABC transporter. *J. Biol. Chem.* **271**, 19351–19357.

Smriti, M., Krishnamurthy, S., Dixit, B.L., Gupta, C.M. and Prasad, R. (2001) ABC transporters Cdr1p, Cdr2p and Cdr3p of a human pathogen *Candida albicans* are general phospholipid translocators. *Yeast (in press)*.

St Georgiev, V. (2000) Membrane transporters and antifungal drug resistance. *Curr. Drug Targets* **1**, 261–284.

Stark, G.R. and Wahl, G.M. (1984) Gene amplification. *Ann. Rev. Biochem.* **53**, 447–491.

Stover, E.P., Loose, D.S., Stevens, D.A. and Feldman, D. (1983) Ketoconazole binds to the intracellular corticosteroid-binding protein in *Candida albicans*. *Biochem. Biophys. Res. Commun.* **117**, 43–50.

Szczypka, M.S., Wemmie, J.A., Moye-Rowley, W.S. and Thiele, D.J. (1994) A yeast metal resistance protein similar to human cystic fibrosis transmembrane conductance regulator (CFTR) and multidrug resistance-associated protein. *J. Biol. Chem.* **269**, 22853–22857.

Taglicht, D. and Michaelis, S. (1998) *Saccharomyces cerevisiae* ABC proteins and their relevance to human health and disease. *Meth. Enzymol.* **292**, 130–162.

Talibi, D. and Raymond, M. (1999) Isolation of a putative *Candida albicans* transcriptional regulator involved in pleiotropic drug resistance by functional complementation of a *pdr1 pdr3* mutation in *Saccharomyces cerevisiae*. *J. Bacteriol.* **181**, 231–240.

Theiss, S., Agabian, N. and Kohler, G. (1999) Molecular characterization of a new ATP-dependent transporter in *Candida albicans*. In: *ASM Conference on Candida and Candidiasis*. Charleston, South Carolina, USA, 1–4 March, 56–57.

Theodoulou, F.L. (2000) Plant ABC transporters. *Biochim. Biophys. Acta.* **1465**, 79–103.

Thornewell, S.J., Peery, R.B. and Skatrud, P.L. (1997) Cloning and characterisation of *CneMDR1*: a *Cryptococcus neoformans* gene encoding a protein related to multidrug resistance proteins. *Gene* **201**, 21–29.

Timpel, C., Strahl-Bolsinger, S., Ziegelbauer, K. and Ernst, J.F. (1998) Multiple functions of Pmt1p-mediated protein *O*-mannosylation in the fungal pathogen *Candida albicans*. *J. Biol. Chem.* **273**, 20837–20846.

Tobin, M.B., Peery, R.B. and Skatrud, P.L. (1997) Genes encoding multiple drug resistance-like proteins in *Aspergillus fumigatus* and *Aspergillus flavus*. *Gene* **200**, 11–23.

Tommasini, R., Vogt, E., Schmid, J., Fromentau, M., Amrheim, N. and Martinoia, E. (1997) Differential expression of genes coding for ABC transporters after treatment of *Arabidopsis thaliana* with xenobiotics. *FEBS Lett.* **411**, 206–210.

Ueda, K., Okamura, N., Hirai, M., Tanigawara, Y., Saeki, T., Kioka, N., Komano, T. and Hori, R. (1992) Human P-glycoprotein transports cortisol, aldosterone and dexamethasone, but not progesterone. *J. Biol. Chem.* **267**, 24248–24252.

Ueda, K., Taguchi, Y. and Morishima, M. (1997) How does P-glycoprotein recognize its substrates? *Sem. Cancer Biol.* **8**, 151–159.

Ueda, K., Yoshida, A. and Amachi, T. (1999) Recent progress in P-glycoprotein research. *Anti-Cancer Drug Design* **14**, 115–121.

Urban, M., Bhargava, T. and Hamer, J.E. (1999) An ATP-driven efflux pump is a novel pathogenicity factor in rice blast disease. *EMBO J.* **18**, 512–521.

Van Den Hazel, H.B., Pichler, H., Do Valle Matta, M.A., Leitner, E., Goffeau, A. and Daum, G. (1999) *PDR16* and *PDR17*, two homologous genes of *Saccharomyces cerevisiae*, affect lipid biosynthesis and resistance to multiple drugs. *J. Biol. Chem.* **274**, 1934–1941.

Van Helvoort, A., Smith, A.J., Sprong, H., Fritzche, I., Schinkel, A.H., Borst, P. and Van Meer, G. (1996) MDR1 P-glycoprotein is a lipid translocase of broad specificity, while MDR3 P-glycoprotein specifically translocates phosphatidylcholine. *Cell* **87**, 507–517.

van Veen, H.W. and Konings, W.N. (1997) Multidrug transporters from bacteria to man: similarities in structure and function. *Sem. Cancer Biol.* **8**, 183–191.

Vanden Bossche, H. and Koymans, L. (1998) Cytochromes P450 in fungi. *Mycoses* **41**, 32–38.

Vanden Bossche, H., Marichal, P., Gorrens, J., Coene, M.-C., Willemsens, G., Bellens, D., Roels, I., Moereels, H. and Janssen, P.A.J. (1989) Biochemical approaches to selective antifungal activity, focus on azole antifungals. *Mycoses* **32**, 35–52.

Vanden Bossche, H., Marichal, P., Odds, F.C., Le Jeune, L. and Coene, M.C. (1992) Characterization of an azole resistant *Candida glabrata* isolate. *Antimicrob. Agents Chemother.* **36**, 2602–2610.

Vanden Bossche, H., Warnock, D.G., Dupont, B., Kerridge, D., Sengupta, S., Improvisi, L., Marichal, P., Odds, F.C., Provost, F. and Ronin, O. (1994) Mechanisms and clinical impact of antifungal drug resistance. *J. Med. Vet. Mycol.* **32**, 189–202.

Vasquez de Aldana, C.R., Marton, M.J. and Hinnebush, A.G. (1995) GCN20, a novel ABC protein and GCN1 reside in a complex that mediates activation of the eIF-2 kinase GCN2 in amino acid starved cells. *EMBO J.* **14**, 3184–3199.

Walker, J.E., Sarsate, M., Runswick, M. and Gay, N.J. (1982) Distantly related sequences in the α- and β-subunits of ATP synthase, myosin, kinases and other ATP-requiring enzymes and a common nucleotide binding fold. *EMBO J.* **1**, 945–951.

Walsh, T.J., Kasai, M., Francesconi, A., Landsman, D. and Chanock, S.J. (1997) New evidence that *Candida albicans* possesses additional ATP-binding cassette MDR-like genes: implications for antifungal azole resistance. *J. Med. Vet. Mycol.* **35**, 133–137.

Wendler, F., Bergler, H., Prutej, K., Jungwirth, H., Zisser, G., Kuchler, K. and Hogenauer, G. (1997) Diazoborine resistance in the yeast *Saccharomyces cerevisiae* reveals a link between *YAP1* and the pleitropic drug resistance genes *PDR1* and *PDR3*. *J. Biol. Chem.* **272**, 27091–27098.

White, T.C. (1997a) Increased mRNA levels of *ERG16, CDR,* and *MDR1* correlate with increased azole resistance in *Candida albicans* isolates from a patient infected with human immunodeficiency virus. *Antimicrob. Agents Chemother.* **41**, 1482–1487.

White, T.C. (1997b) The presence of an R467K amino acid substitution and loss of allelic variation correlate with an azole-resistant lanosterol 14α demethylase in *Candida albicans*. *Antimicrob. Agents Chemother.* **41**, 1488–1494.

White, T.C., Pfaller, M.A., Rinaldi, M.G., Smith, J. and Redding, S.W. (1997) Stable azole drug resistance associated with a substrain of *Candida albicans* from an HIV-infected patient. *Oral Diseases* **3**, 102–109.

White, T.C., Marr, K.A. and Bowden, R.A. (1998) Clinical, cellular, and molecular factors that contribute to antifungal drug resistance. *Clin. Microbiol. Rev.* **11**, 382–402.

Wilkinson, C.F., Hetnarski, K. and Yellin, T.O. (1972) Imidazole derivatives – a new class of microsomal enzyme inhibitors. *Biochem. Phamacol.* **21**, 3187–3192.

Wilkinson, C.F., Hetnarski, K. and Hicks, L.J. (1974) Substituted imidazoles as inhibitors of microsomal oxidation and insecticide synergists. *Pestic. Biochem. Physiol.* **4**, 299–312.

Wirsching, S., Michel, S., Kohler, G. and Morschhauser, J. (2000) Activation of the multidrug resistance gene *MDR1* in fluconazole resistant, clinical *Candida albicans* strains is caused by mutations in a *trans*-regulatory factor. *J. Bacteriol.* **182**, 400–404.

Wolfger, H., Mahe, Y., Parle-McDermott, A., Delahodde, A. and Kuchler, K. (1997) The yeast ATP binding cassette (ABC) protein genes *PDR10* and *PDR15* are novel targets for the Pdr1 and Pdr3 transcriptional regulators. *FEBS Lett.* **418**, 269–274.

Wolfger, H., Mamnun, Y.M. and Kuchler, K. (2001) Fungal ABC proteins: pleiotropic drug resistance, stress response and cellular detoxification. *Res. Microbiol.* **152**, 375–389.

Wu, A.L. and Moye-Rowley, W.S. (1994) GSH1, which encodes γ-glutamylcysteine syn-thetase, is a target gene for yAP-1 transcriptional regulation. *Mol. Cell. Biol.* **14**, 5832–5839.

Yan, C., Lee, L.H. and Davis, L.I. (1998) Crm1p mediates regulated nuclear export of a yeast AP-1-like transcription factor. *EMBO J.* **17**, 7416–7429.

Zhang, X., Collins, K.I. and Greenberger, L.M. (1995) Functional evidence that transmem-brane 12 and the loop between transmembrane 11 and 12 form part of the drug-binding domain in P-glycoprotein encoded by *MDR1*. *J. Biol. Chem.* **270**, 5441–5448.

Zhang, X., Cui, Z., Miyakawa, T. and Moye-Rowley, W.S. (2001) Crosstalk between tran-scriptional regulators of multidrug resistance in *Saccharomyces cerevisiae*. *J. Biol. Chem.* **276**, 8812–8819.

Zhao, X., Malloy, P.J., Ardies, C.M. and Feldman, D. (1995) Oestrogen-binding protein in *Candida albicans*: antibody development and cellular localisation by electron immuno-cytochemistry. *Microbiology* **141**, 2685–2692.

The Physiology and Collective Recalcitrance of Microbial Biofilm Communities

Peter Gilbert[1], Tomas Maira-Litran[2], Andrew J. McBain[1], Alexander H. Rickard[1] and Fraser W. Whyte[1]

[1]*School of Pharmacy and Pharmaceutical Sciences, Coupland III Building, Coupland Street, University of Manchester, Manchester, M13 9PL, UK*
[2]*Channing Laboratory, Harvard Medical School, Longwood Avenue, Boston, Massachusetts, 02115, USA*

ABSTRACT

Microbial biofilms impinge upon all aspects of our lives. Whilst much of this impact is positive, there are many areas in which the presence and activities of biofilms are regarded as problematic and in need of control. It is in this respect that biofilms reveal their recalcitrance towards many of the long-established antibiotics, and industrial and medical treatment strategies. The nature of the resistance of biofilms, in spite of much research, remains an enigma. Whilst it is recognized that reaction–diffusion limitation properties of the biofilm matrix towards the majority of treatment agents will impede access, this cannot be the sole explanation of the observed resistance. Rather, it will delay the death of cells within the community to various extents. Similarly, it is recognized that biofilm communities are phenotypically heterogeneous and that their eradication will reflect the susceptibility of the most resistant phenotype. The nutrient and gaseous gradients that generate this heterogeneity will, however, be destroyed as a result of antimicrobial treatments and cause the phenotype of the survivors to alter from slow-growing resistant cells to fast-growing susceptible ones. Accordingly both explanations can only delay death of the community. In order to explain more fully the long-term recalcitrance of biofilms towards such a wide variety of biocidal agents, more radical hypotheses must be considered. Amongst these are that multidrug efflux pumps could be up-regulated on expression of a biofilm phenotype. Whilst this is an appealing and simple explanation, because of its ability to explain the breadth of agents to which biofilms are resistant, recent work has suggested that this is not the case. Alternative hypotheses attempt to explain the diversity of agents by invoking a common cause of death for which singular resistance mechanisms could be applied. It is therefore suggested that an altruistic majority of sublethally damaged cells in a population commit suicide (apoptosis), thereby providing some protection to the survivors. A proportion of cells (persisters) is suggested to be defective, or repressed, in their suicide response, and survive. The persisters thereby benefit from the self-sacrifice of their compatriots and maintain the gene pool. A second explanation of the presence of persisters is that the general stress response, well known to include the adoption of a viable, nonculturable state of quiescence, is up-regulated in small pockets of the biofilm community, where nutrients are particularly scarce. Such quiescent cells noted for their resistance towards the metabolically acting biocides would potentially have their dormancy broken after treatment by the replenished supply of nutrients caused by the death of the majority. A more recent hypothesis suggests that extracellular signals, 'alarmones', released from killed cells might prime recipients into a state of resistance. Thus, in biofilm communities deep lying cells might be alerted into a resistant state by the premature death of peripheral cells. It is equally possible that 'alarmones', in this context, are merely the post-treatment 'wake-up' call to a previously quiescent subset of cells.

In this review, we attempt to provide a holistic view of the potential mechanisms by which biofilms express resistance. Since these mechanisms are multifaceted then their impact upon resistance will be considered against the context of biofilm formation, growth and maturation.

ABBREVIATIONS

CDAD   *Clostridium difficile*-associated diarrhoea
cfu   Colony-forming units
EPS   Extracellular polysaccharide
GSR   General stress response
HSL   Homoserine lactone
MIC   Minimal inhibitory concentration
RND   Resistance nodulation division
VNC   Viable but non-culturable

1. INTRODUCTION

1.1. Biofilms and Their Impact on the Environment and Man

Almost all surfaces that are found within a moist environment harbour attached, 'sessile' communities of micro-organisms that are termed 'biofilms' (Costerton *et al.*, 1978, 1987; Costerton and Stewart, 2001; Gilbert and McBain, 2001). Whilst the biofilm concept is relatively straightforward, the acceptance of a simple definition of the term has proved difficult. This reflects the structural, metabolic and microbiological diversity of the biofilm phenomenon. Often definitions reflect the specific interests of the individual researcher. At the simplest level, biofilms have been defined as 'functional consortia of cells'. By this definition, any cellular aggregate that possesses properties that are distinct from those of the same cells growing singly should be considered as a biofilm. Further definitions of the term require that the cell(s) must be attached to and immobilized at an interface. In this respect the interface might be a solid inert surface exposed to air or liquids. Such is the case in many biofouling and biocorrosion situations (Cloete et al., 1998; Santo Domingo et al., 1998), in the formation of dental plaque (Marsh and Bradshaw, 1998, 1999; Pratten et al., 1998) and in the infection of indwelling medical devices and implants (Bisno and Waldvogel, 1989). Alternatively, the substratum might be biodegradable and provide nutrients to the microbial community, such as with insoluble polysaccharide particles found within the human large bowel (Macfarlane et al., 1997) and bovine rumen or indeed on and in soft animal tissues (Singh et al., 2000).

Arguably those definitions that hold a consensus of opinion consider biofilms to be, 'functional consortia of cells bound within extracellular polymer matrices (the glycocalyx) and organized at interfaces' (Costerton et al., 1987; Allison and Gilbert, 1992) or 'predominantly two-dimensional microbial communities, which form at solid/liquid interfaces and which may become spatially heterogeneous by virtue of physico-chemical gradients that develop' (Wimpenny, 1993). These further refinements of the definition invoke, as essential biofilm characteristics, the presence of an extracellular polymer matrix, and the generation, through a close proximity of cells, of chemical and physiological gradients. Importantly, regardless as to whether or not substrata are present, the vast majority of biofilm populations interact with other cells, rather than the surface. Also the nature of the extracellular polymer matrix might vary from relatively simple structures involved in cellular aggregation (adhesins, fimbriae, etc.) (Weiss et al., 1988; Klier et al., 1998), insoluble polysaccharide gels (alginate) (Hentzer et al., 2001) to complex gel matrices comprised of colloidal solutions of polysaccharides, proteins, lipids, glycoproteins and nucleic acids (Flemming et al., 2000).

Biofilms that have been implicated in soft tissue infections and in infections of indwelling medical devices are almost exclusively monocultures (Marrie and Costerton, 1984; Gristina and Costerton, 1985; Costerton et al., 1987; Chandra et al., 2001). This reflects the ideal, normal state of microbiological cleanliness for such surfaces, and infections as the 'accidental' arrival of microbes that are able to resist host defences and proliferate. Environmental biofilms, and those associated with exposed surfaces of the body (mouth, gut, skin, urinary genital tract) on the other hand, are usually very heterogeneous communities (Costerton et al., 1987; Allison and Gilbert, 1992; Sibille et al., 1998; Peterson et al., 2001), which might embrace fungi (Elvers et al., 1998), algae (Liehr et al., 1994) and protozoa (Symmons et al., 2001).

What is now indisputable is the ubiquity in nature of the biofilm phenomenon (Costerton *et al.*, 1978, 1987) and the functionality of such interspecies relationships. These communities offer collective microbial activities that are often not possible within axenic cultures (Macfarlane *et al.*, 1997; Bradshaw *et al.*, 1998; Stickler *et al.*, 1998). Indeed, there has recently been much speculation as to whether biofilm communities should be regarded as 'units of proliferation' in their own right that are subject to Darwinian laws of selection and evolution (Caldwell and Costerton, 1996; Caldwell *et al.*, 1997a, b), or whether they simply reflect 'marriages of convenience' between separately evolving species (Wimpenny, 2000). Regardless, biofilm phenotypes and community behaviour have evolved over millions of years and are arguably the major modes of growth for microorganisms. The unique physiological properties of biofilms are as vital to the natural environment as they are in many aspects of life and commerce. Biofilm-associated microorganisms, for example, play crucial roles in terrestrial and benthic nutrient cycling (Paul *et al.*, 1991; Bernhard and Bowser, 1992) and in the biodegradation (and bioremediation) of environmental pollutants (Holden *et al.*, 1997; White *et al.*, 1998). Microbial communities associated with mammalian skin and mucosal surfaces protect the host from pathogenic bacteria, whilst sessile bacteria in the human large intestine play a vital role in host health (McBain and Macfarlane, 1998, 2001; Barcenilla *et al.*, 2000). In natural water systems, biofilm communities degrade organic compounds and detoxify xenobiotics, thereby improving water quality. These metabolic activities have been exploited in commerce, for wastewater management (Yu *et al.*, 1999; Li *et al.*, 2001), sewage treatment (Goncalves *et al.*, 1998) and in biotechnology, through a variety of patented solid-state fermentation processes (Lu *et al.*, 1998).

The rapid expansion of biofilm research that has occurred over the last 20 years, has been characterized by a preoccupation with biofilm-associated problems. Biofilms, for example, are associated with infections that are related to medical implants (Ward *et al.*, 1992; Tunney *et al.*, 1999), that are implicated in cross-contamination in food production (Holah, 1995), with problems of domestic hygiene (Barker and Bloomfield, 2000), in the biofouling of industrial machinery and fluid conduits (Characklis, 1990), and in biodeterioration generally (Webb *et al.*, 1999). Furthermore, external human body surfaces, and the digestive tract are associated with commensal biofilm populations. Pathologies have been suggested to result from the activities of such microfloras, when inimical combinations of cells (dysbioses) occur. In such cases, the 'aetiological agents' are recognized as the associated microbial community. Dental caries (Sissons *et al.*, 1996), ulcerative colitis (Levine *et al.*, 1998), and colon cancer (Roberton, 1993) have all been attributed, at least in part, to dynamic changes in the composition and metabolic activity of the associated microbial community.

Recognition of the problematic nature of biofilm has been compounded by the realization that such communities often display a high degree of recalcitrance towards those chemical and antibiotic treatments that have previously been deployed successfully against free-living planktonic forms (McBain *et al.*, 2000b; Chandra *et al.*, 2001; Stewart and Costerton, 2001). Such resistance is both multifaceted and cumulative with biofilm community formation and maturation. Any consideration of biofilm resistance must therefore be viewed against a background of the formation and maturation process.

1.2. Formation and Development of Biofilm Communities

In most instances in nature, surfaces encountered by micro-organisms will find that they have already been conditioned through the adsorption of molecules and ions from the bulk fluid phase. The conditioning film is, in many instances, of biological origin and may present specific sites for bacterial adhesion (i.e. fibronectin, mucin). Generally the encountered surfaces, rather than being 'sterile' are already colonized by micro-organisms. During the early stages of biofilm formation upon freshly presented or cleaned substrates, all of the planktonic cells that contact the surface have the potential to attach and form microcolonies (primary colonizers). Secondary colonizers will interact either with vacant sites on the surface or with the primary colonizers (Aspiras *et al.*, 2000). There are three possible outcomes to the encounter between a potential immigrant and a newly colonized surface. (1) The surface may be refractory to the potential colonizer due to lack of available/unoccupied binding sites. The immigrant will therefore fail to bind. (2) The immigrant cells may displace one of the early colonizers physically from the surface by possession of a higher binding affinity for a common binding site. This is most likely to occur during the initial attachment phase of film formation and before the deposition of polymer cements. The duration of this phase will be indirectly related to the metabolic potential at each colonized site. (3) Both the immigrant species and the primary colonizer are retained on the surface, either at separate sites, or attached to each other or to the matrix polymers. Where a surface is co-colonized then the degree of interaction between the colonizers will be minimal in the first instance, but will increase as the community grows and adjacent microcolonies come into closer proximity. In some instances aggregates of cells may form in the bulk fluid phase, or be shed from upstream matured biofilm communities (sloughs), and subsequently sequestered to the substratum. Such aggregates will possess some of the properties of mature biofilms, may already include extracellular polymeric matrices, and might be comprised of a single species (Hasman *et al.*, 1999; Roos *et al.*, 1999) or mixtures (Bradshaw *et al.*, 1998; Rickard *et al.*, 2000). Interactions between these surface-bound or free aggregates of cells would be mediated through the production of cell–cell signalling compounds (Davies *et al.*, 1998), specific and nonspecific inhibitors (Gibson and Wang, 1994; BernetCamard *et al.*, 1997), capture of secondary metabolites, and the competition for available nutrients. Alternatively new substrate might become mobilized through co-operative degradation (Macfarlane *et al.*, 1986) and the establishment of cross-feeding partnerships (Degnan and Macfarlane, 1995).

Complimentary microbial activity is responsible for the enhanced catabolism of nutrients and can often provide protection against adverse environmental conditions by the establishment of a more favourable local micro-environment (Caldwell *et al.*, 1997a, b; Marsh and Bradshaw, 1998). The outcome of such interactions might be the elimination of one or other of the initial colonizers, a neutral co-existence of multiple species or the formation of synergistic partnerships (Marsh and Bowden, 2000; Marsh and Bradshaw, 2000).

As the biofilm thickens and population density increases further, the overall heterogeneity of the community becomes more marked through the imposition of nutrient, gaseous and pH gradients generated by the metabolic activity of the enveloped cells (Wimpenny, 1995; Vroom *et al.*, 1999). These will provide a variety of selection pressures, operational at different points in the developing community, each of which influences the outcome of interactions occurring between adjacent species. The biofilm therefore achieves a dynamic equilibrium with respect to its community membership and 'ethnicity'. Throughout this process of maturation, the possibility for microbial succession within the community remains, but as it is dependent upon the successful immigration of allochthonous species, becomes increasingly unlikely (McBain *et al.*, 2000a).

As a consequence of their development in association with intense selection pressures, the climax communities are remarkably resistant to further colonization (colonization resistance/microbial homeostasis) (Alexander, 1971; Marsh and Bradshaw, 1999). Colonization resistance of climax communities has been elegantly demonstrated in gnotobiotic animals. Such animals are devoid of a natural microflora and, as a consequence, are considerably more susceptible to infection by autochthonous bacteria (Boureau *et al.*, 1994). Similarly the increased susceptibility towards infection of animals which possess microflora that have become degraded by antibiotics or radiotherapy, further supports such hypotheses (Brook anc ·dney, 1993; Larson and Welch, 1993). In a similar fashion, biofilm communities that are stressed by chronic exposure to antimicrobial treatments will become destabilized, thereby permitting change and the subsequent immigration of new species. Clonal expansion of the insusceptible, multifunctional community members will occur together with a replacement of the susceptible ones by less susceptible immigrants. These may be less efficient in the displaced community function, but this is complicated by their resistance profile. The potential for biofilm communities to evolve towards tolerance of antimicrobial treatments, yet to maintain their functional integrity, is therefore demonstrated.

1.3. Biofilm Recalcitrance

Some of the earliest reports that link the presence of biofilms with recalcitrance towards chemical treatments stemmed from the clinic in the early 1980s. At this time the use of implanted medical devices was increasing markedly. In some instances, episodes of bacteraemia were reported that were both responsive to antibiotic treatment and also linked to clinical isolates with minimum inhibitory concentrations that were below defined resistance thresholds, but which recurred shortly after the treatment was stopped. Such infectious episodes were only resolved after removal of the implanted devices, which were subsequently found to harbour, as a biofilm, the implicated pathogen (Gristina and Costerton, 1985; Nickel *et al.*, 1985; Costerton *et al.*, 1987; Evans and Holmes, 1987; Gristina *et al.*, 1987, 1989; Bisno and Waldvogel, 1989). At a similar time, reports of microbial recalcitrance towards more aggressive biocidal treatments were being related to surface growth and attachment of micro-organisms in the industrial and environmental sector (Favero *et al.*, 1983; Costerton and Lashen, 1984; McCoy *et al*, 1986a, b). The range of antimicrobial molecules to which biofilm populations appeared to be resistant was considerable and embraced virtually all of the clinically deployed antibiotics, oxidizing biocides, such as the isothiazolones, the halogens, quaternary ammonium compounds, biguanides and established phenolics.

Generally, failure of micro-organisms to succumb to antimicrobial treatments arises through (1) an inherent insusceptibility to the agents employed; (2) the acquisition of resistance, by previously susceptible strains, either by mutation or by transfer of genetic material from another species or genus; or (3) the emergence of pre-existing, but unexpressed, resistance phenotypes. Since the agents that biofilms appeared to be resistant towards encompassed a wide diversity of action mechanism, then the simplest explanation was that the agents were prevented from reaching the target cells. Proposed mechanisms have included reductions in the diffusion characteristics of biofilm relative to water and plasma and diffusion limitation that is augmented by the adsorption or chemical/enzymatic modification of the agent (reaction–diffusion limitation). Such explanations have proved insufficient to explain the whole panoply of resistance (Brown *et al.*, 1988, 1990; Gilbert *et al.*, 1990). Accordingly physiological changes in the biofilm cells mediated through the induction of a slow growth rate and starvation responses, together with induction of separate attachment-specific, drug-resistant physiologies were considered as mediators of biofilm resistance (Allison *et al.*, 2000; Gilbert and Allison, 2000). Whilst there is ample evidence to support all of these mechanistic explanations, it now seems highly unlikely that any single mechanism will account for the general observation of resistance. Rather, these mechanisms are compounded in biofilms to create insusceptibility, and an environment ideally suited for the emergence of tolerant genotypes/communities.

1.4. Chronology of Resistance Development

Since the resistance associated with biofilm communities is multifaceted, the various aspects of it come into play at different stages during formation and maturation. Even before the cells become attached to a substratum, there are often interactions that occur between individual cells within the fluid phase. Such auto-aggregation and co-aggregation may be between individual bacterial cells or initiated by the presence of a small particulate nucleus of coadhesion (i.e. flocculation) (Higgins and Chodoba, 1986). In some instances co-aggregation in the fluid phase is an essential element of community development (Bradshaw et al., 1998). The resistance properties of aggregates differs somewhat from those of single dispersed cells and will be discussed later (Section 2). Equally the susceptibility of single microbial cells becomes altered very rapidly following their attachment to inert surfaces (Ashby et al., 1994; Das et al., 1998; Gilbert et al., 2001). This might be a simple reflection of solution chemistry within the boundary layers leading to localized reductions in the concentration of agent (van Loosdrecht et al., 1990), binding and neutralization of the agent by an organic conditioning film or to phenotypic changes in the cells mediated through some form of 'touch response' (Dagostino et al., 1991). Whilst such changes are rapid, they rarely account for more than a three-fold change in apparent MIC (Das et al., 1998; Gilbert et al., 2001). Exopolymer deposition generally follows rapidly after the initial attachment process (Allison and Sutherland, 1987; Davies et al., 1993, 1998) and brings about further reductions in susceptibility. As the biofilm matures, both its structural and metabolic heterogeneity becomes increasingly marked, as does the lack of susceptibility towards a wide variety of agents. The degree of physiological heterogeneity, marked by the establishment of nutrient, gaseous and pH gradients, continues to increase with time and is positively correlated with the development of drug resistance (Anwar et al., 1990, 1992). It is noteworthy that aggregates of only a few strongly aerobic cells are sufficient to generate anoxic centres within which even obligate anaerobes might survive (Wimpenny, 1993; Marsh and Bowden, 2000).

One of the biological imperatives is the requirement to migrate and colonize new ecological sites (McBain et al., 2000a). If the formation of biofilm were an irreversible process, eventually the world would be devoid of planktonic bacteria and the sessile communities would themselves starve through depletion of their locally derived nutrient. Dispersal of microbial communities is therefore of great importance to biofilm physiology, and might even hold the key to their successful control in problematic situations. Dispersal appears to arise as a consequence of fluid dynamic forces exerted against the communities and a direct result of regulatory events taking place within the community. Both processes lead to the loss of single cells and of whole biofilm. Physical erosion of the surface of the biofilm and dissolution of the exopolymer matrix leads to the shedding of single cells, whereas distortion of the matrix beyond its elastic limit leads to catastrophic sloughing of the community (Allison *et al.*, 1999; Stoodley *et al.*, 2000). The sloughs retain the original community structure and many of the unique properties of the parent biofilm, including resistance. Programmed dispersal of the biofilm community appears to relate to starvation or signal-mediated synthesis of enzymes and molecules capable of dissolving and dispersing the extracellular matrix (polysaccharides, biosurfactants, etc.) (Schooling *et al.*, 2001). This can lead to the liberation of sloughs, which as before retain many of the physiological characteristics of the parent community, but has also been shown to lead to the release of single cells (Allison *et al.*, 1990). The latter grow and divide synchronously when collected and grown in batch culture indicating that the programmed release of single cells is linked into the cellular division cycle. Paradoxically, such released cells are exquisitely sensitive to many antimicrobial agents (Evans *et al.*, 1990a, b; Duguid *et al.*, 1992a, b), but possess surface characteristics that make them unlikely to reattach to a surface (Allison *et al.*, 1990; Gilbert *et al.*, 1993). In infection it is likely that these correspond to bacteraemic cells, and account for the readiness of biofilm-associated bacteraemia to respond to treatment.

1.5. Biofilm Recalcitrance as a Community Phenomenon

Many of the resistance properties of biofilm are lost when the component cells are fully dispersed, separated from their matrix polymers and allowed to grow even for short periods of time in planktonic culture. It is rare, however, for such 'killing experiments' to monitor the complete eradication of the treated population since the detection limits, imposed by dilution and plate count experiments and the need to neutralize residual biocide, are relatively high. Experiments designed to assess the killing of intact biofilm communities, on the other hand, often rely on neutralisation *in situ*, and post-treatment resuspension or incubation of the test piece. The presence of low levels of survival are more likely therefore to be detected in experiments that utilize intact biofilms than in those using suspensions of cells. It is also possible that the local concentration of nutrients and materials released from lysed cells within the intact biofilm experiments facilitates the recovery of sublethally injured bacteria (Gilbert, 1984) that would succumb to the stresses of nutrient starvation, cold and osmotic shock (Gilbert *et al.*, 1995) that so often accompany traditional laboratory experiments (Gilbert and Brown, 1991). Also noteworthy in this respect is that conventional tube-dilution methods to determine the MIC of antimicrobials and antibiotics, limit the inoculum to 10^6 c.f.u. This is intended to minimize the possible effects on the endpoint of 'rogue cells'. Indeed if MICs are determined with inocula that exceed 10^8, then the endpoint is often substantially increased. Experiments with biofilm communities inevitably utilize populations that are greatly in excess of 10^8 and it is therefore not surprising that the outcome sometimes reflects the properties of the least susceptible community member.

Biofilm communities might be inherently more resistant than single cells, because the genetic diversity within the population increases the likelihood of the survival and detection of individual clones that possess inherently lower levels of susceptibility than the majority. For example, dental plaque communities exhibit a high degree of species complexity that renders the plaque itself comparatively resistant towards antibacterial agents. This is because no single agent possesses a broad enough spectrum of activity to eliminate all of the resident species (Kinniment et al., 1996; Pratten et al., 1998). Furthermore, even biofilms that are comprised of a single microbial species will exhibit sufficient genetic and phenotypic diversity for such effects to be evident. It is in this context that we must consider the true nature of biofilm survival from inimical treatments, since in many instances the vast majority of cells (circa 1–3 log reductions = 99.9%) within the target population succumb and die. Victory over the biocide is indeed often a Pyrrhic one (Gilbert and McBain, 2000).

2. CHANGES IN SUSCEPTIBILITY ASSOCIATED WITH CELLULAR AGGREGATES

It is widely acknowledged that the shapes of survival curves and in the associated rate constants for activation are greatly affected by the presence of clumps or clusters of cells. In such instances, singular reductions in colony-forming units reflect the death of the last survivor within one clump. In such a manner, experiments performed with aggregates comprised of two to three cells will generate rate constants that are substantially reduced from those of single-cell suspensions. It is for such reasons that many of the test strains advocated for the testing of disinfectants and preservatives are ones that do not auto-aggregate, and many of the early tests for disinfection additionally advocate a coarse prefiltration of the test inocula to remove clumps. In theory, if a population consists of single cells of identical phenotype then the inactivation of that population, by physical and chemical agents, ought to obey the laws of probability and thereby approximate to first-order kinetics. In practice, variance from this is related to distributions of susceptibility with the target populations, depletion of bioavailable agent through adsorption, reaction with cellular materials and dead cells, and to aggregation (Hoff, 1978; LeChevallier *et al.*, 1981, 1988), coaggregation, association with particulates (Ngian *et al.*, 1977; Hoff and Akin, 1986) and/or other surfaces (LeChevallier *et al.*, 1981; Daly *et al.*, 1998; Mah and O'Toole, 2001). Whilst some of these characteristics can be explained arithmetically as an effect of cell clustering, there appears to be some survival advantage to individual cells within aggregates. Whilst this can be readily explained in terms of some physical agencies (i.e. ultra-violet light) (Blatchley *et al.*, 2001) that will be absorbed and therefore fail to penetrate to the core of an aggregate, similar protective effects are apparent even towards some oxidative antimicrobial chemicals. Since autoaggregation, coaggregation, coadhesion and the attachment of individual cells to surfaces are integral components of the biofilm formation process then it is therefore essential to consider their separate contributions to the development of resistance.

2.1. Autoaggregation

Autoaggregation is the self-recognition and adherence to one another of genetically identical bacterial cells. Autoaggregates are commonly observed in nature (Hirsch, 1974; LeChevallier *et al.*, 1981; Sly and Cahill, 1997) and are often seen in cultures of common laboratory bacteria. For example, motile *E. coli* cells often form organized patterns in semi-solid media which is due to chemotactically mediated autoaggregation (Shapiro, 1995). Similarly, *Sphingomonas natatoria* produces aggregates in liquid culture and very adhesive colonies on agar (Hugenholtz and Fuerst, 1992). There have been a large number of publications that identify the surface polymers that mediate autoaggregation (Ziebuhr *et al.*, 1997; Kjaergaard *et al.*, 2000), but relatively few studies that compare the susceptibilities to antibiotics and biocides of the aggregated cells to those of disassociated ones. Mir *et al.* (1997) have shown, however, that the autoaggregation of *Bacillus mycoides*, as well as other species of wild-type freshwater bacteria, enhances their survival in chlorine-treated water. In addition, Olson and Stewart (1990) have demonstrated that microbial susceptibility to various potable water treatment agents could be reduced two- to 30-fold by autoaggregation. They further showed that insusceptibility towards one of these agents, chloramine, developed by *Klebsiella pneumoniae* under low nutrient growth conditions, was jointly functions of cellular aggregation and localized protection of sulphydryl groups within the cell by the presence of an extracellular polymer matrix (Stewart and Olson, 1992).

From a physicochemical perspective, autoaggregation may also protect cells from chemical attack. Intra- and extracellular components of the aggregate, including EPS, LPS and proteins will limit the mass transfer of antimicrobials into the aggregate to create a reaction–diffusion gradient towards the core (Ngian *et al.*, 1977; Olson and Stewart, 1990; Stewart and Olson, 1992). Bacteria at the core would thereby have a heightened chance of survival and the larger the aggregate became, then the greater would be the proportionate survival of component cells. Mass transfer resistance would be greater in those instances that involve biocidal molecules that possess high-affinity (Giles *et al.*, 1974a, b; Broxton *et al.*, 1984) binding sites on the cell. High-affinity binding is particularly prevalent amongst the cationic biocides and, provided the interactive sites remain unsaturated, leads to negligible residual concentrations of agent (Giles *et al.*, 1974b).

In addition to the advantage afforded the cells through mass transfer resistance, autoaggregated communities have been shown to exchange plasmids at much higher frequencies than do compacts of cells that do not have the ability to autoaggregate (Walsh and McKay, 1981; Licht *et al.*, 1999; Ghigo, 2001). Autoaggregates are likely to be comprised of communities where each individual contains the 'population library' of transposon-mediated and plasmid-mediated resistance genes. This, in turn, could enhance the resistance characteristics of the survivors of inimical treatments.

2.2. Coaggregation

Coaggregation is the specific recognition and adherence of genetically distinct bacteria (Kolenbrander and London, 1993) and is usually mediated by lectin–saccharide interactions (Rickard *et al.*, 1999; Kolenbrander, 2000). Although coaggregation was initially reported and studied within the human oral microflora (Gibbons and Nygaard, 1970), evidence is now emerging that it is a universal phenomenon and occurs between bacteria within a wide variety of diverse environments, including mammalian intestinal tracts (Reid *et al.*, 1988; Vandevoorde *et al.*, 1992) and freshwater ecosystems (Rickard *et al.*, 2000). It is probable that exactly the same protection against biocides will be afforded to coaggregated communities as is afforded to autoaggregated ones. When more than one morphological type is present, however, the effects of mass transfer resistance could become amplified. This is particularly the case when partnerships result in the close proximity of sensitive cells to ones that possess a high binding/reaction capacity for the treatment agent. Since the absorption of lipophilic biocides such as the quaternary ammonium compounds by coryneform bacteria greatly exceeds that by Gram-negatives, then the presence of coryneform bacteria within a coaggregated cluster of cells would form a dissolution sink. Equally, if susceptible cells were present within coaggregate, that included organisms capable of enzymically neutralizing the treatment agent, protection would be afforded to all (i.e. catalase, β-lactamase, and aldehyde lyase producers might protect sensitive cells against hydrogen peroxide, penicillin and formaldehyde treatments, respectively).

2.3. Coadhesion

Coadhesion is an extension of coaggregation, where one of the partner organisms is already attached to a surface, or where one partner mediates the attachment of the coaggregate (Busscher and Mei, 2000). Coadhesion contributes, at least in part, to the development of multispecies biofilms (Kolenbrander, 1989), and has been described within the oral cavity (Bos *et al.*, 1998) and the urogenital tract (Reid *et al.*, 1988). Mechanistically, the only difference between coaggregation and coadhesion is the presence of a surface. There are numerous reports in the literature, however, of changes in cellular physiology and gene expression that occur immediately upon the immobilization of a cell at a surface (discussed below). Whilst it is obvious that such changes will operate for a coadhered partnership, it is not altogether obvious whether such changes will also occur when the attachment is to another cell. It is our belief that such touch responses and the generation of specific phenotypes will not be unique to attachment on inert surfaces.

2.4. Association with Inert and Other Surfaces

The ability of bacterial cells to respond to their attachment to an inert surface has been long postulated. Indeed, Zobell (1943) was the first to describe the stimulation of bacterial metabolism at solid–liquid interfaces. This was termed the 'bottle effect' (Fletcher, 1991). Similarly, McFeters *et al.* (1990) observed an enhancement of nitriloacetate degradation when the degradative organisms were attached, but where nitriloacetate was unable to adsorb to the test surface. Gliding bacteria lack extracellular polymer biosynthesis when they are grown in suspension culture, but rapidly initiate it following attachment (Humphrey *et al.*, 1979; Abbanat *et al.*, 1988). Lateral flagella gene transcription in *Vibrio parahaemolyticus* is found during growth on solid, but not in liquid media. This has been attributed to a restriction of their movement of polar flagella when at a surface brought about through an increased fluid viscosity (Belas *et al.*, 1984, 1986; McCarter *et al.*, 1988).

Such observations led to the suggestion that bacterial cells were able to sense the proximity of surfaces (Dagostino *et al.*, 1991). Whether or not this can be attributed to specific touch receptors or to chemical and electrochemical differences between the surface boundary layer and the bulk phase is still subject to debate (van Loosdrecht *et al.*, 1990). With the advent of modern molecular tools, it has however, become unquestioned that bacteria can respond to the close proximity of surfaces (Armitage, 1999; Danese *et al.*, 2000), and that such sensing is more likely to relate to the accumulation of chemical signals than to the presence of a sense of touch (Davies *et al.*, 1998; Parsek and Greenberg, 1999). There is no reason to suppose that similar responses will not occur between autoaggregated and coaggregated bacteria. Amongst those changes that are immediate upon attachment are the up-regulation of a number of regulons associated with the production of EPS, the transcription of proteins that are specific to the 'biofilm phenotype', and the suppression of others (Cochran *et al.*, 2000; Xu *et al.*, 2001). Consequently, cells that have become adhered to surfaces display markedly changed suscep-tibilities toward antibiotics (Ashby *et al.*, 1994) and biocides (Das *et al.*, 1998). Using novel spectrophotometric methods, which allowed for simultaneous monitoring of the growth of planktonic and biofilm bacteria within the wells of a microtitre plate, Das *et al.* (1998) showed that the susceptibility of *P. aerug-inosa* and *Staphylococcus aureus* to a range of different biocides, changed rapidly after cellular attachment and biofilm formation. In some instances three to five-fold decreases in susceptibility occurred immediately upon attach-ment and could occur in the presence of biocide concentrations that exceeded the MIC for planktonic populations. These observed changes in susceptibility occurred before there had been any significant growth of the micro-organisms on the surface and before exopolymers had been accumulated. In a similar study, Fujiwara *et al.* (1998) investigated the immediate effect of adherence upon the antimicrobial susceptibilities of *P. aeruginosa*, *Serratia marcescens* and *Proteus mirablis*. Their results demonstrated that, after 1 h of incubation, involving adherence of bacteria to the surface of a plastic tissue culture plate, the MIC for adherent bacteria became markedly elevated even before biofilm formation had occurred. The magnitude of the decreases in susceptibility observed after bacterial attachment, but before biofilm formation, are however in general far less marked than those observed for mature biofilms. They are therefore insufficient to account for the reported levels of resistance in biofilm communities.

3. SUSCEPTIBILITY CHANGES ASSOCIATED WITH THE SYNTHESIS AND PRESENCE OF THE GLYCOCALYX

We have so far considered only the effects on antimicrobial susceptibility of attachment of single cells to one another and to surfaces. It is generally accepted, however, that the most profound changes to antimicrobial suscepti-bility occur when these small aggregates and sedentary cells initiate the formation of microcolonies. With few exceptions, such communities are enveloped within an extensive extracellular polymeric matrix (glycocalyx). This not only cements and immobilizes the cells with respect to one another, but it also traps molecules and ions of environmental and microbial origin. It is not surprising therefore that many of the early explanations of biofilm resistance towards antimicrobial substances intuitively invoked the properties of 'slime'.

3.1. What is the Glycocalyx?

For biofilms, the term 'glycocalyx' refers to the matrix of polymers that sur-round and envelop biofilm communities. The proportion that such extracellular matrix occupies within biofilm communities can vary between 50 and 90% of the total organic matter (Flemming and Wingender, 2001). Whilst virtually all defi-nitions of biofilm include reference to substrata it is pertinent to note that it will be inaccessible to the majority of biofilm cells. These interact with one another and the matrix. Direct light and electron microscopy show the exopolymeric matrix in biofilms to be comprised of ordered arrays of fine fibres. These arrays provide a relatively thick coating to the cells that, in Gram-negative bacteria, are predominantly composed of gelled and highly hydrated exopolysaccharides (Sutherland, 1997). Glycocalices should not, however, be thought of as being exclusively polysaccharide in nature. Rather, they contain varying proportions of polysaccharide, nucleic acids (Nielsen *et al.*, 1997), proteins, globular glyco-proteins (Flemming and Wingender, 2001), lipids (Goodwin and Forster, 1989), phospholipids (Sand and Gehrke, 1999), humic acids, and sequestered ions (Sutherland, 1985; Flemming *et al.*, 2000). Whilst alginate slimes might well dominate pure culture *Pseudomonas* biofilms, within these proteins often com-prise a substantial proportion of the matrix (Flemming and Wingender, 2001). Indeed, a number of studies concerned with wastewater biofilms have shown proteins to be the dominant component (Rudd *et al.*, 1983; Nielsen *et al.*, 1997; Jorand *et al.*, 1998). The molecules that comprise the glycocalyx are only rarely electrically neutral (i.e. dextran and levan); more commonly these are polyelec-trolytes and present both negative and positive charged sites to the aqueous environment. Anionic groups are often presented as carboxyl, sulphate and phos-phate groups (Uhlinger and White, 1983), and as nucleic acids and proteins (Dignac *et al.*, 1998), whereas cationic groups are generally provided by amino-sugars (Hejzlar and Chudoba, 1986; Veiga *et al.*, 1997). Extracellular proteins contribute to the hydrophobic properties of the matrix (Jorand *et al.*, 1998) through their deposition of amino acids such as alanine, leucine and glycine (Higgins and Novak, 1997). Clearly in mixed species biofilms, where the matrix properties of each component species are different, then the overall matrix might be represented as a mosaic comprising negatively and positively charged micro-zones together with hydrophobic pockets (Jorand *et al.*, 1998). The implications for the distribution and pharmacodynamic effects of added antimicrobials are profound (below). Whether the matrix polymers differ from those associated with planktonically grown cells, and also whether the matrix polymers, which bind cells to other cells, differ from those 'footprint polymers' which cement the primary colonizers to the substratum, is uncertain at present (Sutherland, 1997; Allison *et al.*, 1998). There is, however, evidence that extracellular polysaccha-ride synthesis is up-regulated within minutes of the irreversible attachment of a cell to a surface (Allison and Sutherland, 1987; Davies *et al.*, 1993). It seems likely that this is necessary for the subsequent formation of an extant biofilm. Deposition of EPS within the developing microcolony then proceeds over a period of hours. Matrix polymers ultimately determine the physical properties and binding and reaction capacities of the biofilm community.

3.2. Regulation and Dynamics of the Biofilm Matrix

The synthesis of matrix polymers appears to be regulated by a variety of factors, of which surface attachment appears to be of particular importance. Davies *et al.* (1993) have shown that the up-regulation of alginate synthesis in *P. aeruginosa* occurs within minutes of attachment to the surface of a glass flow cell. This was well before there had been any accumulation of biomass. Exopolysaccharide (alginate) production was also derepressed in the biofilms, compared with planktonic populations. These authors suggest that, in most Gram-negative organisms, such up-regulation of EPS is under the control of signal substances in the form of *N*-acyl homoserine lactone (HSL) (Davies *et al.*, 1998). These global regulators of transcriptional activation are responsible for cell–cell signalling in Gram-negative bacteria through cell density-mediated events and as such, may also act as regulators of biofilm-specific physiology (Williams *et al.*, 1992; Gambello *et al.*, 1993; Cooper *et al.*, 1995; Heys *et al.*, 1997).

In Gram-positive bacteria, such as *Staphylococcus epidermidis*, the exopolysaccharide is a β-1,6-linked glucosaminoglycan (Mack *et al.*, 1996) and is encoded by the *icaABCD* operon (Gerke *et al.*, 1998). Evans *et al.* (1994) had previously shown that exopolysaccharide production, and the synthesis of various extracellular virulence factors by *S. epidermidis* increases progressively with reductions in specific growth rate and, as for Gram-negative bacteria, is substantially higher for biofilm than for planktonic cells. Such effects of growth rate would provide for increased exopolymer production within the slow-growing heart of a thick aerobic microcolony/biofilm and would alter the distribution and density of cells throughout the matrix. In this respect, it is notable that with the exception of the alginates, the majority of the extracellular polysaccharides are soluble to some degree in water. Biofilm structure is therefore dynamic with a gradual solubilization of polymers occurring at the periphery being compensated by an increased production of polymers within the depths.

In biofilms, signal substances such as HSL would become concentrated within the geometric centre of the microcolonies/biofilm, thereby increasing exopolymer production. This would alter the distribution and density of cells throughout the matrix, and once again confer some structural organization upon the community to provide customized microniches at various points within the biofilm (Costerton *et al.*, 1994). The extent and nature of exopolymer production is also dependent upon physiological factors such as the relative availability of carbon and nitrogen (Sutherland, 1985), and the properties of the matrix responsive to sequestered ions.

3.3. Implications of an Extracellular Polymeric Matrix for Communities

The glycocalyx, particularly in mixed community biofilms, cannot be considered as being homogeneous, rather it will vary its degree of hydration with depth into the community. The physical-chemical properties of the matrix will vary spatially as does the mosaic of enveloped microcolonies of individual community members. At the interface of such mosaics, the physicochemical properties of the mixed colloidal solutions, which comprise the juxtaposed polymers, will be substantially different from that of the component polymers. Adsorption sites within the matrix serve to anchor extracellular enzymes from the producer organisms, and will actively concentrate ionic materials from the bulk fluid phase and secondary metabolites (including some cell–cell signalling substances) from the community itself. Immobilized enzymes are not only capable of mobilizing complex nutrients captured from the fluid phase, but they are also capable of degrading many antibacterial substances. Thus, the glycocalyx is able to moderate the microenvironments of each of the individual community members.

Such molecular organization within biofilm communities has led some workers to compare them with primitive eukaryotic tissues and with the evolution of homeostasis (Costerton *et al.,* 1994). These polymer mosaics will not only bind the community together, but will also form compartments, of differing hydrophobicity and diffusivity throughout which antimicrobial substances must interact. Such interactions must be considered in terms of partitioning and diffusion limitation, the provision of interactive targets for the antimicrobial that compete locally with those associated with the target organisms (reaction–diffusion limitation), and the retention and organization of drug-inactivating enzymes (enzyme-mediated reaction–diffusion limitation). Such considerations are nowadays inherent in the pharmacokinetic modelling of the bioavailability of drugs in animals and humans and should perhaps now be formally applied to biofilm communities.

3.3.1. Resistance to Chemical Agents Through Diffusion Limitation

Most, but not all, of the resistance characteristics of biofilm communities are lost when those communities are resuspended and separated from their extracellular products. Indeed, intuitively one would suggest that the presence of a charged, hydrated, exopolymer matrix surrounding individual cells and microcolonies would influence profoundly the access of molecules and ions, including biocides and antibiotics, to the cell walls and membranes. It is therefore not surprising that many groups of workers have suggested that the glycocalyx acts as a protective umbrella that physically prevents the access of antimicrobials to the cell surface. Many of the early studies of antibiotic action on biofilms attributed their recalcitrance, purely and simply, to exclusion (Slack and Nichols, 1981, 1982; Costerton *et al.*, 1987; Suci *et al.*, 1994). Such universal explanations have been refuted (Gordon *et al.*, 1988; Nichols *et al.*, 1988, 1989), since reductions in the diffusion coefficients of antibiotics, such as tobramycin and cefsulodin, within biofilms or microcolonies, are insufficient to account for the observed changes in susceptibility. In this context, Gristina *et al.* (1987) found no difference in the susceptibility towards antibiotics of slime-producing and non-slime-producing strains of *Staphylococcus epidermidis*. Evans *et al.* (1991), in a similar study, assessed susceptibilities to the quinolone antibiotic ciprofloxacin, of mucoid and non-mucoid strains of *P. aeruginosa* grown as biofilms. In stark contrast to Gristina *et al.* (1987), however, these workers showed that, whilst the resuspended biofilm of each were equally susceptible, the biofilm form of the mucoid strains was substantially less susceptible than those of the wild-type. The results from collected studies such as these demonstrate that, whilst the possession of a mucoid phenotype may be associated with decreases in susceptibility, reductions in the diffusion coefficient across polymeric matrices relative to liquid media, are insufficient to account solely for it. This is because the equilibration of antibiotic from the bulk phase across the biofilm will be rapid (circa 1 min), and at equilibrium the concentrations at the cell surfaces and in the surrounding bulk aqueous phase would be equal. Such ready access of antimicrobial agent throughout the community is further aided by the presence of 'water channels'. These permeate not only the established laboratory monoculture biofilms (Costerton *et al.*, 1994), but also those associated with dental plaque (J.W. Costerton, personal communication).

Diffusion limitation studies have generally focused on antibiotics rather than biocides, and upon medically relevant biofilm populations rather than biofouling situations. The dimensions of biofilms *in vivo* are in the order of tens of micrometres, but for industrial biofilms they may be in the order of tens of centimetres thick (Nichols *et al.*, 1989). Whilst thickness of the matrix will not affect diffusion properties *per se*, it will however greatly increase the extent to which access of agent, to the underlying cells, is retarded. This retardation might possibly be to such an extent that, where treatments are periodic and short-lived, the deeper lying cells might escape lethal exposure (Nichols, 1993). Stewart (1996) has also reached similar conclusions, albeit using a theoretical model, by investigating antibiotic penetration into microbial biofilms.

On the basis of the data available in the literature, the extent of retardation of antibiotic diffusion does not appear to be sufficient to account for reduced biofilm susceptibility. If, however, the antimicrobial agents are strongly charged (i.e. tobramycin) or highly reactive (i.e. halogens/peroxygens), then they will be chemically quenched within the matrix during diffusion, either by adsorption to matrix-associated charged sites or by direct neutralization reactions.

3.3.2. Resistance to Chemical Agents Through Adsorptive Losses

Adsorptive losses of antimicrobial to the matrix might occur for highly charged drug molecules, such as the glycopeptides, bisbiguanides and quaternary ammonium compounds. Many of such agents would become irreversibly bound to the predominant anionic sites within the matrix (Hoyle and Costerton, 1991; Hoyle *et al.,* 1992), and would have to saturate each and every available binding site as diffusion through the matrix occurred. Thus, the matrix would act in an analogous fashion to an ion exchange column. Availability of cationic sites within many glycocalices would extend this phenomenon to anionic drugs. Curiously, macrolide antibiotics, which are positively charged, but also very hydrophobic, are relatively unaffected by the presence of exopolymers (Ichimiya *et al.,* 1994). Poor penetration through anionic matrices might therefore be a phenomenon restricted to the more hydrophilic, positively charged agents. Clearly, whether or not the exopolymeric matrix constitutes a physical barrier to antimicrobial penetration depends greatly upon the nature of the agent, the binding capacity of the polymeric matrix towards it, the levels of agent used therapeutically (Nichols, 1993), the distribution of biomass and local hydrodynamics (DeBeer *et al.,* 1994), together with the rate of turnover of the microcolony relative to antibiotic diffusion rate (Kumon *et al.,* 1994). For antibiotics such as tobramycin and cefsoludin, such effects are therefore likely to be minimal (Nichols *et al.,* 1988, 1989), but will be high for positively charged antibiotics, such as the aminoglycosides (Nichols *et al.,* 1988) and biocides, such as polymeric biguanides (Gilbert *et al.,* 2001). In all instances, diffusion would be slowed, but not halted. If the underlying cells were to be protected from the actions of an antimicrobial through adsorptive losses, then the number of adsorption sites must be sufficient to deplete the available drug within the bulk phase. This is an unlikely situation in soft tissue infections, but is possible in confined body compartments, such as the lung and brain and in industrial situations.

3.3.3. Resistance to Chemical Agents Through Reaction–Diffusion Limitation

In addition to a potential action as an adsorption column, matrix polymers and cellular materials at the periphery of a biofilm community may react chemically with, and neutralize, the treatment agents. This thereby further reduces their bioavailability. Such effects will be most pronounced with oxidative biocides, such as iodine and iodine–polyvinylpyrollidone complexes (Favero *et al.*, 1983), and for chlorine and peroxides (Huang *et al.*, 1995), which react nonspecifically in a consumptive manner with general exopolymers and cellular materials. Isothiazolone, and other thiol interactive agents, might equally be quenched by the presence of thiol-containing amino acids within the exopolymers. In all instances, the greater the activity of the antimicrobial agent (i.e. effective and therefore deployed at lower concentration) then the greater will be the susceptibility to such neutralization (Gilbert *et al.*, 2001). The resistance of *Klebsiella pneumoniae* and *P. aeruginosa* biofilms towards monochloramine treatment has been explained as depletion of the biocide within the interior of the biofilm through reaction–diffusion interactions at the periphery (Huang *et al.*, 1995). Huang *et al.* grew biofilms of the two organisms together on stainless steel surfaces using a continuous-flow annular reactor. Biofilms were treated with monochloramine (2 mg L^{-1}) for 2 h and stained using a fluorogenic redox indicator that could differentiate respiring from nonrespiring cells. Epifluorescent micrographs of frozen cross-sections, taken at regular time intervals, revealed gradients of respiratory activity within biofilms in response to monochloramine treatment. Cells near the biofilm–bulk fluid interface lost respiratory activity early in the treatment, whereas residual respiratory activity persisted near the substratum or in the centre of small, viable cell clusters, even after 2 h treatment. A further study, using both oxidizing and non-oxidizing biocides (Stewart *et al.*, 1998), used an artificial, biofilm construct of alginate-entrapped *Enterococcus faecalis* to demonstrate a similar lack of penetration, and action, against the entrapped cells by chlorine, glutaraldehyde, isothiazolone and quaternary ammonium biocides. In a real-life situation, however, the volume and reactive capacity of a biofilm would be insufficient to deplete the bulk availability of biocide, and interactive sites within the matrix polymers would become saturated with adsorbed/reacted biocide. The net effect once again would be to delay, rather than to prevent, the inhibitory process (Huang *et al.*, 1995). Provided that the exposure to biocide were brief, or that the biocide were not in a vast excess of requirements, then reaction–diffusion limitation could allow the survival of cells at the base of the biofilm which would flourish once the biocide were removed or depleted. If, however, the bulk phase concentration of antimicrobial were not reduced to sublethal levels by adsorptive or consumptive losses within the glycocalyx, then neither of these processes could account for the observation of long-term recalcitrance towards biocidal molecules.

3.3.4. Resistance to Chemical Agents Through Enzyme-Mediated Reaction–Diffusion Limitation

The reaction–diffusion limitation properties of the glycocalyx could be significantly enhanced if it contained extracellular enzymes that were capable of degrading the diffusing substrate. A catalytic (i.e. enzymatic) reaction could lead to severe antibiotic penetration failure (Stewart, 1996), provided that the turnover of substrate by the enzyme was sufficiently rapid. It has been noted that hydrolytic enzymes, such as β-lactamases, are induced/derepressed in adherent populations and in those exposed to sublethal concentrations of imipenem and/or piperacillin (Giwercman et al., 1991; Lambert et al., 1993). These enzymes become trapped and concentrated within the biofilm matrix and are able to further impede the penetration and action of susceptible antibiotics. Inactivation of formaldehyde by the enzymes formaldehyde lyase and formaldehyde dehydrogenase (Sondossi et al., 1985) has also been observed within biofilms of P. aeruginosa and is sufficient to confer some degree of resistance towards aldehyde biocides and preservatives. Similarly the presence within mixed community biofilms of bacteria such as P. aeruginosa, which are strong producers of the enzymes catalase and superoxide dismutase, will confer community protection against peroxide and superoxide treatment (Elkins et al., 1999; Hassett et al., 1999; Stewart et al., 2000).

4. PHENOTYPIC HETEROGENEITY WITHIN BIOFILM COMMUNITIES AND ITS EFFECT UPON ANTIMICROBIAL SUSCEPTIBILITY

Adherent microcolonies form functional consortia and influence their microenvironment through the localized concentration of enzymes and metabolic products, and the relative exclusion of gases such as oxygen (Costerton et al., 1987). A major consequence of growth within a dense community, regardless of the presence of an extracellular matrix, is that the rate of growth of individual cells will depend on their spatial location within the community (Brown et al., 1988). Cells on the periphery would have ready access to nutrients and oxygen, and would rapidly lose secondary metabolites. With positions progressively further into the community, nutrients would become scarce, there would be an increased tendency for the build-up of toxic waste products, and oxygen would become rate-limiting. Under such conditions many facultative anaerobes would switch to a fermentative mode of growth, and the active growth of more fastidious anaerobes would be encouraged. Accordingly one can postulate that mature biofilm is comprised of a plethora of spatially arranged phenotypes and that each will possess slightly different susceptibilities towards a diverse range of antimicrobial treatments. Brown et al. (1988) pointed out that in many of the pioneer studies of biofilm resistance to antibiotics the susceptibility of intact biofilms had been compared with that of fast-growing planktonic populations. They pointed out that differences in susceptibility that were being attributed to the biofilm mode of growth could be replicated either in continuous culture comparing fast and slow growth rates, or in batch culture with log phase and stationary phase cells.

5. EFFECTS OF GROWTH RATE AND NUTRIENT DISPOSITION UPON ANTIMICROBIAL SUSCEPTIBILITY

It is well established that the susceptibilities of bacterial cells towards antibiotics and biocides are profoundly affected by their nutrient status and growth rate, as well as by temperature, pH and prior exposure to subeffective concentrations of antimicrobial agents (Brown and Williams, 1985; Williams, 1988; Brown et al., 1990; Gilbert et al., 1990). Such effects relate not only to growth rate-dependent changes in a variety of cellular components that include membrane fatty acids, phospholipids and envelope proteins (Gilbert and Brown 1978a, b, 1980; Klemperer et al., 1979; Al-Hiti and Gilbert, 1980; Wright and Gilbert, 1987a, b, c), but also to the production of extracellular enzymes (Evans et al., 1990a, b; Giwercman et al., 1991) and polysaccharides (Govan and Fyfe, 1978).

At any particular time within a biofilm community, a plethora of phenotypes is represented that reflects the chemical heterogeneity of the biofilm environment and the imposition of chemical and gaseous gradients. Whilst this has long been considered to be the case for biofilm communities (Brown et al., 1988, 1990), such gradients and spatial mosaics have only recently been visualized. Wentland et al. (1996) used acridine orange staining of K. pneumoniae biofilms to show that the regions of fastest growth occurred in the outer 30 microns of the biofilm, closest to the bulk liquid. Dispersed regions of slow growth, tending towards the substratum, were also noted. Similar observations of gradients of growth rate were also reported recently by Sternberg et al. (1999). Kinniment and Wimpenny (1992) observed directly that the distribution of adenylate and of adenylate charge was unequal across a model Pseudomonas aeruginosa biofilm grown in the constant depth biofilm fermenter. Huang et al. (1998) and Xu et al. (1998) noted that patterns and rates of protein biosynthesis, and of phosphatase activity in response to phosphate starvation were asymmetrically distributed across biofilm communities. A major contributor towards the observed resistance of biofilm is therefore associated with physiological gradients of growth rate and nutritional status. It should be noted that even within monoculture biofilms, grown in laboratory systems, the established physiological gradients are non-uniform and take on the appearance of a mosaic, with pockets of very slow-growing cells juxtaposed with relatively fast growing ones (Xu et al., 2000).

A distinction can be made between those effects related to the nature of the least available nutrient (nutrient limitation/depletion) and the cellular growth rate. Within the depths of a biofilm, growth rates will be generally suppressed relative to planktonic cells growing in the same environment. Ashby *et al.* (1994) used biofilm:planktonic ratios of isoeffective concentration (growth inhibition and bactericidal activity), determined for a wide range of antibiotics against cells grown either in broth or on urinary catheter discs, to indicate the extent of biofilm resistance. They noted that such ratios followed closely those generated between non-growing and actively growing cultures. With the exception of ciprofloxacin, antibiotic agents that were most effective against non-growing cultures (i.e. imipenem, meropenem) were also the most active against these biofilms. Other workers have used perfused biofilm fermenters (Gilbert *et al.*, 1989; Hodgson *et al.*, 1995) to directly control and study the effects of growth rate within biofilms. Using control populations that comprised planktonic cells grown in a chemostat, the separate contributions of growth rate, and association within a biofilm were evaluated. Decreased susceptibility of *S. epidermidis* to tobramycin (Duguid *et al.*, 1992a) and of *Escherichia coli* to tobramycin (Evans *et al.*, 1990b) and cetrimide (Evans *et al.*, 1990a) could be explained largely in terms of growth rate. Cells resuspended from growth rate controlled biofilms and planktonic cells grown at the same growth rate possessed virtually identical susceptibilities to these agents. When intact biofilms were treated however, susceptibility was decreased somewhat from that of planktonic and resuspended biofilm cells, indicating the potential benefit to the cells of organization within an exopolymeric matrix.

6. WHY CHEMICAL AND PHYSIOLOGICAL GRADIENTS PROVIDE INCOMPLETE EXPLANATIONS OF BIOFILM RESISTANCE

Stewart (1994) developed a mathematical model, which incorporated the concepts of metabolism-driven oxygen gradients and growth rate-dependent killing, together with reaction–diffusion limitation, to examine the susceptibility of *S. epidermidis* biofilms to various antibiotics. The model accurately predicted that susceptibility would be reduced in thicker biofilms due to oxygen limitation. Oxygen gradients within the biofilm may also directly influence the activity of some antibacterials (Shepherd *et al.*, 1988; Zabinski *et al.*, 1995). Since nutrient and gaseous gradients will increase in extent as biofilms thicken and mature, then growth-rate effects upon susceptibility, such as these, will become particularly marked in aged biofilms (Anwar *et al.*, 1989, 1990). Such changes probably contribute to reports that aged biofilms are more recalcitrant to antibiotic and biocide treatment than are younger ones (Anwar *et al.*, 1989).

Whilst the contribution of reduced growth rate within biofilm communities cannot be denied, as with diffusion limitation it cannot be the sole explanation of resistance (Xu *et al.*, 2000). Physiological gradients across biofilms depend on the growth and metabolism of cells at the periphery consuming nutrients before they permeate to the more deeply placed cells. The peripheral cells will have growth rates and nutrient profiles that are not very dissimilar to those of planktonic cells. They will therefore be relatively sensitive to the treatments imposed and will quickly cease to metabolize. The lysis products of the killed cells would in turn feed the cells within the depths of the biofilm, which as a consequence would step up their metabolism and growth rate, adopt a more susceptible phenotype and die (McBain *et al.*, 2000b). Indeed the growth rate of cells within this layer might even exceed that of untreated peripheral cells through a mobilization of nutrients from the dead biomass. This phenomenon would occur throughout the biofilm, proceeding inwards from the outside, until the biofilm was completely killed. Should the supply of antimicrobial agent cease, then the biofilm could re-establish almost as fast as it was destroyed because of the relative abundance of nutrients. Whilst such growth rate-related processes might delay the onset of killing in the recesses of the film, they could not confer resistance against a sustained exposure to antimicrobial agents. While reaction–diffusion limitation of the access of agent and the existence of physiological gradients within biofilms provide explanations for their reduced susceptibility, neither explanation, separately or together, explains the long-term tolerance displayed towards a variety of chemically and physiologically unrelated antimicrobial agents. For long-term tolerance towards antimicrobial agents to occur, the biofilm population must either contain cells which express a physiology that has not yet been taken into account, or the surviving portion of the population must adapt to a resistant phenotype during the 'time-window' of opportunity provided by the buffering effect (i.e. a rapid response to sublethal treatment).

7. DRUG-RESISTANT PHYSIOLOGIES

Since neither of the general hypotheses concerning the resistance of biofilms towards such diversely acting agents can account for its sustained nature during prolonged treatments, or in the presence of a vast excess of treatment agent, alternative explanations must be sought. Since the biofilm victory over antibiotics is Phyrric, then long-term survival might be related to the presence of a small fraction of the population that expresses a highly recalcitrant physiology. Candidate physiologies include dormant 'quiescent' cells, expression of efflux pumps, and suicide-less mutants.

7.1. Quiescence

There has been much speculation over the last 10 years over the ability of non-sporulating forms of bacteria to adopt spore-like status through the adoption of a quiescent state. In this state, specific growth rates of the cells approximates to zero (Moyer and Morita, 1989), as they undergo a reductive division in order to complete the segregation of initiated rounds of chromosome replication (Novitsky and Morita, 1976, 1977; Moyer and Morita, 1989). Such cells have generally been associated with marine biofilms where there is poor availability of nutrient (Kjelleberg *et al.*, 1982) and have been suggested to be the dominant form of bacteria within natural environments of low nutrient availability (Morita, 1986). Whilst mainly associated with aquatic Gram-negative bacteria, such quiescence has recently been reported in Gram-positives (Lleo *et al.*, 1998), and would appear to be a universal response to extreme nutrient stress (Matin *et al.*, 1989). This has recently been termed the general stress response (GSR). GSR leads to populations of cells that, by virtue of their much reduced metabolism, and the synthesis of highly phosphorylated nucleotides ppApp and pppApp (Rhaese *et al.*, 1975; Piggot and Coote, 1976), are resistant to a wide range of physical and chemical agents (Matin *et al.*, 1989; Hengge-Aronis, 1996). The GSR is now thought to account for much of the resistance observed in stationary phase cultures and is induced under conditions of extreme starvation. Various terms have been adopted to describe such phenotypes, including quiescent (Trainor *et al.*, 1999), dormant (Amy *et al.*, 1983; Lim *et al.*, 1999), resting (Munro *et al.*, 1989) ultra-microbacteria (Novitsky and Morita, 1976) and somnicells (Roszak and Colwell, 1987a). It is highly probable that the same phenomena describe the state of viable but non-culturable (VNC) (Barer and Harwood, 1999), since such bacterial cells often fail to produce colonies when plated directly on to nutrient-rich media. Indeed, when bacteria are collected and plated from natural oligotrophic environments, and from infections, then there is often a great disparity between the viable and total cell counts obtained (Roszak and Colwell, 1987a, b; Defives *et al.*, 1999). Even within biofilms that have an abundant supply of nutrient, regions exist where nutrients are scarce and even absent. Under such circumstances, a small proportion of the cells present within a mature biofilm will be expressing the GSR regulator and will be relatively recalcitrant to inimical treatments. Similar mechanisms have been proposed for the hostile take-over of batch cultures by killer phenotypes during the stationary phase (Zambrano and Kolter, 1995), induced as part of the GSR in *Escherichia coli*. This can lead to a phenotype that is not only more competitive in its growth than non-stressed cells, but which can also directly bring about the death of non-stressed ones (Zambrano *et al.*, 1993). Whilst such phenomena have been reported in variously grown batch cultures, adoption of a killer phenotype by either the biofilm or émigrés from a biofilm would facilitate colonization resistance or invasiveness, respectively. In this respect, the general stress response can be associated with the separate regulation of at least 30 distinct proteins (Zambrano and Kolter, 1995) some of which might be assigned as binding-receptors/cell-bound bacteriocins.

The *rpoS*-encoded sigma factor σ^s is a master regulator in a complex network of stationary phase responsive genes in *E. coli* (Hengge-Arronis, 1996) whereas in *P. aeruginosa* it appears that at least two sigma factors RpoS and AlgU, and also the density-dependent cell–cell signalling systems orchestrate such responses (Foley *et al.*, 1999). Indeed there is a hierarchical link between *n*-acyl homoserine lactones and rpoS expression (Latifi *et al.*, 1996) which might specifically induce the quiescent state at locations within a biofilm where signals accumulate and where nutrients are most scarce. It was elegantly demonstrated by Foley *et al.* (1999) that the GSR response regulator rpoS was highly expressed in all 19 *P. aeruginosa*-infected sputum samples taken from cystic fibrosis patients.

7.2. Efflux Pumps

An increasingly observed resistance mechanism is the expression and over-production of multidrug efflux pumps (Nikaido, 1996). Expression of such pumps is induced, in Gram-negative and Gram-positive bacteria, through sublethal exposure to a plethora of agents (George and Levy, 1983; Ma *et al.*, 1993). These include not only small hydrophilic antibiotics, but also other xenobiotics such as pine oil and salicylate (Miller and Sulavick, 1996; Moken *et al.*, 1997). Efflux pumps are operational in a wide variety of Gram-negative organisms, and may be plasmid or chromosomally encoded (Nikaido, 1996). Multidrug efflux pumps encoded by *qacA-G* also contribute to biocide tolerance in *Staphylococcus aureus* (Rouch *et al.*, 1990). Sublethal exposure to many antimicrobials can select for cells that hyperexpress an alternative efflux pump *mex*CD, in *mex*AB-deleted mutants (Chuanchuen *et al.*, 2000, 2001). This indicates the multiplicity of efflux genes and their highly conserved nature.

Membrane-associated efflux pumps are a common feature of both prokaryotic and eukaryotic cells. Several attempts have been made to group them into families (Griffith *et al.*, 1992) and to predict structure and function of the proteins themselves (Saier, 1994; Johnson and Church, 1999). In bacteria, efflux pumps belonging to four superfamilies have been recognized (Saier and Paulsen, 2001). The ATP binding cassette (ABC) family transporters utilize ATP hydrolysis to power drug efflux, whilst the major facilitator superfamily (MFS), resistance nodulation division (RND), drug metabolite transporter (DMT) and multi-antimicrobial extrusion (MATE) families use H^+ or Na^+ antiport mechanisms. Although the families share no significant sequence identity, substrate specificity is often shared between them (Paulsen *et al.*, 1996).

All the efflux superfamilies contain, to a greater or lesser extent, pumps specific for a single agent and broader specificity multidrug efflux pumps, capable of removing a wide range of structurally unrelated antibiotics and disinfectants. Any type of efflux pump may be involved primarily with the extrusion of endogenous metabolites or, alternatively, may be involved primarily with the efflux of chemotherapeutic agents. Indeed, it is probable that these exporters were originally developed to extrude endogenous metabolites, but that a coincident ability to exclude harmful substances has proven to be a desirable survival strategy which has been selected for, and incorporated, by almost every known genus and species of bacterium. In other cases, expression of efflux pumps is necessary for the intrinsic resistance profile of an organism. For example, *E. coli* is an autochthonous inhabitant of the human intestine, an environment rich in bile salts. Deletion of the *acrB* efflux pump gene renders it incapable of growth in medium containing bile salts, thus destroying a facility considered characteristic of the organism (Nikaido and Zgurskaya, 1999).

The nature of the cell envelope of Gram-negative bacteria has allowed certain transporters, particularly RND pumps, to function in harmony with accessory proteins. In these bacteria, pump proteins can often increase their efficacy by working in parallel with periplasmic and outer-membrane factors to form efflux pump complexes.

Studies with planktonic cultures have shown that expression of AcrAB is enhanced under conditions of general stress, such as exposure to ethanol or simply entry into stationary phase (Ma *et al.*, 1995). This is interesting as the conditions found within a biofilm more closely parallel the stationary phase than any other mode of planktonic growth and raise the question of whether efflux is, in general, up-regulated in biofilms.

Notable amongst the multidrug resistance operons are *mar* and efflux pumps such as *acrAB* (George and Levy, 1983; Ma *et al.*, 1993). The *mar* locus of *E. coli* was the first mechanism found to be involved in the chromosomally encoded, intrinsic resistance of Gram-negative bacteria to multiple drugs. Homologues have since been described in many Gram-negative bacteria. Moken *et al.* (1997) and McMurry *et al.* (1998b) have shown that mutations causing overexpression of *mar*A or *acr*AB are associated with exposure and reduced susceptibility towards a wide range of chemicals and antibiotics. The importance of *mar* and efflux systems generally would be far greater, however, if induced by growth as a biofilm *per se*. In such instances, a generalized efflux of toxic agents would provide explanation of the ubiquitous observation of resistance, regardless of the treatment agent, and would be conferred upon the cells prior to exposure.

Ciprofloxacin exposure does not induce the expression of *mar* or *acrAB* in *E. coli*, but such expression will confer limited protection against this agent.

Exposure to ciprofloxacin of biofilms comprised of wild-type, constitutive and *mar*-deleted strains ought to evaluate whether or not such genes were up-regulated in unexposed biofilm communities. Maira-Litran *et al.* (2000a) perfused biofilms of such *E. coli* strains for 48 h with various concentrations of ciprofloxacin. These experiments, whilst demonstrating reduced suscepti-bility in the *mar* constitutive strain showed little or no difference between wild-type and *mar*-deleted strains (Maira-Litran *et al.*, 2000a). Similar exper-iments using biofilms constructed from strains in which the efflux pump *acrAB* was either deleted or constitutively expressed (Maira-Litran, 1998; Maira-Litran *et al.*, 2000b) showed the *acrAB* deletion did not significantly affect susceptibility over that of the wild-type strain. Clearly neither *mar* nor *acrAB* is induced by sublethal treatment of biofilms, other than with inducer sub-stances. On the other hand, constitutive expression of *acrAB* protected the biofilm against low concentrations of ciprofloxacin. Studies conducted in con-tinuous culture with a *lacZ* reporter gene fused to $marO_{II}$ showed *mar* expression to be inversely related to specific growth rate (Maira-Litran *et al.*, 2000b). Hence, following exposure of biofilms to sublethal levels of β-lactams, tetracyclines and salicylates, *mar* expression will be greatest within the depths of the biofilm, where growth rates are suppressed, and might account for the long-term survival of the community when exposed to inducer molecules. Another recent study of efflux in biofilms showed that expression of the major RND multidrug efflux pumps of *P. aeruginosa* actually decreased as the biofilm developed. In addition, although expression was greatest in the depths of the biofilm, experiments with deletion mutants showed that none of the mul-tidrug efflux pumps were contributing to the general, increased resistance to antibiotics exhibited by the biofilm (DeKievet *et al.*, 2001).

7.3. Suicide-less Mutants

It is becoming increasingly recognized that many species of bacteria are capa-ble of undergoing programmed cell death (Jensen and Gerdes, 1995; Naito *et al.*, 1995; Yarmolinsky, 1995; Franch and Gerdes, 1996; Boutibonnes, 1997; Hochman, 1997; Cellini *et al.*, 1998; Engelberg-Kulka and Glaser, 1999; Lewis, 2000) and related autolytic processes. Programmed cell death in bacteria cannot be viewed as advantageous if these organisms exist primarily in a planktonic mode. Indeed, such suicide is not beneficial to the individual cell, but is often a highly evolved mechanism displayed within tissues. Since biofilm populations are arguably functional tissues (Costerton *et al.*, 1994), and possibly proliferat-ing entities (Caldwell and Costerton, 1996), then it is highly probable that programmed cell death relates to the biofilm mode of growth.

A recent and novel hypothesis for the considerable recalcitrance of biofilm relates to the potential of damaged bacterial cells to undergo apoptosis or programmed cell death. Lewis (2000, 2001) suggested that death of cells following treatments with bactericidal agents results not from direct action of the agent, but from a programmed suicide mechanism and cellular lysis (Moyed and Bertrand, 1983a, b; Black et al., 1991). If this is the case, cells subjected to different treatment agents with different mechanisms of action may well die from a common process. A singular mechanism of death allows us to speculate that singular mechanisms exist for the resistance towards and survival from inimical treatments.

If an entire population of cells underwent programmed cell death simultaneously, as the result of a sublethal exposure of an antimicrobial agent, then little benefit would be derived. It is imperative that a small proportion of the population be able to avoid such a response and ultimately be responsible for the survival and recovery of the community. It is equally important that this trait of selfishness is not retained in the resultant clones.

The biocide literature of the last 50 years has been punctuated with reports of low-level, persistent survival of antimicrobial treatments (tailing), where the agent has not been quenched and where the survivors do not demonstrate resistance when recultured or cloned (Bigger, 1944). Whilst this might relate to a subpopulation of cells that are quiescent (see above), recent evidence suggests that such cells, rather than being resistant to the agent (Koch, 1987) might be defective in programmed cell death (Brooun et al., 2000). Following removal of an inimical stress, these damaged persisters would grow rapidly in the presence of nutrients released from their lysed community partners and the community would become restored. It is also postulated that biofilm populations are enriched in 'persister' cells, possibly as a biofilm-specific phenotype. Due to their protection from immune responses or predation within the biofilm, these cells would survive treatment phases and proliferate in the post-treatment phase (Lewis, 2000, 2001). This would engender considerable recalcitrance within the biofilm community.

8. SELECTION/INDUCTION OF RESISTANT PHYSIOLOGIES THROUGH SUBEFFECTIVE TREATMENTS

The basic tenet of Darwinian theories of evolution is that all populations are genetically diverse and that continued exposure to any environmental stress will lead to an expansion of the most suited genotype/phenotype (survival of the fittest). This is particularly the case for mixed community biofilms, but applies equally to planktonic cultures of bacteria. Exposure to sublethal concentrations of biocides and antibiotics will therefore enrich in the population the least susceptible clones. Exposure to such conditions may also directly cause the induction of less susceptible phenotypes.

8.1. Selection of Less Susceptible Clones

It has long been demonstrated that pure cultures of bacteria can be 'trained' to become more tolerant of antibiotics (Brown *et al.*, 1969) and biocides (Brozel and Cloete, 1993; McMurry *et al.*, 1998a; Maclehose *et al.*, 2001). In such experiments cultures are grown either in liquid medium which contains concentrations of agent that are below the MIC, or they are streaked on to gradient plates that incorporate the agent. At each step in the process, the MIC is redetermined and the process repeated. In such a fashion, it is relatively easy to select for populations of bacteria that have significantly reduced MIC values towards the selected agents. In some instances, the change in MIC is sufficient to render the cells resistant to normal treatment regimes. Where groups of agents have common biochemical targets, it is possible for selection by one agent to confer cross-resistance to a third-party agent (Chuanchuen *et al.*, 2001). Such 'resistance training' has for many years been regarded as artefactual, since it is difficult to imagine a set of circumstances in natural environments where bacteria will be exposed to gradually increasing concentrations of an inhibitory agent over a prolonged period. Repeated, sublethal treatment of biofilms, in the environment and in infection, however, provide one situation where this might happen (McBain *et al.*, 2000b). As with any process involving changes in susceptibility to inimical agents the nature of the genotype/phenotype selected reflects changes in the biocidal/inhibitory targets, the adoption of alternative physiologies that circumvent the target and changes in drug access to those targets. The last might be through modifications in the cell envelope (Brown *et al.*, 1990; Gilbert *et al.*, 1990) or it might reflect active efflux mechanisms (Levy, 1992). Generally there is a fitness cost associated with such adaptation, but this appears to decrease with continued exposure to the stress (Levin *et al.*, 2000). It has recently been shown that sublethal exposure of Gram-negative bacteria to the commonly deployed antibacterial agent triclosan selects for cells that are mutated in the *fabI* gene (McMurry *et al.*, 1998a). This encodes the enoyl reductase associated with fatty acid biosynthesis (Heath and Rock, 1995). Similarly exposure of pseudomonads to sublethal concentrations of isothiazolones causes the repression of an outer membrane protein T, thought to facilitate uptake of this biocide in normal cells (Brozel and Cloete, 1994).

Mutations that increase the expression of multidrug efflux pumps result in elevated levels of resistance to a wide range of agents. Thus, mutations in the *mar* operon increase the expression of the *acrAB* efflux pump in *E. coli* (McMurry *et al.*, 1998b) and mutations in the mexAB operon of *P. aeruginosa* leads to significant overexpression (Rella and Haas, 1982). It must be borne in mind that the primary function of energetic efflux is to defend the cell against naturally occurring environmental toxicants (Miller and Sulavick, 1996). Efflux is often nonspecific and equivalent to an emetic 'vomit' response. Cells that efflux permanently will be poor competitors in heterogeneous communities and will not prosper in the absence of the selection stress. Treatment with antimicrobials that act as substrates, but are not themselves inducers (Maira-Litran et al., 2000a, b; Chuanchuen et al., 2001), might lead to a clonal expansion of mutant cells that are constitutive in efflux pump expression. Treatment with agents that are both strong inducers and also substrates (Thanassi et al., 1995; Moken et al., 1997; Sundheim et al., 1998) will confer no selective advantage upon the efflux mutants, but it must be borne in mind that induction of efflux by one agent will confer a broad spectrum of resistance.

8.2. Alarmones

Alarmones have recently been subject to a number of excellent reviews (Rowbury 2001a, b, c). Intracellular alarmones (Bochner et al., 1984) enable cells to respond to the intracellular accumulation of chemicals to toxic levels. These are well documented and include the sensing by SoxR and OxyR of superoxide and peroxides and superoxide (Kullik et al., 1995). Even if these inducers were released from those cells that had been targeted and killed by chemical agents, they are unlikely to modify cells within the vicinity of the damage since they only poorly penetrate healthy cell walls and membranes. There is, however, a growing body of literature that shows bacteria to produce constitutively extracellular sensing components that can be converted (activated) into extracellular inducers by certain types of stress. Unlike the intracellular alarmones, these small, readily diffusible molecules can easily reach other unstressed bacteria and induce the expression of tolerance before the recipients of the signal become exposed (Rowbury, 2001b). Such inducible expression of tolerance has been related to the exposure of bacteria to a variety of physical and chemical agents (Rowbury, 2000a) and manifests itself in reduced susceptibility towards acids, alkali, alkylating agents, electrophiles, oxidizing agents and heat (Samson and Cairns, 1977; Demple and Halbrook, 1983; Mackey and Derrick, 1986). To our knowledge there have been no publications relating extracellular alarmones to the resistance of biofilm communities. It is, however, tempting to postulate that such extracellular sensors are retained within the biofilm matrix and act as an early warning system of the arrival of toxic environmental agents. Cells deep within the biofilm would sense their impending doom during chemical treatments and adopt a resistant phenotype. Such phenotypes might include the GSR, adoption of a suicide-less behaviour or the expression of efflux pumps. It is also tempting to suggest however, that rather than induce resistance during an inimical treatment, alarmones actually awaken quiescent cells post-treatment.

8.3. Efflux Pump Induction by Sublethal Exposure to Antibiotics and Biocides

Neither reaction–diffusion limitation nor the existence of physiological gradients and mosaics can account for the extreme tolerance towards antibiotics and biocides observed in the field. Each of these can delay the onset of a lethal effect but cannot, unless turnover of the matrix polymers and cells is very rapid, account for survival during chronic long-term treatments. One effect of a delayed action with respect to the underlying cells within a biofilm is that they will be exposed, for an extended period, to the presence of subinhibitory levels of treatment agent. These might well act as inducers/transcriptional activators of a more tolerant phenotype.

It is interesting to note in this light that the expression of multidrug resistance operons, such as *mar* and efflux pumps such as *acrAB*, have been shown to be up-regulated by exposure to subeffective concentrations of antibiotics such as tetracycline and chloramphenicol (George and Levy, 1983; Ma *et al.*, 1993), and xenobiotics such as salicylate. The *mar* locus is chromosomal, variously induced and represented within a wide range of Gram-negative bacteria. Induction of operons such as *mar*, during the delayed onset of the action of inducer-antibiotics directed at biofilms is a tempting explanation of the biofilms' long-term resistance. Similar systems, under the regulation of different inducer agents might extend this explanation of biofilm tolerance to include other treatment agents. It is noteworthy that resistance operons such as *mar* have also been reported to be up-regulated in response to exposures to many commonly occurring, natural inhibitory substances such as garlic, chilli peppers, herbs and spices (Whyte *et al.*, 2001). Possible selection pressures towards efflux mutants are not therefore restricted to man-made chemicals.

9. COMMUNITY RESPONSES TO CHRONIC SUBLETHAL CHEMICAL STRESS

Once biofilm communities have reached a certain level of maturity, and provided all significant extrinsic factors remain constant, then the population will generally achieve a dynamic steady state (McBain *et al.*, 2000a). Although such 'climax communities' are subject to a gradual turnover of indigenous species, they achieve a degree of temporal and spatial stability. In nature such conditions are rare, since growth and division will inevitably lead to increased genetic diversity, and even a constant selection pressure will always favour the most competitive morphotype arising from an inherent phenotypic and genotypic plasticity in living systems. Even in the most controlled of laboratory experiments, such as those involving chemostats inoculated with single clones, then under steady-state conditions the cultures are subject to a natural genetic drift towards a more competitive phenotype/genotype (Atwood et al., 1951; Dykhuizen and Hartl, 1983).

Nethertheless, one consequence of their development under intense selection pressures is that mature biofilm communities are remarkably colonization-resistant (Alexander, 1971; Marsh and Bradshaw, 1999). Such resistance of microbial communities has been illustrated using gnotobiotic animals, which lack a protective natural, commensal microflora and are therefore considerably more susceptible to infection by allochthonous bacteria (Boureau et al., 1994).

Biofilm communities will respond to sustained antimicrobial exposure through adaptation of their phenotypic and genotypic profile. This will probably include the expression of multidrug efflux pumps, together with a clonal expansion of less susceptible individuals or species within the community. Treatment might well result in the death and loss from the community, of particular species and groups of organism. Since they had been a part of a colonization-resistant climax community, their function within it will have become lost, and the colonization resistance will be temporarily broken. This will provide ecological niches that may become filled by fresh immigration into the community. Equally, the less susceptible clones of species that initially survive the treatment might be less competitive in their community function than potential immigrants that are inherently tolerant of the agents. In this fashion, sublethal antimicrobial treatments will destabilize the community. Continued exposure will lead to the establishment of a new climax community that reflects the clonal expansion of pre-existing, less susceptible phenotypes and the displacement of some community members by émigrés.

A key aspect of biofilm community dynamics relates to the immigration of planktonic cells. Since a majority of the species represented within a climax community are unlikely to have been present within the primary colonizing group, new species must be sequestered into the biofilms to enable community structures to evolve (McBain et al., 2000a). Although selection pressures will apply equally to the residual biofilm community and to the immigrant cells, it is likely that a successful immigration would be more probable immediately after a cull event, such as would occur during antimicrobial treatment. This is shown by the increased susceptibility towards infection in animals when their intestinal microflora have become degraded by antibiotics or radiotherapy (Brook and Ledney, 1993; Larson and Welch, 1993). Failure of colonization resistance may be used as an index of microbial community perturbation. Perhaps the most notable example of such effects is Clostridium difficile-associated diarrhoea (CDAD). This bacterium is the cause of approximately 25% of all cases of antibiotic-associated diarrhoea (Bartlett, 1992). The risk for CDAD after clindamycin therapy for hospitalized individuals has been estimated to range from one in 10 to one in 10,000 (Lusk et al., 1977).

The initial step in development of CDAD is disruption of the normal flora of the colon, usually caused by antibiotics frequently clindamycin, but also ampicillin, amoxicillin and the cephalosporins (Johnson *et al.*, 1999). Antineoplastic or immunosuppressive drugs have been less frequently associated with the condition (Anand and Glatt, 1993). Colonization usually occurs by the faecal–oral route, where ingested spores of *C. difficile* are able to survive the gastric acid barrier and germinate in the colon (Kelly *et al.*, 1994; Kelly and LaMont, 1998). Symptoms of CDAD may occur at any stage during antibiotic therapy, or up to 6 weeks or longer after cessation of treatment (Anand *et al.*, 1994).

The assembly of altered community profiles and ultimately distinct climax communities during inimical conditions has been elegantly demonstrated by Watanabe *et al.* (1998), using activated sludge exposed to phenol. In this study, culture-dependent and molecular ecological techniques were employed to characterize activated sludge communities obtained from a municipal sewage treatment plant before and during a 20-day *in vitro* exposure to phenol. Following addition of phenol, community profiles were markedly altered, as shown by PCR/temperature gradient gel electrophoresis and reached quasi-steady states after day 10. Three major bands, corresponding to the PCR products of the dominant phenol-degrading populations, were visualized. Sequence analysis showed that specialized phenol-degrading communities, dominated by novel proteobacteria (γ and β subclass) became established, but that a species distantly related to *Psychroserpens burtonensis* was probably the most functionally active community member.

Caldwell's proliferation hypothesis (Caldwell and Costerton, 1996; Caldwell *et al.*, 1997a, b) is particularly pertinent when considering the endurance of a biofilm community following chemical degradation. On such occasions, the community itself is a dynamic entity which may persist, albeit with an altered composition, despite the extinction of component lineages. Where one or more clones become uncompetitive, due to altered selection pressures (in this case, chemical stresses), they may either be replaced by the immigration of new species or replaced by the clonal expansion of an existing community member. In either case, where the chemical treatment is subeffective, the community as an entity will persist.

10. OVERVIEW

In moist environments, micro-organisms are able to survive and grow, as sessile biofilm communities, in close association with surfaces. Whilst the unique physiological properties of microbial biofilms are essential to many aspects of life and commerce, they are generally regarded as problematic. This is particularly the case in relation to periodontal disease, dental caries, infections of indwelling medical devices, industrial biofouling, biocorrosion and in public health, where biofilms often form the *nidus* of infection. As a consequence, much research has been directed towards prevention and control of biofilms. Most antibacterial agents have been developed and optimized for their activity against fast-growing, planktonic populations of individual species. It is therefore not surprising that these agents have failed to control biofilm communities, which are reportedly up to 1000 times less susceptible than their free-living counterparts. Since many of the resistance characteristics associated with biofilms are lost when the communities are physically disrupted, descriptions of resistance mechanisms frequently relate the localized high concentration of extracellular enzymes and cells within an extracellular polyanionic polymer matrix (slime) to the imposition of a reaction–diffusion barrier towards antimicrobials. Such barriers impede the penetration of chemically reactive and oxidizing biocides, cationic agents and also those agents that can be degraded enzymically. The duration of the protection afforded to the community must depend upon the reaction capacity of the cells and their surrounding matrices, the rate at which such materials are replaced, and the spatial arrangement of individual species. Less attention has been given to the effects that spatial heterogeneity of individual species and phenotypes within the biofilm matrix might have upon the response to antibacterial treatments than have been given to the matrix itself. At its most simplistic level, spatial location relates to starvation of the more deeply lying cells, exclusion of oxygen, and the accumulation and retention of secondary metabolites. Accordingly, deep-lying cells within biofilms will not only be partially occluded with respect to the treatment agents, but will also express phenotypes that are substantially different from those of the same cells growing in suspension. Where these reductions in susceptibility are associated with reductions in growth rate, and the expression of starvation or shock responses, multidrug efflux pumps (i.e. *acrAB*, *mex*) or further increases in exopolymer synthesis, then susceptibility will be markedly reduced, even after the community has been physically disrupted.

Further to these physiological responses to growth and starvation within dense bacterial communities, it is postulated that the attachment of microorganisms to surfaces per se causes the expression of biofilm-specific phenotypes that are intrinsically less susceptible to conventional treatment agents than planktonic cells. The spatial organization of individual species within the community will ultimately affect the pharmacodynamics of the community response.

A major problem associated with these accepted mechanisms of biofilm recalcitrance is that they account for only a delay in the lethality of chemical treatments. Recently, a number of concepts have been introduced that might profoundly affect our understanding of the resistance of biofilm communities.

The first recognizes that materials lost from damaged cells may act as signals or 'alarmones' that induce the adoption of a less susceptible phenotype in the vicinity of the inimical stress. The second recognizes that it is generally only a small group of 'persister cells' that survive inimical treatments, and that this subset of cells is present in all microbial populations, but has remained unrecognized outside the biofilm context. It has been proposed that the 'persister phenotype' represents cells that are unable to undergo apoptosis before the levels of chemically induced damage achieve lethality in their own right. Alternatively, persisters might represent 'quiescent' cells that are normally dormant, but which become awakened post-treatment. If any of these mechanisms were globally operational, they would represent a common resistance mechanism that may be of particular advantage within biofilm communities.

In the present article, we have attempted to present a holistic view of biofilm physiology and resistance towards inimical treatments. Since susceptibility of microbial cells towards chemical antimicrobials alters from the moment that they become associated with one another and/or surfaces, through to the formation and maturation of the biofilm community, resistance has been viewed in the context of biofilm formation, maturation and population dynamics.

11. CONCLUSIONS

Resistance of microbial biofilms to a wide variety of antimicrobial agents is clearly associated with the organization of cells within an extensive exopolymer matrix. Such organization is able to moderate the concentrations of antimicrobial agents and antibiotics to which the more deeply lying members of the biofilm community are exposed. Such cells are coincidentally slow-growing, starved and express stressed phenotypes that may include the up-regulation of efflux pumps. The expressed phenotype of the deeply seated biofilm community reduces their susceptibility to the treatment agents and exacerbates the likelihood of their being exposed sublethally. The deeper-lying cells will out-survive those at the surface and multiply if the bulk of the treatment agent is depleted or the exposure transient. At the fringes of action, selection pressures will enrich the populations with the least susceptible genotype. It is possible under such circumstances for repeated chronic exposure to sublethal treatments to select for a more resistant population.

Alternative explanations of the resistance of biofilm communities lies with their expression of biofilm-specific phenotypes, which are so different to those of planktonic cells, that the agents developed against the latter fail to operate. Whilst such phenotypes are known to be expressed and might be regulated through quorum-sensing mechanisms, they do not appear to contribute greatly to the susceptibility pattern of individual biofilm cells. Such processes do, however, offer the possibility of novel agents that might prevent the formation of dense, polymer-encased communities and thereby circumvent the problem of killing intact, mature biofilms. More probably, biofilms provide an environment that, unlike traditional laboratory cultures, causes the expression of a general stress response.

Current strategies for the control of microbial biofilms involve the design of antimicrobial agents that are specifically targeted towards the biofilm-specific phenotype or which chemically degrade the glycocalyx. These include molecules with high diffusion–reaction ratios and agents targeted at slow or non-growing cells. Such approaches have, to date, met with only limited success and the need to develop efficient, low-cost hygienic cleansing systems remains as urgent as ever. To aid the search for novel antimicrobial targets, there is a need not only to develop our knowledge of biofilm physiology, but also to examine the various mechanisms associated with resistance of biofilms towards antimicrobial agents.

ACKNOWLEDGEMENTS

We wish to acknowledge the support given towards the biofilm research conducted within the Manchester group by the following organizations: Unilever, Procter and Gamble, S.C. Johnson, Rohm & Haas, The Royal Society and the BBSRC.

REFERENCES

Abbanat, D.R., Godchaux, W. and Leadbetter, E.R. (1988) Surface-induced synthesis of new sulfonolipids in the gliding bacterium *Cytophaga johnsonae*. *Arch. Micobiol.* **149**, 358–364.

Alexander, M. (1971). *Microbial Ecology*. John Wiley, New York.

Al-Hiti, M.M. and Gilbert, P. (1980) Changes in preservative sensitivity for the USP antimicrobial agents effectiveness test micro-organisms. *J. Appl. Bacteriol.* **49**, 119–126.

Allison, D.G. and Gilbert, P. (1992) Bacterial biofilms. *Sci. Prog.* **76**, 305–321.

Allison, D.G. and Sutherland, I.W. (1987) The role of exopolysaccharides in adhesion of fresh-water bacteria. *J. Gen. Microbiol.* **133**, 1319–1327.

Allison, D.G., Evans, D.J. and Brown, M.R.W. (1990) Possible involvement of the division cycle in dispersal of *Escherichia coli* from biofilms. *J. Bacteriol.* **172**, 1667–1669.

Allison, D.G., Ruiz, B., SanJose, C., Jaspe, A. and Gilbert, P. (1998) Extracellular products as mediators of the formation and detachment of *Pseudomonas fluorescens* biofilms. *FEMS Microbiol. Lett.* **167**, 179–184.

Allison, D.G., Heys, S., Willcock, L., Holah, J. and Gilbert, P. (1999) Cellular detachment and dispersal from bacterial biofilms: a role for quorum sensing? In: *Biofilms: The Good, the Bad and the Ugly* (J. Wimpenny, P. Gilbert, M. Brading, J. Walker, R. Bayston, ed.), pp. 279–286. Bioline, Cardiff.

Allison, D.G., McBain, A.J. and Gilbert, P. (2000) Microbial biofilms: Problems of control. In: *Community Structure and Cooperation in Biofilms* (D. Allison, P. Gilbert, H. Lappin-Scott, M. Wilson, eds), pp. 309–327. Cambridge University Press, Cambridge.

Amy, P.S., Pauling, C. and Morita, R.Y. (1983) Recovery from nutrient starvation by a marine *Vibrio* sp. *Appl. Environ. Microbiol.* **45**, 1685–1690.

Anand, A. and Glatt, A.E. (1993) *Clostridium difficile* infection associated with antineoplastic chemotherapy: a review. *Clin. Infect. Dis.* **17**, 109–113.

Anand, A., Bashey, B., Mir, T. and Glatt, A.E. (1994) Epidemiology, clinical manifestations, and outcome of *Clostridium difficile*-associated diarrhea. *Am. J. Gastroenterol.* **89**, 519–523.

Anwar, H., Dasgupta, M., Lam, K. and Costerton, J.W. (1989) Tobramycin resistance of mucoid *Pseudomonas aeruginosa* biofilm grown under iron limitation. *J. Antimicrob. Chemother.* **24**, 647–655.

Anwar, H., Dasgupta, M.K. and Costerton, J.W. (1990) Testing the susceptibility of bacteria in biofilms to antibacterial agents. *Antimicrob. Agents Chemother.* **34**, 2043–2046.

Anwar, H., Strap, J.L., Chen, K. and Costerton, J.W. (1992) Dynamic interactions of biofilms of mucoid *Pseudomonas aeruginosa* with tobramycin and piperacillin. *Antimicrob. Ag. Chemother.* **36**, 1208–1214.

Armitage, J.P. (1999) Bacterial tactic responses. *Adv. Microb. Physiol.* **41**, 229–289.

Ashby, M.J., Neale, J.E., Knott, S.J. and Critchley, I.A. (1994) Effect of antibiotics on non-growing cells and biofilms of *Escherichia coli*. *J. Antimicrob. Chemother.* **33**, 443–452.

Aspiras, M.B., Kazmerzak, K.M., Kolenbrander, P.E., McNab, R., Hardegen, N. and Jenkinson, H.F. (2000) Expression of green fluorescent protein in *Streptococcus gordonii* DL1 and its use as a species-specific marker in coadhesion with *Streptococcus oralis* 34 in saliva-conditioned biofilms in vitro. *Appl. Environ. Microbiol.* **66**, 4074–4083.

Atwood, K.C., Schneider, L.K. and Ryan, F.J. (1951) Periodic selection in *Escherichia coli*. *Proc. Natl Acad. Sci. USA* **37**, 146–155.

Barcenilla, A., Pryde, S.E., Martin, J.C., Duncan, S.H., Stewart, C.S., Henderson, C. and Flint, H.J. (2000) Phylogenetic relationships of butyrate-producing bacteria from the human gut. *Appl. Environ. Microbiol.* **66**, 1654–1661.

Barer, M.R. and Harwood, C.R. (1999) Bacterial viability and culturability. *Adv. Microb. Physiol.* **41**, 93–137.

Barker, J. and Bloomfield, S.F. (2000) Survival of salmonella in bathrooms and toilets in domestic homes following salmonellosis. *J. Appl. Microbiol.* **89**, 137–144.

Bartlett, J.G. (1992) Antibiotic-associated diarrhea. *Clin. Infect. Dis.* **15**, 573–581.

Belas, R., Mileham, A., Simon, M., Silverman, M. (1984) Transposon mutagenesis of marine *Vibrio* spp. *J. Bacteriol.* **158**, 890–896.

Belas, R., Simon, M. and Silverman, M. (1986) Regulation of lateral flagella gene transcription in *Vibrio parahaemolyticus*. *J. Bacteriol.* **167**, 210–218.

BernetCamard, M.F., Lievin, V., Brassart, D., Neeser, J.R., Servin, A.L. and Hudault, S. (1997) The human *Lactobacillus acidophilus* strain LA1 secretes a nonbacteriocin antibacterial substance(s) active *in vitro* and *in vivo*. *Appl. Environ. Microbiol.* **63**, 2747–2753.

Bernhard, J.M. and Bowser, S.S. (1992) Bacterial biofilms as a trophic resource for certain benthic foraminifera. *Mar. Ecol-Prog. Ser.* **83**, 263–272.

Bigger, J.W. (1944) Treatment of staphylococcal infections with penicillin. *Lancet* **ii**, 497–500.

Bisno, A.L. and Waldvogel, F.A. (1989) *Infections Associated with Indwelling Medical Devices*. ASM, Washington.

Black, D.S., Kelly, A.J., Mardis, M.J. and Moyed, H.S. (1991) Structure and organisation of htp, an operon that affects lethality due to inhibition of peptidoglycan or DNA synthesis. *J. Bacteriol.* **173**, 5732–5739.

Blatchley, E.R., Dumoutier, N., Halaby, T.N., Levi, Y. and Laine, J.M. (2001) Bacterial responses to ultraviolet irradiation. *Water Sci. Technol.* **43**, 179–186.

Bochner, B.R., Lee, P.C., Wilson, S.W., Cutler, C.W. and Ames, B.N. (1984) AppppA and related adenylated nucleotides are synthesised as a result of oxidation stress. *Cell* **37**, 227–232.

Bos, R., van der Mei, H.C. and Busscher, H.J. (1998) On the role of coaggregation and co-adhesion in dental plaque formation. In: *Oral biofilms and plaque control.* (H.J Busscher and L.V Evans, eds). pp. 163–173. Harwood Academic Publishers, Amsterdam.

Boureau, H., Salanon, C., Decaens., C. and Bourlioux, P. (1994) Caecal localization of the specific microbiota resistant to *Clostridium difficile* colonization in gnotobiotic mice. *Microb. Ecol. Health. Dis.* **7**, 1111–1117.

Boutibonnes, P. (1997) La mort des bacteries: provoquée ou program-mée? Subie ou voulue? *Med. Sci.* **13**, 73–80.

Bradshaw, D.J., Marsh, P.D., Watson, G.K. and Allison, C. (1998) Role of *Fusobacterium nucleatum* and coaggregation in anaerobe survival in planktonic and biofilm oral microbial communities during aeration. *Infect. Immun.* **66**, 4729–4732.

Brook, I. and Ledney, G.D. (1993) Quinolone and glycopeptide therapy for infection in mice following exposure to mixed-field neutron-y-irradiation. *Int. J. Rad. Biol.* **64**, 771–777.

Brooun, A., Liu, S. and Lewis, K. (2000) A dose–response study of antibiotic resistance in *Pseudomonas aeruginosa* biofilms. *Antimicrob. Agents Chemother.* **44**, 640–646.

Brown, M.R.W. and Williams, P. (1985) Influence of substrate limitation and growth phase on sensitivity to antimicrobial agents. *J. Antimicrob. Chemother.* **15** (Suppl. A), 7–14.

Brown, M.R.W., Watkins, W.M. and Foster, J.H. (1969) Step-wise resistance to polymyxin and other agents by *Pseudomonas aeruginosa. J. Gen. Microbiol.* **55**, 17–18.

Brown, M.R.W., Allison, D.G. and Gilbert, P. (1988) Resistance of bacterial biofilms to antibiotics: a growth rate related effect. *J. Antimicrob. Chemother.* **22**, 777–789.

Brown, M.R.W., Collier, P.J. and Gilbert, P. (1990) Influence of growth rate on the susceptibility to antimicrobial agents: modification of the cell envelope in batch and continuous culture. *Antimicrob. Agents Chemother.* **34**, 1623–1628.

Broxton, P., Woodcock, P.M. and Gilbert, P. (1984) Binding of some polyhexamethylene biguanides to the cell envelope of *Escherichia coli* ATCC 8739. *Microbios.* **41**, 15–22.

Brozel, V.S. and Cloete, T.E. (1993) Adaptation of *Pseudomonas aeruginosa* to 2,2′-methylenebis (4-chlorophenol). *J. Appl. Bacteriol.* **74**, 94–99.

Brozel, V.S. and Cloete, T.E. (1994) Resistance of *Pseudomonas aeruginosa* to isothiazolone. *J. Appl. Bacteriol.* **76**, 576–582.

Busscher, H.J. and van der Mei, H.C. (2000) Initial microbial adhesion events: mechanisms and implications. In: *Community Structure and Co-operation in Biofilms* (D. Allison, P. Gilbert, H. Lappin-Scott, M. Wilson, eds). pp. 25–36. Cambridge University Press, UK.

Caldwell, D.E. and Costerton, J.W. (1996) Are bacterial biofilms constrained to Darwin's concept of evolution through natural selection? *Microbiologia* **12**, 347–358.

Caldwell, D.E., Atuku, E., Wilkie, D.C., Wivcharuk, K.P., Karthikeyan, S., Korber, D.R., Schmid, D.F. and Wolfaardt, G.M. (1997a) Germ theory *vs* community theory in understanding and controlling the proliferation of biofilms. *Adv. Dent. Res.* **11**, 4–13.

Caldwell, D.E, Wolfaardt, G.M., Korber, D.R. and Lawrence, J.R. (1997b) Do bacterial communities transcend Darwinism? *Adv. Microb. Ecol.* **15**, 105–191.

Cellini, L., Robuffo, I., Di Campli, E., Di Bartolomeo, S., Taraborelli, T. and Dainelli, B. (1998) Recovery of *Helicobacter pylori* ATCC43504 from a viable but not culturable state: regrowth or resuscitation? *APMIS* **106**, 571–579.

Chandra, J., Kuhn, D.M. Mukherjee, P.K., Hoyer, L.L., McCormick, T. and Ghannoum,

M.A. (2001) Biofilm formation by the fungal pathogen *Candida albicans*: development, architecture, and drug resistance. *J. Bacteriol.* **183**, 5385–5394.

Characklis, W.G. (1990) Microbial fouling. In: *Biofilms* (W.G. Characklis and K.C. Marshall, eds). pp. 523–584. Wiley, New York.

Chuanchuen, R., Beinlich, K. and Schweitzer, H.P. (2000) Multidrug efflux pumps in *Pseudomonas aeruginosa. Abstr. Ann. Meet. Am. Soc. Microbiol.* Los Angeles, A31.

Chuanchuen, R., Beinlich, K., Hoang, T.T., Becher, A., Karkhoff-Schweizer, R.R. and Schweizer, H.P. (2001) Cross-resistance between triclosan and antibiotics in *Pseudomonas aeruginosa* is mediated by multidrug efflux pumps: exposure of a susceptible mutant strain to triclosan selects nfxB mutants overexpressing MexCD-OprJ. *Antimicrob. Agents Chemother.* **45**, 428–432.

Cloete, T.E., Jacobs, L. and Brozel, V.S. (1998) The chemical control of biofouling in industrial water systems. *Biodegradation* **9**, 23–37.

Cochran, W.L., Suh, S.J., McFeters, G.A. and Stewart, P.S. (2000) Role of RpoS and AlgT in *Pseudomonas aeruginosa* biofilm resistance to hydrogen peroxide and monochloramine. *J. Appl. Microbiol.* **88**, 546–553.

Cooper, M., Batchelor, S.M. and Prosser, J.I. (1995) Is cell density signalling applicable to biofilms? In: *The Life and Death of Biofilm* (J. Wimpenny, P. Handley, P. Gilbert and H. Lappin-Scott, eds). pp. 93–97. Bioline Press, Cardiff.

Costerton, J.W. and Lashen, E.S. (1984) Influence of biofilm on the efficacy of biocides on corrosion-causing bacteria. *Materials Perform.* **23**, 34–37.

Costerton, J.W. and Stewart, P.S. (2001) Battling biofilms. *Sci. Am.* **285**, 74–81.

Costerton, J.W., Geesey, G.G. and Cheng, K.J. (1978) How bacteria stick. *Sci. Am.* **238**, 86–95.

Costerton, J.W., Cheng, K.J., Geesey, G.G., Ladd, T.I., Nickel, J.C., Dasgupta, M. and Marrie, T.J. (1987) Bacterial biofilms in nature and disease. *Ann. Rev. Microbiol.* **41**, 435–464.

Costerton, J.W., Lewandowski, Z., Caldwell, D.E., Korber, D.R. and Lappin-Scott, H.M. (1994) Biofilms: the customised microniche. *J. Bacteriol.* **176**, 2137–2142.

Dagostino, L., Goodman, A.E. and Marshall, K.C. (1991) Physiological responses induced in bacteria adhering to surfaces. *Biofouling* **4**, 113–119.

Daly, B., Betts, W.B., Brown, A.P. and O'Neill, J.G. (1998) Bacterial loss from biofilms exposed to free chlorine. *Microbios* **96**, 7–21.

Danese, P.N., Pratt, L.A., Dove, S.L. and Kolter, R. (2000) The outer membrane protein, antigen 43, mediates cell-to-cell interactions within *Escherichia coli* biofilms. *Mol. Microbiol.* **37**, 424–432.

Das, J.R., Bhakoo, M., Jones, M.V. and Gilbert, P. (1998) Changes in the biocide susceptibility of *Staphylococcus epidermidis* and *Escherichia coli* cells associated with rapid attachment to plastic surfaces. *J. Appl. Microbiol.* **84**, 852–858.

Davies, D.G., Chakrabarty, A.M. and Geesey, G.G. (1993) Exopolysaccharide production in biofilms: substratum activation of alginate gene expression by *Pseudomonas aeruginosa. Appl. Environ. Microbiol.* **59**, 1181–1186.

Davies, D.G., Parsek, M., Pearson, J., Iglewski, B., Costerton, J. and Greenberg, E. (1998) The involvement of cell-to-cell signals in the development of a bacterial biofilm. *Science* **280**, 295–298.

DeBeer, D., Srinivasan, R. and Stewart, P.S. (1994) Direct measurement of chlorine penetration into biofilms during disinfection. *Appl. Environ. Microbiol.* **60**, 4339–4344.

Defives, C., Guyard, S., Oulare, M.M., Mary, P. and Hornez, J.P. (1999) Total counts, culturable and viable, and non-culturable microflora of a French mineral water: a case study. *J. Appl. Microbiol.* **86**, 1033–1038.

Degnan, B.A. and Macfarlane, G.T. (1995) Arabinogalactan utilization in continuous cultures of *Bifidobacterium longum* – Effect of co-culture with *Bacteroides thetaiotaomicron. Anaerobe* **2**, 103–112.

DeKievet, T.R., Parkins, M.D., Gillis, R.J., Srikumar, R., Ceri, H., Poole, K, Iglewski, B.H. and Storey, D.G. (2001) Multidrug efflux pumps: expression patterns and contribution to antibiotic resistance in *Pseudomonas aeruginosa* biofilms. *Antimicrob. Agents Chemother.* **45**, 1761–1770.

Demple, B. and Halbrook, J. (1983) Inducible repair of oxidative damage in *Escherichia coli. Nature* **304**, 466–468.

Dignac, M.F., Urbain, V., Rybacki, D., Bruchet, A., Snidaro, D. and Scribe, P. (1998) Chemical disruption of extracellular polymers: implication on activated sludge floc structure. *Water Sci. Technol.* **38**, 45–53.

Duguid, I.G., Evans, E., Brown, M.R. and Gilbert, P. (1992a) Effect of biofilm culture upon the susceptibility of *Staphylococcus epidermidis* to tobramycin. *J. Antimicrob. Chemother.* **30**, 803–810.

Duguid, I.G., Evans, E., Brown, M.R.W. and Gilbert, P. (1992b) Growth rate-dependent killing by ciprofloxacin of biofilm-derived *Staphylococcus epidermidis* – evidence for cell-cycle dependency. *J. Antimicrob. Chemother.* **30**, 791–802.

Dykhuizen, D.E. and Hartl, D.L. (1983) Selection in chemostats. *Microbiol. Rev.* **47**, 150–168.

Elkins, J.G., Hassett, D.J., Stewart, P.S., Schweizer, H.P. and McDermott, T.R. (1999) Protective role of catalase in *Pseudomonas aeruginosa* biofilm resistance to hydrogen peroxide. *Appl. Environ. Microbiol.* **65**, 4594–4600.

Elvers, K.T., Leeming, K., Moore, C.P. and Lappin-Scott, H.M. (1998) Bacterial–fungal biofilms in flowing water photo-processing tanks. *J. Appl. Microbiol.* **84**, 607–618.

Engelberg-Kulka, H. and Glaser, G. (1999) Addiction modules and programmed cell death and anti-death in bacterial cultures. *Ann. Rev. Microbiol.* **53**, 43–70.

Evans, R.C. and Holmes, C.J. (1987) Effect of vancomycin hydrochloride on *Staphylococcus epidermidis* biofilm associated with silicone elastomer. *Antimicrob. Agents Chemother.* **31**, 889–894.

Evans, D.J., Allison, D.G., Brown, M.R. and Gilbert, P. (1990a) Effect of growth-rate on resistance of gram-negative biofilms to cetrimide. *J. Antimicrob. Chemother.* **26**, 473–478.

Evans, D.J., Brown, M.R., Allison, D.G. and Gilbert, P. (1990b) Susceptibility of bacterial biofilms to tobramycin: role of specific growth rate and phase in the division cycle. *J. Antimicrob. Chemother.* **25**, 585–591.

Evans, D.J., Allison, D.G., Brown, M.R.W. and Gilbert, P. (1991) Susceptibility of *Escherichia coli* and *Pseudomonas aeruginosa* biofilms towards ciprofloxacin: Effect of specific growth rate. *J. Antimicrob. Chemother.* **27**, 177–184.

Evans, E., Brown, M.R.W. and Gilbert, P. (1994) Iron chelator, exopolysaccharide and protease production of *Staphylococcus epidermidis*: a comparative study of the effects of specific growth rate in biofilm and planktonic culture. *Microbiology* **140**, 153–157.

Favero, M.S., Bond, W.W., Peterson, N.J. and Cook, E.H. (1983) Scanning electron microscopic observations of bacteria resistant to iodophor solutions. In: *Proc. Int. Symp. Povidone*, pp.158–166. University of Kentucky, Lexington, USA.

Flemming, H-C. and Wingender, J. (2001) Relevance of microbial extracellular polymeric substances (EPSs) – Part I: Structural and ecological aspects. *Water Sci. Technol.* **43**, 1–8.

Flemming, H-C., Wingender, J., Mayer, C., Korstgens, V. and Borchard, W. (2000) Cohesiveness in biofilm matrix polymers. In: *Community Structure and Cooperation in Biofilms* (D. Allison, P. Gilbert, H. Lappin-Scott and M. Wilson, eds), pp. 87–105. Cambridge University Press, Cambridge.

Fletcher, M. (1991) The physiological activity of bacteria attached to solid surfaces. *Adv. Microb. Physiol.* **32**, 53–85.

Foley, I., Marsh, P., Wellington, E.M., Smith, A.W. and Brown, M.R. (1999) General stress response master regulator rpoS is expressed in human infection: a possible role in chronicity. *J. Antimicrob. Chemother.* **43**, 164–165.

Franch, T. and Gerdes, K. (1996) Programmed cell death in bacteria: translational repression by mRNA end-pairing. *Mol. Microbiol.* **21**, 1049–1060.

Fujiwara, S., Miyake, Y., Usui, T. and Suginaka, H. (1998). Effect of adherence on antimicrobial susceptibility of *Pseudomonas aeruginosa*, *Serratia marcescens*, and *Proteus mirabilis*. *J. Med. Sci.* **47**, 1–5.

Gambello, M.J., Kaye, S. and Inglewski, B.H. (1993) *LasR* of *Pseudomonas aeruginosa* is a transcriptional activator of the line protease gene (*apr*) and an enhancer of exotoxin A expression. *Infect. Immun.* **61**, 1180–1184.

George, A.M. and Levy, S.B. (1983) Amplifiable resistance to tetracycline, chloramphenicol, and other antibiotics in *Escherichia coli*: involvement of a non-plasmid-determined efflux of tetracycline. *J. Bacteriol.* **155**, 531–540.

Gerke, C., Kraft, A., Sussmuth, R., Schweitzer, O. and Gotz, F. (1998) Characterization of the N-acetylglucosaminyltransferase activity involved in the biosynthesis of the *Staphylococcus epidermidis* polysaccharide intercellular adhesin. *J. Biol. Chem.* **273**, 18586–18593.

Ghigo, J.M. (2001) Natural conjugative plasmids induce bacterial biofilm development. *Nature* **26**, 442–445.

Gibbons, R.J. and Nygaard, M. (1970) Inter-bacterial aggregation of plaque bacteria. *Arch. Oral. Biol.* **15**, 1397–1400.

Gibson, G. and Wang, X. (1994) Regulatory effects of bifidobacteria on the growth of other colonic bacteria. *J. Appl. Bacteriol.* **77**, 412–420.

Gilbert, P. (1984) The revival of microorganisms sublethally injured by chemical inhibitors. *Soc. Appl. Bacteriol. Symp. Ser.* **12**, 175–197.

Gilbert, P. and Allison, D. (2000) Biofilms and their resistance towards antimicrobial agents. In: *Dental Plaque Revisited* (H. Newman and M. Wilson, eds). Bioline, Cardiff.

Gilbert, P. and Brown, M.R. (1978a) Influence of growth rate and nutrient limitation on the gross cellular composition of *Pseudomonas aeruginosa* and its resistance to 3- and 4-chlorophenol. *J. Bacteriol.* **133**, 1066–1072.

Gilbert, P. and Brown, M.R. (1978b) Effect of R-plasmid RP1 and nutrient depletion on the gross cellular composition of *Escherichia coli* and its resistance to some uncoupling phenols. *J. Bacteriol.* **133**, 1062–1065.

Gilbert, P. and Brown, M.R. (1980) Cell wall-mediated changes in sensitivity of *Bacillus megaterium* to chlorhexidine and 2-phenoxyethanol, associated with growth rate and nutrient limitation. *J. Appl. Bacteriol.* **48**, 223–230.

Gilbert, P. and Brown, M.R. (1991) Out of the test tube into the frying pan: post-growth, pre-test variables. *J. Antimicrob. Chemother.* **27**, 859–860.

Gilbert, P. and McBain, A.J. (2000) Resistance development by biofilm communities. *ASM, Biofilm Symp. 2000*, p. 46.

Gilbert, P. and McBain, A.J. (2001) Biofilms: Their impact on health and their recalcitrance toward biocides. *Am. J. Infect. Control.* **29**, 252–255.

Gilbert, P., Allison, D.G., Evans, D.J., Handley, P.S. and Brown, M.R.W. (1989) Growth rate control of adherent microbial populations. *Appl. Environ. Microbiol.* **55**, 1308–1311.

Gilbert, P., Collier, P.J. and Brown, M.R.W. (1990) Influence of growth rate on susceptibility to antimicrobial agents: Biofilms, cell cycle, dormancy and stringent response. *Antimicrob. Agents Chemother.* **34**, 1865–1868.

Gilbert, P., Evans, D.J. and Brown, M.R.W. (1993) Formation and dispersal of bacterial biofilms *in vivo* and *in situ*. *J. Appl. Bacteriol.* **74** (Suppl), 67S–78S.

Gilbert, P., Collier, P.J., Andrews, J. and Brown, M.R.W. (1995) Influence of post-growth procedures on the properties of microorganisms. In: *Microbial Quality Assurance: A Guide to Relevance and Reproducibility of Inocula.* (Brown, M.R.W. and P. Gilbert, eds). pp. 125–136, CRC Press, Boca Raton, USA.

Gilbert, P., Das, J.R., Jones, M. and Allison, D.G. (2001) Assessment of the biocide activity upon various bacteria following their attachment to and growth on surfaces. *J. Appl. Microbiol.* **91**, 248–255.

Giles C.H., Smith, D. and Huitson, A. (1974a) A general treatment and classification of the solute adsorption isotherm. Part I. Theoretical. *J. Colloid. Interface Sci.* **47**, 755–765.

Giles, C.H., D'Silva, A.P. and Easton, I.A. (1974b) A general treatment and classification of the solute adsorption isotherm. Part II. Experimental interpretation. *J. Colloid. Interface Sci.* **47**, 766–778.

Giwercman, B., Jensen, E.T., Hoiby, N., Kharazmi, A. and Costerton, J.W. (1991) Induction of ß-lactamase production in *Pseudomonas aeruginosa* biofilms. *Antimicrob. Agents Chemother.* **35**, 1008–1010.

Goncalves, R.F., de Araujo, V.L. and Chernicharo, C.A.L. (1998) Association of a UASB reactor and a submerged aerated biofilter for domestic sewage treatment. *Water Sci. Technol.* **38**, 189–195.

Goodwin, J.A.S. and Forster, C.F.A.N. (1989) Examination of the extracellular polymers produced by activated-sludge. *Microbios* **57**, 2179–2185.

Gordon, C.A., Hodges, N.A. and Marriot, C. (1988) Antibiotic interaction and diffusion through alginate and exopolysaccharide of cystic fibrosis derived *Pseudomonas aeruginosa. J. Antimicrob. Chemother.* **22**, 667–674.

Govan, J.R. and Fyfe, J.A. (1978) Mucoid *Pseudomonas aeruginosa* and cystic fibrosis: resistance of the mucoid to carbenicillin, flucloxacillin and tobramycin and the isolation of mucoid variants *in vitro. J. Antimicrob. Chemother.* **4**, 233–240.

Griffith, J.K., Baker, M.E., Rouch, D.A., Page, M.G.P., Skurray, R.A., Paulsen, I.T., Chater, K.F., Baldwin, S.A. and Henderson, P.J.F. (1992) Membrane transport proteins: implications of sequence comparisons. *Curr. Opin. Cell Biol.* **4**, 684–695.

Gristina A.G. and Costerton J.W. (1985) Bacterial adherence to biomaterials and tissue. The significance of its role in clinical sepsis. *J. Bone Joint Surg. Am.* **67**, 264–273.

Gristina, A.G., Hobgood, C.D., Webb, L.X. and Myrvik, Q.N. (1987) Adhesive colonisation of biomaterials and antibiotic resistance. *Biomaterials* **8**, 423–426.

Gristina, A.G, Jennings, R.A., Naylor, P.T., Myrvik, Q.N. and Webb, L.X. (1989) Comparative *in vitro* antibiotic-resistance of surface-colonizing coagulase-negative staphylococci. *Antimicrob. Agents Chemother.* **33**, 813–816.

Hasman, H., Chakrabarty, T. and Klemm, P. (1999) Antigen-43-mediated autoaggregation of *Escherichia coli* is blocked by fimbriation. *J. Bacteriol.* **181**, 4834–4841.

Hassett, D.J., Elkins, J.G., Ma, J.F. and McDermott, T.R. (1999) *Pseudomonas aeruginosa* biofilm sensitivity to biocides: use of hydrogen peroxide as model antimicrobial agent for examining resistance mechanisms. *Meth. Enzymol.* **310**, 599–608.

Heath, R.I. and Rock, C.O. (1995) Enoyl-acyl carrier protein reductase *(fabI)* plays a determinant role in completing cycles of fatty-acid elongation in *Escherichia coli. J. Biol. Chem.* **270**, 26538–26542.

Hejzlar, J. and Chudoba, J. (1986) Microbial polymers in the aquatic environment. I. Production by activated sludge microorganisms under different conditions. *Water Res.* **20**, 1209–1216.

Hengge-Aronis R. (1996) Back to log phase: sigma S as a global regulator in the osmotic control of gene expression in *Escherichia coli. Mol. Microbiol.* **21**, 887–893.

Hentzer, M., Teitzel, G.M., Balzer, G.J., Heydorn, A., Molin, S., Givskov, M. and Parsek

M.R. (2001) Alginate overproduction affects *Pseudomonas aeruginosa* biofilm structure and function. *J. Bacteriol.* **183**, 5395–5401.

Heys, S.J.D., Gilbert, P. and Allison, D.G. (1997) Homoserine lactones and bacterial biofilms. In: *Biofilms: Community Interactions and Control* (J.W.T. Wimpenny, P. Handley, P. Gilbert and H.M. Lappin-Scott, eds). pp. 103–112. Bioline, Cardiff.

Higgins, M.J. and Chodoba, J. (1986) Characterization of extracellular protein and its role in bioflocculation. *J. Environ. Eng.* **123**, 479–485.

Higgins, M.J. and Novak, J.T. (1997) Characterization of extracellular protein and its role in bioflocculation. *J. Environ. Eng.* **123**, 479–485.

Hirsch, P. (1974) Budding bacteria. *Ann. Rev. Microbiol.* **28**, 391–442.

Hochman, A. (1997) Programmed cell death in prokaryotes. *Crit. Rev. Microbiol.* **23**, 207–214.

Hodgson, A.E., Nelson, S.M., Brown, M.R. and Gilbert, P. (1995) A simple *in vitro* model for growth control of bacterial biofilms. *J. Appl. Bacteriol.* **79**, 87–93.

Hoff, J.C. (1978) The relationship of turbidity to disinfection of potable water. In: *Evaluation of the Microbiology Standards for Drinking Water.* (Hendricks, C.W., ed.). EPA 5700/9-78-00C. US Environmental Protection Agency, Washington, DC.

Hoff, J.C. and Akin, E.W. (1986) Microbial resistance to disinfectants: mechanisms and significance. *Environ. Health Perspect.* **69**, 7–13.

Holah, J.T. (1995). Disinfection of food production areas. *Rev. Sci. Tech.* **14**, 343–363.

Holden, P.A., Hunt, J.R., and Firestone, M.K. (1997) Toluene diffusion and reaction in unsaturated *Pseudomonas putida* biofilms. *Biotechnol. Bioeng.* **56**, 656–670.

Hoyle, B.D., and Costerton, J.W. (1991) Bacterial resistance to antibiotics: the role of biofilms. *Prog. Drug. Res.* **37**, 91–105.

Hoyle, B.D., Wong. C.K.W. and Costerton, J.W. (1992) Disparate efficacy of tobramycin on Ca^{2+}-treated, Mg^{2+}-treated, and Hepes-treated *Pseudomonas aeruginosa* biofilms. *Can. J. Microbiol.* **38**, 1214–1218.

Huang, C.T., Yu, F.P., McFeters, G.A. and Stewart, P.S. (1995) Non-uniform spatial patterns of respiratory activity within biofilms during disinfection. *Appl. Environ. Microbiol.* **61**, 2252–2256.

Huang, C.T., Xu, K.D., McFeters, G.A. and Stewart, P.S. (1998) Spatial patterns of alkaline phosphatase expression within bacterial colonies and biofilms in response to phosphate starvation. *Appl. Environ. Microbiol.* **64**, 1526–1531.

Hugenholtz, P. and Fuerst, J.A. (1992) Heterotrophic bacteria in an air-handling system. *Appl. Environ. Microbiol.* **58**, 3914–3920.

Humphrey, B.A., Dickson, M.R. and Marshall, K.C. (1979) Physiological and *in situ* observations on the adhesion of gliding bacteria to surfaces. *Arch. Microbiol.* **120**, 231–238.

Ichimiya, T., Yamaski, T. and Nasu, M. (1994) *In vitro* effects of antimicrobial agents on *Pseudomonas aeruginosa* biofilm formation. *J. Antimicrob. Chemother.* **34**, 331–341.

Jensen, R.B. and Gerdes, K. (1995) Programmed cell death in bacteria: proteic plasmid stabilised systems. *Mol. Microbiol.* **17**, 205–210.

Johnson, J.M. and Church, G.M. (1999) Alignment and structure prediction of divergent protein families: periplasmic and outer membrane proteins of bacterial efflux pumps. *J. Mol. Biol.* **287**, 695–715.

Johnson, S., Samore, M.H., Farrow, K.A., Killgore, G.E., Tenover, F.C., Lyras, D., Rood, J.I., DeGirolami, P., Baltch, A.L., Rafferty, M.E., Pear, S.M. and Gerding, D.N. (1999) Epidemics of diarrhea caused by a clindamycin-resistant strain of *Clostridium difficile* in four hospitals. *N. Engl. J. Med.* **341**, 1645–1651.

Jorand, F., Boué-Bigne, F., Block, J.C. and Urbain, V. (1998) Hydrophobic/hydrophilic properties of activated sludge exopolymeric substances. *Water Sci. Technol.* **37**, 307–315.

Kelly, C.P. and LaMont, J.T. (1998) *Clostridium difficile* infection. *Ann. Rev. Med.* **49**, 375–390.

Kelly, C.P., Pothoulakis, C. and LaMont, J.T. (1994) *Clostridium difficile* colitis. *N. Engl. J. Med.* **330**, 257–262.

Kinniment, S.L. and Wimpenny, J.W. (1992) Measurements of the distribution of adenylate concentrations and adenylate energy charge across *Pseudomonas aeruginosa* biofilms. *Appl. Environ. Microbiol.* **58**, 1629–1635.

Kinniment, S.L., Wimpenny, J.W., Adams, D. and Marsh, P.D. (1996) The effect of chlorhexidine on defined, mixed culture oral biofilms grown in a novel model system. *J. Appl. Bacteriol.* **81**, 120–125.

Kjaergaard, K., Schembri, M.A., Ramos, C., Molin, S. and Klemm, P. (2000) Antigen 43 facilitates formation of multispecies biofilms. *Environ. Microbiol.* **2**, 695–702.

Kjelleberg, S., Humphrey, B.A. and Marshall, S.C. (1982) Effects of interphases on small, starved marine bacteria. *Appl. Environ. Microbiol.* **43**, 1166–1172.

Klemperer, R.M., Gilbert, P., Meier, A.M., Cozens, R.M. and Brown, M.R. (1979) Influence of suspending media upon the susceptibility of *Pseudomonas aeruginosa* NCTC 6750 and its spheroplasts to polymyxin B. *Antimicrob. Agents Chemother.* **15**, 147–151.

Klier, C.M., Roble, A.G. and Kolenbrander, P.E. (1998) Actinomyces serovar WVA963 coaggregation-defective mutant strain PK2407 secretes lactose-sensitive adhesin that binds to coaggregation partner *Streptococcus oralis* 34. *Oral Microbiol. Immunol.* **13**, 337–340.

Koch A.L. (1987) Similarities and differences of individual bacteria within a clone. In: *Escherichia coli and Salmonella: Cellular and Molecular Biology* (F.C. Neidhardt, R.I. Curtiss, J.L. Ingraham, C.C.I. Lin, K.B. Low, B. Magasanik, W.S. Reznikoff, M. Riley, M. Scgaechter and H.E. Umbarger, eds) pp. 1640–1651, ASM Press, Washington DC.

Kolenbrander, P.E. (1989) Surface recognition among oral bacteria: multigeneric coaggregations and their mediators. *Crit. Rev. Microbiol.* **17**, 137–159.

Kolenbrander, P.E. (2000) Oral microbial communities: biofilms, interactions, and genetic systems. *Annu. Rev. Microbiol.* **54**, 413–437.

Kolenbrander, P.E. and London, J. (1993) Adhere today, here tomorrow: oral bacterial adherence. *J. Bacteriol.* **175**, 3247–3252.

Kullik, I., Toledano, M.B., Tartaglia, L.A. and Storz, G. (1995) Mutation analysis of the redox sensitive transcriptional regulator OxyR: regions important for oxidation and transcriptional activation. *J. Bact.* **177**, 1275–1284.

Kumon, H., Tomochika, K-I., Matunaga, T., Ogawa, M. and Ohmori, H. (1994) A sandwich cup method for the penetration assay of antimicrobial agents through *Pseudomonas* exopolysaccharides. *Microbiol. Immunol.* **38**, 615–619.

Lambert, P.A., Giwercman, B. and Hoiby, N. (1993) Chemotherapy of *Pseudomonas aeruginosa* in cystic fibrosis. In: *Bacterial Biofilms and their Control in Medicine and Industry* (J.T. Wimpenny, W.W. Nichols, D. Stickler and H.M. Lappin-Scott, eds) pp. 151–153, Bioline, Cardiff.

Larson, H.E. and Welch, A. (1993) *In vitro* and *in vivo* characterization of resistance to colonization with *Clostridium difficile*. *J. Med. Microbiol.* **38**, 103–108.

Latifi, A., Foglino, M., Tanaka, K., Williams, P. and Lazdunski, A. (1996) A hierarchical quorum-sensing cascade in *Pseudomonas aeruginosa* links the transcriptional activators LasR and RhlR (VsmR) to expression of the stationary-phase sigma factor RpoS. *Mol. Microbiol.* **21**, 1137–1146.

LeChevallier M.W., Evans, T.M. and Seidler, R.J. (1981) Effect of turbidity on chlorination efficiency and bacterial persistence in drinking water. *Appl. Environ. Microbiol.* **42**, 159–167.

LeChevallier M.W., Cawthon, C.D. and Lee, R.G. (1988) Factors promoting survival of bacteria in chlorinated water supplies. *Appl. Environ. Microbiol.* **54**, 649–654.

Levin, B.R., Perrot, V. and Walker, N. (2000) Compensatory mutations, antibiotic resistance and the population genetics of adaptive evolution in bacteria. *Genetics* **154**, 985–997.

Levine, J., Ellis, C.J., Furne, J.K., Springfield, J. and Levitt, M.D. (1998) Fecal hydrogen sulfide production in ulcerative colitis. *Am. J. Gastroenterol.* **93**, 83–87.

Levy, S.B. (1992) Active efflux mechanisms for antimicrobial resistance. *Antimicrob. Agents Chemother.* **36**, 695–703.

Lewis, K. (2000) Programmed cell death in bacteria. *Microbiol. Mol. Biol. Rev.* **64**, 503–514.

Lewis, K. (2001) Riddle of biofilm resistance. *Antimicrob. Agents Chemother.* **45**, 999–1007.

Li, Y., Gu, G., Zhao, J. and Yu, H. (2001) Anoxic degradation of nitrogenous heterocyclic compounds by acclimated activated sludge. *Process Bioch.* **37**, 81–86.

Licht, T.R., Christensen, B.B., Krogfelt, K.A. and Molin, S. (1999) Plasmid transfer in the animal intestine and other dynamic bacterial populations: the role of community structure and environment. *Microbiology* **145**, 2615–2622.

Liehr, S., Chen, H. and Lin, S. (1994) Metals removed by algal biofilms. *Water Sci. Technol.* **30**, 59–68.

Lim, A., Eleuterio, M., Hutter, B., Murugasu-Oei, B. and Dick T. (1999) Oxygen depletion-induced dormancy in *Mycobacterium bovis* BCG. *J. Bacteriol.* **181**, 2252–2256.

Lleo, M.D., Tafi, M.C. and Canepari, P. (1998) Nonculturable *Enterococcus faecalis* cells are metabolically active and capable of resuming active growth. *Syst. Appl. Microbiol.* **21**, 333–339.

Lu, M.Y., Maddox, I.S. and Brooks, J.D. (1998) Application of a multi-layer packed-bed reactor to citric acid production in solid-state fermentation using *Aspergillus niger.* *Process Biochem.* **33**, 117–123.

Lusk, R.H., Fekety, F.R. Jr., Silva, J. Jr., Bodendorfer, T., Devine, B.J., Kawanishi, H., Korff, L., Nakauchi, D., Rogers, S. and Siskin, S.B. (1977) Gastrointestinal side effects of clindamycin and ampicillin therapy. *J. Infect. Dis.* **135** (Suppl), S111–S119.

Ma, D., Cook, D.N., Alberti, M., Pon, N.G., Nikaido, H. and Hearst, J.E. (1993) Molecular cloning and characterization of acrAB and acrE genes of *Escherichia coli. J. Bacteriol.* **175**, 6299–6313.

Ma, D., Cook, D.N., Alberti, M., Pon, N.G., Nikaido, H., Hearst, J.E. (1995) Genes *acrA* and *acrB* encode a stress-induced efflux system of *Esherichia coli. Mol. Microbiol.* **16**, 45–55.

Macfarlane, G.T., Cummings, J.H. and Allison, C. (1986) Protein-degradation by human intestinal bacteria. *J. Gen. Microbiol.* **132**, 1647–1656.

Macfarlane, S., McBain A.J. and Macfarlane, G.T. (1997) Consequences of biofilm and sessile growth in the large intestine. *Adv. Dent. Res.* **11**, 59–68.

Mack, D., Fischer, W., Krokotsch, A., Leopold, K., Hartmann, R., Egge, H. and Laufs, R. (1996) The intercellular adhesin involved in biofilm accumulation of *Staphylococcus epidermidis* is a linear β-1,6-linked glucosaminoglycan: purification and structural analysis. *J. Bacteriol.* **178**, 175–183.

Mackey, B.M. and Derrick, C.M. (1986) Changes in the heat resistance of *Salmonella typhimurium* during heating at rising temperatures. *Lett. Appl. Microbiol.* **4**, 13–16.

Maclehose, H.G., Allison, D.G. and Gilbert, P. (2001) Susceptibility of *Pseudomonas* species to biocides and antibiotics following chronic sub-inhibitory exposure. *Abst. 101st Ann. Gen. Meet. Am. Soc. Gen. Microbiol.* A-153, p. 36.

Mah, T.F. and O'Toole, G.A. (2001) Mechanisms of biofilm resistance to antimicrobial agents. *Trends Microbiol.* **9**, 34–39.

Maira-Litran, T. (1998) An investigation into the potential of the *mar* operon to moderate the antibiotic resistance of biofilms. PhD Thesis, University of Manchester, Manchester.

Maira-Litran, T., Allison, D.G., and Gilbert, P. (2000a) An evaluation of the potential role of the multiple antibiotic resistance operon (*mar*) and the multi-drug efflux pump *acr*AB in the resistance of *E. coli* biofilms towards ciprofloxacin. *J. Antimicrob. Chemother.* **45**, 789–795.

Maira-Litran, T., Allison, D.G. and Gilbert, P. (2000b) Expression of the multiple resistance operon (mar) during growth of *Escherichia coli* as a biofilm. *J. Appl. Microbiol.* **88**, 243–247.

Marrie, T.J. and Costerton, J.W. (1984) Morphology of bacterial attachment to cardiac pacemaker leads and power packs. *J. Clin. Microbiol.* **19**, 911–914.

Marsh, P. and Bowden, G. (2000) Community structure and cooperation in biofilms. In: *Community Structure and Cooperation in Biofilms* (D. Allison, P. Gilbert, H. Lappin-Scott and M. Wilson, eds). Cambridge University Press, Cambridge.

Marsh, P. and Bradshaw, D. (1998) Dental plaque: Community spirit in action. In: *Microbial Pathogenesis: Current and Emerging Issues* (D. LeBlanc, M. Lanz and L. Switalski, eds), pp. 41–53. University of Indiana, Indianapolis.

Marsh, P. and Bradshaw, D. (1999) Microbial community aspects of dental plaque. In: *Dental Plaque Revisited: Oral Biofilms in Health and Disease* (H. Newman and M. Wilson, eds), pp. 237–253, Bioline, Cardiff.

Marsh, P. and Bradshaw, D. J. (2000) Microbial community aspects of dental plaque. In: *Dental Plaque Revisited* (H. Newman and M. Wilson, eds). Bioline, Cardiff.

Matin, A., Auger, E.A.., Blum, P.H. and Schultz, J.E. (1989) Genetic basis of starvation survival in non-differentiating bacteria. *Annu. Rev. Microbiol.* **43**, 293–316.

McBain, A.J. and Macfarlane, G.T. (1998) Ecological and physiological studies on large intestinal bacteria in relation to production of hydrolytic and reductive enzymes involved in formation of genotoxic metabolites. *J. Med. Microbiol.* **47**, 407–416.

McBain, A.J. and Macfarlane, G.T. (2001) Non-digestible oligosaccharide metabolism modulates genotoxic enzyme activities in *in vitro* human gut bacterial ecosystems. *J. Med. Microbiol.* **50**, 832–841.

McBain, A.J., Allison, D.G. and Gilbert, P. (2000a) Population dynamics in microbial biofilms. In: *Community Structure and Cooperation in Biofilms.* (D.G. Allison, P. Gilbert, H. Lappin-Scott, M. Wilson, eds), pp. 257–278, Cambridge University Press, Cambridge.

McBain, A.J., Allison, D.G. and Gilbert, P. (2000b) Emerging strategies for the chemical treatment of microbial biofilms. *Biotechnol. Genet. Eng. Rev.* **17**, 267–279.

McCarter, L., Hilmen, M. and Silverman, M. (1988) Flagellar dynamometer controls swarmer cell differentiation of *V. parahaemolyticus. Cell* **54**, 345–351.

McCoy, W., Ridge, J. and Lashen, E. (1986a) Efficacy of biocides in a laboratory model cooling tower. *Material Perform.* Aug, 9–14.

McCoy, W., Wireman, J. and Lashen, E. (1986b) Efficacy of methylchloroisothiazolone biocide against *Legionella pneumophila* in cooling tower water. *Chimica. Oggi.* **4**, 79–83.

McFeters, G.A., Egil, T., Wilberg, E., Adler, A., Schneider, R., Snozzi, M. and Giger, M. (1990) Activity and adaptation of nitriloacetate (NTA) degrading bacteria: field and laboratory studies. *Water Res.* **24**, 875–881.

McMurry, L.M., Oethinger, M. and Levy, S.B. (1998a) Triclosan targets lipid synthesis. *Nature* **394**, 531–532.

McMurry, L.M., Oethinger, M. and Levy, S.B. (1998b) Overexpression of *mar*A, *sox*S, or *acr*AB produces resistance to triclosan in laboratory and clinical strains of *Escherichia coli. FEMS Microbiol. Lett.* **166**, 305–309.

Miller, P.F. and Sulavick, M.C. (1996) Overlaps and parallels in the regulation of intrinsic multiple antibiotic resistance in *Escherichia coli. Mol. Microbiol.* **21**, 441–448.

Mir, J., Morato, J. and Ribas, F. (1997) Resistance to chlorine of freshwater bacterial strains. *J. Appl. Microbiol.* **82**, 7–18.

Moken, M.C., McMurry, L.M. and Levy, S.B. (1997) Selection of multiple antibiotic-resistant (mar) mutants of *Escherichia coli* by using the disinfectant pine oil: roles of the mar and acrAB loci. *Antimicrob. Agents Chemother.* **41**, 2770–2772.

Morita, R.Y. (1986) Starvation survival: the normal mode of most bacteria in the ocean. *Proc. 4th Int. Symp. Microbiol. Ecol. Slovene Soc. Microbiol.* 242–248.

Moyed, H.S. and Bertrand, K.P. (1983a) *hlp*A, a newly recognised gene of *E. coli* that affects frequency of persistence after inhibition of murein synthesis. *J. Bacteriol.* **155**, 768–775.

Moyed, H.S. and Bertrand, K.P. (1983b) Mutations in multicopy Tn10 tet plasmids that confer resistance to inhibitory effects of inducers of tet gene expression. *J. Bacteriol.* **155**, 557–564.

Moyer, C.L. and Morita, R.Y. (1989) Effect of growth rate and starvation survival on the viability and stability of a psychrophylic marine bacterium. *Appl. Environ. Microbiol.* **55**, 1122–1127.

Munro, P.M., Gauthier, M.J., Breittmayer, V.A. and Bongiovanni, J. (1989) Influence of osmoregulation processes on starvation survival of *Escherichia coli* in seawater. *Appl. Environ. Microbiol.* **55**, 2017–2024.

Naito, T., Kusano, K. and Kobayashi, I. (1995) Selfish behaviour of restriction modification systems. *Science* **267**, 897–899.

Ngian, K.F., Lin, S.H. and Martin, W.R. (1977). Effect of mass transfer resistance on the Lineweaver-Burk plots for flocculating microorganisms. *Biotechnol. Bioeng.* **19**, 1773–1784.

Nichols, W.W. (1993) Biofilm permeability to antibacterial agents. In: *Bacterial Biofilms and their Control in Medicine and Industry* (J. Wimpenny, W.W. Nichols, D. Stickler and H. Lappin-Scott, eds). pp. 141–149. Cardiff, Bioline Press.

Nichols, W.W., Dorrington, S.M., Slack, M.P.E. and Walmsley, H.L. (1988) Inhibition of tobramycin diffusion by binding to alginate. *Antimicrob. Agents Chemother.* **32**, 518–523.

Nichols, W.W., Evans, M.J., Slack, M.P.E. and Walmsley, H.L. (1989) The penetration of antibiotics into aggregates of mucoid and non-mucoid *Pseudomonas aeruginosa*. *J. Gen. Microbiol.* **135**, 1291–1303.

Nickel, J.C., Ruseska, I., Wright, J.B. and Costerton, J.W. (1985) Tobramycin resistance of *Pseudomonas aeruginosa* cells growing as a biofilm on urinary catheter material. *Antimicrob. Agents Chemother.* **27**, 619–624.

Nielsen, P-H., Jahn, A. and Palmgren, R. (1997) Conceptual model for production and composition of exopolymers in biofilms. *Water Sci. Technol.* **36**, 11–19.

Nikaido, H. (1996) Multidrug efflux pumps of Gram-negative bacteria. *J. Bacteriol.* **178**, 5853–5859.

Nikaido, H. and Zgurskaya, H.I. (1999) Antibiotic efflux mechanisms. *Curr. Opin. Infect. Dis.* **12**, 529–536.

Novitsky, J.A. and Morita, R.Y. (1976) Morphological characterisation of small cells resulting from nutrient starvation of a psychrophilic marine *Vibrio*. *Appl. Environ. Microbiol.* **32**, 617–622.

Novitsky, J.A. and Morita, R.Y. (1977) Survival of a psychrophilic marine *Vibrio* under long-term nutrient starvation. *Appl. Environ. Microbiol.* **33**, 635–641.

Olson, B.H. and Stewart, M.H. (1990) Factors that change bacterial resistance to disinfection. *Water Chlorination* **6**, 885–904.

Parsek, M.R. and Greenberg, E.P. (1999) Quorum-sensing signals in development of *Pseudomonas aeruginosa* biofilms. *Meth. Enzymol.* **310**, 43–55.

Paul B.J., Duthie, H.C. and Taylor, W.D. (1991) Nutrient cycling by biofilms in running waters of differing nutrient status. *J.N. Am. Benthol. Soc.* **10**, 31–41.

Paulsen, I.T., Brown, M.H. and Skurray, R.A. (1996) Proton-dependent multidrug efflux systems. *Microbiol. Rev.* **60**, 575–608.

Peterson, B.J., Wollheim, W.M., Mulholland, P.J., Webster, J.R., Meyer, J.L., Tank, J.L., Marti, E., Bowden, W.B., Valett, H.M., Hershey, A.E., McDowell, W.H., Dodds, W.K., Hamilton, S.K., Gregory, S. and Morrall, D.D. (2001) Control of nitrogen export from watersheds by headwater streams. *Science* **292**, 86–90.

Piggot, P.J. and Coote, J.G. (1976) Genetic aspects of bacteria enospore formation. *Bacterial Rev.* **40**, 908–962.

Pratten, J., Smith, A.W. and Wilson, M. (1998) Response of single species biofilms and micro-cosm dental plaques to pulsing with chlorhexidine. *J. Antimicrob. Chemother.* **42**, 453–459.

Reid, G., McGroarty, J.A., Angotti, R. and Cook, R.L. (1988) Lactobacillus inhibitor production against *Escherichia coli* and coaggregation ability with uropathogens. *Can. J. Microbiol.* **34**, 344–351.

Rella, M. and Haas, D. (1982) Resistance of *Pseudomonas aeruginosa* PAO to nalidixic acid and low levels of beta-lactam antibiotics: mapping of chromosomal genes. *Antimicrob. Agents Chemother.* **22**, 242–249.

Rhaese, H., Dichtelmüller, R., Grade, R. and Groscurth, R. (1975). High phosphorylated nucleotides involved in regulation of sporulation in *Bacillus subtilis*. In: *Spores, VI* (P. Gerhardt and R. Costliow, eds), pp. 335–340. American Society for Microbiology, Washington, DC.

Rickard, A.H., Leach, S.A., Buswell, C.M., High, N.J. and Handley, P.S. (1999) Coaggregation amongst aquatic and oral bacteria is mediated by lectin–saccharide interactions. In: *Biofilms: The good, the Bad and the Ugly* (J. Wimpenny, P. Gilbert, J. Walker, M. Brading, R. Bayston, eds) pp. 343–354. Bioline, Cardiff.

Rickard, A.H., Leach, S., Buswell, C., High, N. and Handley, P. (2000) Coaggregation between aquatic bacteria is mediated by specific-growth-phase-dependent lectin–saccharide interactions. *Appl. Environ. Microbiol.* **66**, 431–434.

Roberton, A.M. (1993) Roles of endogenous substances and bacteria in colorectal cancer. *Mut. Res.* **290**, 71–78.

Roos, S., Lindgren, S. and Jonsson, H. (1999) Autoaggregation of *Lactobacillus reuteri* is mediated by a putative DEAD-box helicase. *Mol. Microbiol.* **32**, 427–436.

Roszak, D.B. and Colwell, R.R. (1987a) Metabolic activity of bacterial cells enumerated by direct viable count. *Appl. Environ. Microbiol.* **53**, 2889–2893.

Roszak, D.B. and Colwell, R.R. (1987b) Survival strategies of bacteria in the natural environment. *Microbial. Rev.* **51**, 365–379.

Rouch, D.A., Cram, D.S., Dibernadino, D., Littlejohn, T.G. and Skurray, R.A. (1990) Efflux-mediated antiseptic gene *qacA* from *Staphylococcus aureus*: common ancestry with tetracycline and sugar transport proteins. *Mol. Microbiol.* **4**, 2051–2062.

Rowbury, R.J. (2000a). Killed cultures of *Escherichia coli* can protect living organisms from acid stress. *Microbiology* **146**, 1759–1760.

Rowbury, R.J. (2001b) Cross-talk involving extracellular sensors and extracellular alarmones gives early warning to unstressed *Escherichia coli* of impending lethal chemical stress and leads to induction of tolerance responses. *J. Appl. Microbiol.* **90**, 677–695.

Rowbury, R.J. (2001c) Extracellular sensing components and extracellular induction component alarmones give early warning against stress in *Escherichia coli. Adv. Microb. Physiol.* **44**, 215–257.

Rudd, T., Sterrit, R.M. and Lester, J.N. (1983) Extraction of extracellular polymers from activated sludge. *Biotechnol. Lett.* **5**, 327–332.

Saier, M.H. (1994) Computer-aided analyses of transport protein sequences: gleaning evidence concerning function, structure, biogenesis and evolution. *Microbiol. Rev.* **58**, 71–93.

Saier, M.H. and Paulsen, I.T. (2001) Phylogeny of multidrug transporters. *Semin. Cell Dev. Biol.* **12**, 205–213.

Samson, L. and Cairns, J. (1977) A new pathway for DNA repair in *Escherichia coli.* *Nature* **267**, 281–283.

Sand, W. and Gehrke, T. (1999) Analysis and function of the EPS from the strong acidophile *Thiobacillus ferrooxidans.* In: *Bacterial Extracellular Polymeric Substances* (J. Wingender, T. Neu and H.C. Flemming, eds), pp. 127–141. Springer, Berlin.

Santo Domingo, J.W., Berry, C.J., Summer, M. and Fliermans, C.B. (1998) Microbiology of spent nuclear fuel storage basins. *Curr. Microbiol.* **37**, 387–394.

Schooling, S., Gilbert, P. and Allison, D. (2001) Is there a role for homoserine lactones in the biofilm phenotype. In: *Biofilm Communities: Chance or necessity?"* (P. Gilbert, D. Allison, M. Brading, J. Verran and J. Walker, eds). pp. 201–210. Bioline, Cardiff.

Shapiro, J.A. (1995) The significances of bacterial colony patterns. *Bioessays* **17**, 597–607.

Shepherd, J.A., Waigh, R.D. and Gilbert, P. (1988) Antibacterial action of 2-bromo-2-nitropropane-1,3-diol (bronopol). *Antimicrob. Agents Chemother.* **32**, 1693–1698.

Sibille, I., Sime-Ngando, T., Mathieu, L. and Block, J.C. (1998) Protozoan bacterivory and *Escherichia coli* survival in drinking water distribution systems. *Appl. Environ. Microbiol.* **64**, 197–202.

Singh, P.K, Schaefer, A.L, Parsek, M.R., Moninger, T.O., Welsh, M.J. and Greenberg, E.P. (2000) Quorum-sensing signals indicate that cystic fibrosis lungs are infected with bacterial biofilms. *Nature* **407**, 762–764.

Sissons, C.H., Wong, L. and Cutress, T.W. (1996) Inhibition by ethanol of the growth of biofilm and dispersed microcosm dental plaques. *Arch. Oral Biol.* **41**, 27–34.

Slack, M.P.E. and Nichols, W.W. (1981) The penetration of antibiotics through sodium alginate and through the exopolysaccharide of a mucoid strain of *Pseudomonas aeruginosa. Lancet* **11**, 502–503.

Slack, M.P.E. and Nichols, W.W. (1982) Antibiotic penetration through bacterial capsules and exopolysaccharides. *J. Antimicrob. Chemother.* **10**, 368–372.

Sly, L.I. and Cahill, M.M. (1997) Transfer of *Blastobacter natatorius* (Sly 1985) to the genus *Blastomonas* gen. nov. as *Blastomonas natatoria* comb. nov. *Int. J. Syst. Bacteriol.* **47**, 566–568.

Sondossi, M., Rossmore, H.W. and Wireman, J.W. (1985) Observation of resistance and cross-resistance to formaldehyde and a formaldehyde condensate biocide in *Pseudomonas aeruginosa. Int. Biodet. Biodeg.* **21**, 105–106.

Sternberg, C., Christensen, B.B., Johansen, T., Toftgaard Nielsen, A., Andersen, J.B., Givskov, M. and Molin, S. (1999) Distribution of bacterial growth activity in flow-chamber biofilms. *Appl. Environ. Microbiol.* **65**, 4108–4117.

Stewart, M.H. and Olson, B.H. (1992) Physiological studies of chloramine resistance developed by *Klebsiella pneumoniae* under low-nutrient growth conditions. *Appl. Environ. Microbiol.* **58**, 2918–2927.

Stewart, P.S. (1994) Biofilm accumulation model that predicts antibiotic resistance of *Pseudomonas aeruginosa* biofilms. *Antimicrob. Agents Chemother.* **38**, 1052–1058.

Stewart, P.S. (1996) Theoretical aspects of antibiotic diffusion into microbial biofilms. *Antimicrob. Agents Chemother.* **40**, 2517–2522.

Stewart, P.S. and Costerton, J.W. (2001) Antibiotic resistance in bacterial biofilms. *Lancet* **358**, 135–138.

Stewart, P.S., Grab, L. and Diemer, J.A. (1998) Analysis of biocide transport limitation in an artificial biofilm system. *J. Appl. Microb.* **85**, 495–500.

Stewart, P.S., Roe, F., Rayner, J., Elkins, J.G., Lewandowski, Z., Ochsner, U.A. and Hassett, D.J. (2000) Effect of catalase on hydrogen peroxide penetration into *Pseudomonas aeruginosa* biofilms. *Appl. Environ. Microbiol.* **66**, 836–838.

Stickler, D.J., Morris, N.S., McLean, R.J. and Fuqua, C. (1998) Biofilms on indwelling urethral catheters produce quorum-sensing signal molecules *in situ* and *in vitro*. *Appl. Environ. Microbiol.* **64**, 3486–3490.

Stoodley, P., Hall-Stoodley, L., Boyle, J., Jorgensen, F. and Lappin-Scott, H. (2000) Environmental factors and genetic factors influencing biofilm structure. In: *Community Structure and Cooperation in Biofilms* (D. Allison, P. Gilbert, H. Lappin-Scott and M. Wilson, eds). Cambridge University Press, Cambridge.

Suci, P.A., Mittelman, M.W., Yu F.P. and Geesey, G.G. (1994) Investigation of ciprofloxacin penetration into *Pseudomonas aeruginosa* biofilms. *Antimicrob. Agents Chemother.* **38**, 2125–2133.

Sundheim, G., Langsrud, S., Heir, E. and Holck, A.L. (1998) Bacterial resistance to disinfectants containing quaternary ammonium compounds. *Int. Biodet. Biodeg.* **41**, 235–239.

Sutherland, I.W. (1985) Biosynthesis and composition of Gram-negative bacterial extracellular and wall polysaccharides. *Ann. Rev. Microbiol.* **39**, 243–270.

Sutherland, I.W. (1997) Microbial biofilm exopolysaccharides – superglues or velcro? In: *Biofilms:Community Interactions and Control* (J. Wimpenny, P. Handley, P. Gilbert, H. Lappin-Scott, M. Jones, ed.) pp. 33–39. Bioline, Cardiff.

Symmons, S.A., McBain, A.J. and Gilbert, P. (2001) In: *Biofilm Community Interactions: Chance or Necessity?* (P. Gilbert, D.G. Allison, M. Brading, J. Walker and J. Verran, eds). pp. 241–253, Bioline, Cardiff.

Thanassi, D.G., Suh, G.S. and Nikaido, H. (1995) Role of outer membrane barrier in efflux-mediated tetracycline resistance of *Escherichia coli*. *J. Bacteriol.* **177**, 998–1007.

Trainor, V.C., Udy, R.K., Bremer, P.J. and Cook, G.M. (1999) Survival of *Streptococcus pyogenes* under stress and starvation. *FEMS Microbiol. Lett.* **176**, 421–428.

Tunney, M.M., Patrick, S., Curran, M.D., Ramage, G., Anderson, N., Davis, R.I., Gorman, S.P. and Nixon, J.R. (1999) Detection of prosthetic joint biofilm infection using immunological and molecular techniques. *Meth. Enzymol.* **310**, 566–576.

Uhlinger, D.J. and White, D.C. (1983) Relationship between physiological status and formation of extracellular polysaccharide glycocalyx in *Pseudomonas atlantica*. *Appl. Environ. Microbiol.* **45**, 64–70.

Vandevoorde, L., Christiaens, H. and Verstraete, W. (1992) Prevalence of coaggregation reactions among Chicken lactobacilli. *J. Appl. Bacteriol.* **2**, 214–219.

Van Loosdrecht, M.C., Lyklema, J., Norde, W. and Zehnder, A.J. (1990) Influence of interfaces on microbial activity. *Microbiol. Rev.* **54**, 75–87.

Veiga, M.C., Jain, M.K., Wu, W.-M., Hollingsworth, R. and Zeikus, J.G. (1997) Composition and role of extracellular polymers in methanogenic granules. *Appl. Environ. Microbiol.* **63**, 403–407.

Vroom, J.M., de Grauw, K.J., Gerritsen, H.C., Bradshaw, D.J, Marsh, P.D, Watson, G.K., Birmingham, J.J. and Allison, C. (1999) Depth penetration and detection of pH gradients in biofilms by two-photon excitation microscopy. *Appl. Environ. Microbiol.* **65**, 3502–3511.

Walsh, P.M. and McKay, L.L. (1981) Recombinant plasmid associated cell aggregation and high-frequency conjugation of *Streptococcus lactis* ML3. *J. Bacteriol.* **146**, 937–944.

Ward, K.H., Olson, M.E., Lam, K. and Costerton, J.W. (1992) Mechanism of persistent infection associated with peritoneal implants. *J. Med. Microbiol.* **36**, 406–413.

Watanabe, K., Teramoto, M., Futamata, H. and Harayama, S. (1998) Molecular detection, isolation, and physiological characterization of functionally dominant phenol-degrading bacteria in activated sludge. *Appl. Environ. Microbiol.* **64**, 4396–4402.

Webb, J.S., Van der Mei, H.C., Nixon, M., Eastwood, I.M., Greenhalgh, M., Read, S.J., Robson, G.D. and Handley, P.S. (1999) Plasticizers increase adhesion of the deteriogenic fungus *Aureobasidium pullulans* to polyvinyl chloride. *Appl. Environ. Microbiol.* **65**, 3575–3581.

Weiss, E.I., London, J., Kolenbrander, P.E., Hand, A.R. and Siraganian, R. (1988) Localization and enumeration of fimbria-associated adhesins of *Bacteroides loescheii*. *J. Bacteriol.* **170**, 1123–1128.

Wentland, E.J., Stewart, P.S., Huang, C.T. and McFeters, G.A. (1996) Spatial variations in growth rate within *Klebsiella pneumoniae* colonies and biofilm. *Biotechnol. Prog.* **12**, 316–321.

White, C., Sharman, A.K. and Gadd, G.M. (1998) An integrated microbial process for the bioremediation of soil contaminated with toxic metals. *Nat. Biotechnol.* **16**, 572–575.

Whyte, F.W., Allison, D.G., Jones, M.V. and Gilbert, P. (2001) *In vitro* detection of multiple antibiotic resistance (Mar) phenotype induction by miscellaneous groceries. *Proc. 101st Ann. Gen. Meet. Am. Soc. Microbiol.* A99, p 24.

Williams, P. (1988) Role of the cell envelope in bacterial adaptation to growth *in vivo* in infections. *Biochemie* **70**, 987–1011.

Williams, P., Bainton, N.J., Swift, S., Chhabra, S.R., Winson, M.K., Stewart, G.S.A.B., Salmond, G.P.C. and Bycroft, B.W. (1992) Small molecule-mediated density dependent control of gene expression in prokaryotes: bioluminescence and the biosynthesis of carbapenem antibiotics. *FEMS Microbiol. Lett.* **100**, 161–168.

Wimpenny, J. (1993). Bacterial biofilms and their control in medicine and industry. In: *Bacterial Biofilms and their Control in Medicine and Industry* (J. Wimpenny, W. Nicols, D. Stickler and H. Lappin-Scott, eds). Bioline, Cardiff.

Wimpenny, J. (1995). How thick must a biofilm be for gradients to determine its physiology? In: *The Life and Death of Biofilm* (J. Wimpenny, P. Handley, P. Gilbert and H. Lappin-Scott, eds), pp. 109–112. Bioline, Cardiff.

Wimpenny, J. (2000) An overview of biofilms as functional communities. In: *Community Structure and Co-operation in Biofilms* (H. Lappin-Scott, P. Gilbert, M. Wilson and D. Roberts, eds). Cambridge University Press, Cambridge.

Wright, N.E. and Gilbert, P. (1987a) Antimicrobial activity of n-alkyltrimethylammonium bromides: influence of specific growth rate and nutrient limitation. *J. Pharm. Pharmacol.* **39**, 685–690.

Wright, N.E. and Gilbert, P. (1987b) Influence of specific growth rate and nutrient-limitation upon the sensitivity of *Escherichia coli* towards polymyxin B. *J. Antimicrob. Chemother.* **20**, 303–312.

Wright, N.E. and Gilbert, P. (1987c) Influence of specific growth rate and nutrient limitation upon the sensitivity of *Escherichia coli* towards chlorhexidine diacetate. *J. Appl. Bacteriol.* **62**, 309–314.

Xu, K.D., Stewart, P.S., Xia, F., Huang, C.T. and McFeters, G.A. (1998) Spatial physiological heterogeneity in *Pseudomonas aeruginosa* biofilm is determined by oxygen availability. *Appl. Environ. Microbiol.* **64**, 4035–4039.

Xu, K.D., McFeters, G.A. and Stewart, P.S. (2000) Biofilm resistance to antimicrobial agents. *Microbiology* **146**, 547–549.

Xu, K.D., Franklin, M.J., Park, C.H., McFeters, G.A. and Stewart, P.S. (2001) Gene expression and protein levels of the stationary phase sigma factor, RpoS, in continuously-fed *Pseudomonas aeruginosa* biofilms. *FEMS Microbiol. Lett.* **199**, 67–71.

Yarmolinsky, M.B. (1995) Programmed cell death in bacterial populations. *Science* **267**, 836–837.

Yu, J.A., Ji, M. and Yue, P.L. (1999) A three-phase fluidized bed reactor in the combined anaerobic/aerobic treatment of wastewater. *J. Chem. Technol. Biot.* **74**, 619–626.

Zabinski, R.A., Walker, K.J., Larsson, A.J., Moody, J.A., Kaatz, G.W. and Rotschafer, J.C. (1995) Effect of aerobic and anaerobic environments on antistaphylococcal activities of five fluoroquinolones. *Antimicrob. Agents Chemother.* **39**, 507–512.

Zambrano, M. and Kolter, R. (1995) Changes in bacterial cell properties in going from exponential growth to stationary phase. In: *Microbial Quality Assurance: A Guide Towards Relevance and Reproducibility of Inocula* (M.R.W. Brown and P. Gilbert, eds) pp. 21–30. CRC Press, Boca Raton.

Zambrano, M., Siegele, D.A., Almiron, M., Tormo, A. and Kolter, R. (1993) Microbial competition: *Escherichia coli* mutants that take over stationary phase culture. *Science* **259**, 1757–1760.

Ziebuhr, W., Heilmann, C., Gotz, F., Meyer, P., Wilms, K., Straube, E. and Hacker, J. (1997) Detection of the intercellular adhesion gene cluster (ica) and phase variation in *Staphylococcus epidermidis* blood culture strains and mucosal isolates. *Infect. Immun.* **65**, 890–896.

Zobell, C.E. (1943) The effect of solid surfaces upon bacterial activity. *J. Bacteriol.* **43**, 39–56.

Biochemistry, Regulation and Genomics of Haem Biosynthesis in Prokaryotes

Mark R. O'Brian[1] and Linda Thöny-Meyer[2]

[1]*Department of Biochemistry, State University of New York at Buffalo,
Buffalo, NY 14214, USA*
[2]*Institute of Microbiology, ETH Zürich, Schmelzbergstrasse 7,
CH-8092 Zürich, Switzerland*

ABSTRACT

Haems are involved in many cellular processes in prokaryotes and
eukaryotes. The biosynthetic pathway leading to haem formation is,
with few exceptions, well-conserved, and is controlled in accordance
with cellular function. Here, we review the biosynthesis of haem and
its regulation in prokaryotes. In addition, we focus on a modification of
haem for cytochrome *c* biogenesis, a complex process that entails both
transport between cellular compartments and a specific thioether
linkage between the haem moiety and the apoprotein. Finally, a whole
genome analysis from 63 prokaryotes indicates intriguing exceptions
to the universality of the haem biosynthetic pathway and helps define
new frontiers for future study.

ADVANCES IN MICROBIAL PHYSIOLOGY VOL 46
ISBN 0-12-027746-8

ABBREVIATIONS

ALA	δ-Aminolaevulinic acid
GSA	Glutamate 1-semialdehyde
GTR	Glutamyl-tRNA reductase
PBG	Porphobilinogen
RF1	Peptide release factor 1

1. INTRODUCTION

Haems are the prosthetic groups or active moieties of proteins involved in the metabolism of molecular oxygen and other diatomic gases, in electron transfer, and oxidation-reduction reactions. For aerobic metabolism, haem is the prosthetic group of cytochromes, haemoglobins, catalases and peroxidases. Whereas respiration and oxidative phosphorylation are synonymous with O_2-based metabolism in eukaryotes, many prokaryotes carry out oxidative respiration in the absence of O_2, using nitrate, sulphate or other oxidants. Here, cytochromes are required nevertheless for electron transfer and the final reduction of the terminal electron acceptor as occurs with O_2-based respiration. Haem proteins can serve as sensors for diatomic gases such as O_2, CO or NO in signal transduction pathways (Rodgers, 1999). Exogenous haem serves as a nutritional iron source to some pathogenic bacteria that invade and colonize eukaryotic hosts. Finally, haem is a substrate for synthesis of phytochrome and bilins, which serve as light-sensing and light-harvesting molecules, respectively, in photosynthetic organisms.

Haem refers to protohaem, the prosthetic group of haemoglobins and *b*-type cytochromes, and also to modified protohaem derivatives found in cytochrome oxidases and the covalently bound prosthetic group of cytochromes *c* (Fig. 1). These modifications of haem, in conjunction with the context within the protein to which they associate, increase their versatility by altering their oxidation-reduction potential and ligand-binding capability. Protohaem is ubiquitous in organisms that make haem, but the other haems are more specialized.

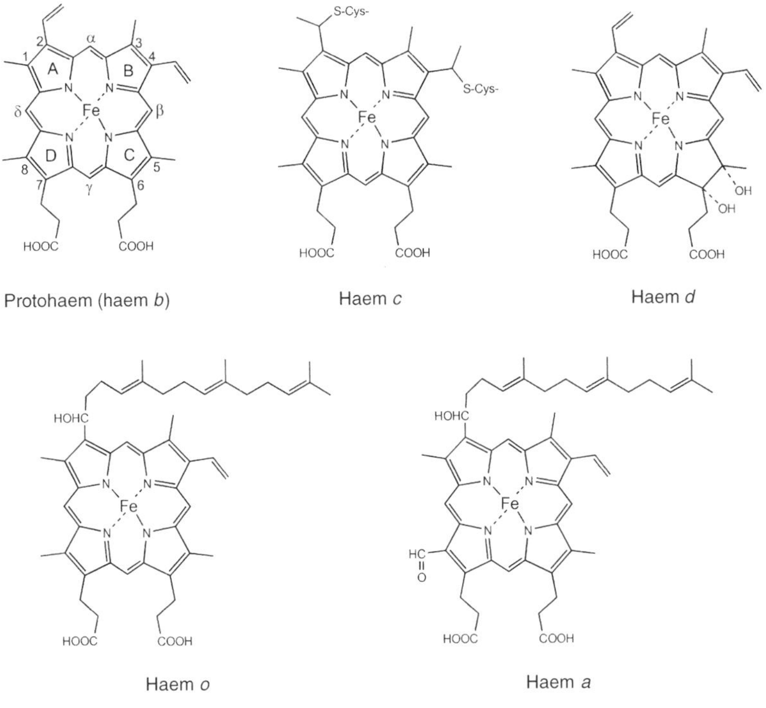

Protohaem (haem *b*)   Haem *c*   Haem *d*

Haem *o*   Haem *a*

Figure 1   Protohaem and its modifications. The numbering of protohaem carbons is by the Fischer system.

Cytochrome c is particularly interesting because, unlike the other haems, haem c is covalently linked to the apoprotein via thioether linkages. Thus, cytochrome c biogenesis requires both the translocation of haem from the cytoplasmic membrane, where it is synthesized to the periplasmic space concomitant with the ligation reactions.

Haem (protohaem) is the end product of a biosynthetic pathway and is a major focus of this review. For the most part, this pathway is highly conserved in terms of the intermediate precursors and the enzymes that catalyse each step. Cytochrome-based respiratory chains, hence haem, apparently arose early in the evolution of life (Schütz et al., 2000), and its presence in eukaryotes, Archaea and Bacteria is predominantly a reflection of a common ancestry rather than lateral gene transfer (Castresana and Moreira, 1999). However, the information explosion arising from whole genome sequencing has allowed us to examine haem biosynthesis genes in many organisms. As described here, this information reveals possible exceptions to the generally held view that the pathway is conserved, which needs to be resolved in order to understand more completely haem biosynthesis.

Regulation of biosynthetic pathways is essential for normal cell function, and contributes to the adaptive success of prokaryotes to diverse environmental conditions. Accordingly, haem biosynthesis is highly regulated to accommodate the qualitative and quantitative changes in the haem protein profile. In addition, control assures that toxic haem precursors do not accumulate in cells.

2. OVERVIEW OF TETRAPYRROLE SYNTHESIS

Haems belong to a larger class of molecules called tetrapyrroles, which include chlorophylls, chlorins, corrins and bilins. The first universal tetrapyrrole precursor δ-aminolaevulinic acid (ALA) is synthesized either from glutamate by

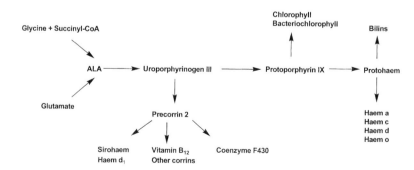

Figure 2   Outline of tetrapyrrole synthesis.

the C_5 pathway or from glycine and succinyl coenzyme A via ALA synthase (Fig. 2). The synthesis of uroporphyrinogen III from ALA requires three enzymatic steps with conserved intermediates. Uroporphyrinogen III is the final common precursor for all tetrapyrroles.

Uroporphyrinogen III is methylated to form precorrin 2, a precursor for vitamin B_{12}, sirohaem, haem d_1 and coenzyme F430 (Fig. 2). Vitamin B_{12} is a cofactor in numerous methyl transfer reactions. Sirohaem is a prosthetic group of sulphite reductase and nitrite reductase, and haem d_1 is also a prosthetic group of nitrite reductase in some bacteria. Thus, sirohaem and haem d_1 are not derived from protoporphyrin, and will not be described further. Coenzyme F430 is a nickel tetrapyrrole which serves as the prosthetic group of methyl coenzyme M reductase, an enzyme required for methanogenesis.

For haem and chlorophyll synthesis, uroporphyrinogen III is metabolized by three successive enzymatic steps that modify the side groups of the macrocycle to yield protoporphyrin. Protoporphyrin chelates either magnesium or iron catalysed by different chelatases for chlorophyll and protohaem formation, respectively. Protohaem is covalently modified further for c-, d-, o- and a-type cytochromes (Fig. 1), or is linearized for bilin formation. (We note here that the letter designation for a type of haem does not correlate with a gene designation. For example, haem c refers to a specific haem in cytochrome c ultimately derived from protohaem, whereas the *hemC* gene encodes an enzyme of the biosynthetic pathway leading to protohaem formation.)

3. STEPS OF THE HAEM SYNTHETIC PATHWAY

3.1. ALA Synthase

ALA synthase is a pyridoxal phosphate-dependent enzyme that catalyses the condensation of glycine and succinyl coenzyme A to form ALA (Fig. 3). The carboxyl carbon of glycine is lost as CO_2, and the remainder is incorporated into ALA. ALA synthase is found in non-photosynthetic eukaryotes, where it is located in mitochondria, and in some bacteria. ALA formation by the C_5 pathway occurs in plants, algae and in most prokaryotes. The photosynthetic phytoflagellate *Euglena gracilis* is the only organism reported to have both the C_5 pathway and ALA synthase (Weinstein and Beale, 1983). In that case, ALA derived by ALA synthase is incorporated into mitochondrial tetrapyrroles, whereas plastid pigments are made from ALA synthesized by the C_5 pathway (Weinstein and Beale, 1983).

Amongst prokaryotes, ALA synthase is restricted to the α-proteobacteria, a taxonomic group that includes photosynthetic bacteria, obligate intracellular pathogens of eukaryotes, plant symbionts and in general it encompasses a

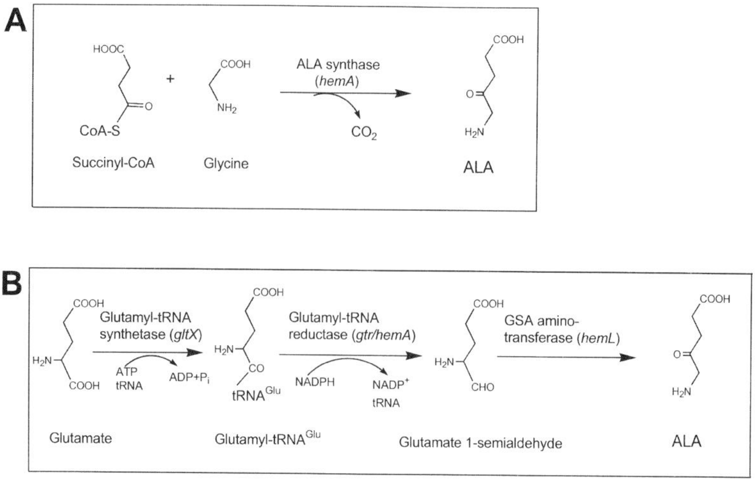

Figure 3   Two routes of ALA formation. (A) ALA synthesis from succinyl-coenzyme A and glycine via ALA synthase. (B) The C5 pathway for ALA synthesis from glutamate. The gene designations are shown in parentheses. *hemA* denotes genes encoding both glutamyl tRNA reductase and ALA synthase, but they are unrelated proteins that catalyse different reactions. We suggest *gtr* to denote the gene encoding glutamyl-tRNA reductase.

metabolically diverse group of organisms. The strict retention of ALA synthase and exclusion of the C_5 pathway within the α-proteobacterial group despite the diversity of tetrapyrroles and ecological niches that it represents, suggests that the distribution of ALA synthase reflects primarily a common ancestry rather than an advantageous evolutionary adaptation in extant organisms. ALA synthase must have arisen in an organism that makes succinyl-CoA. However, the distribution of the synthase is restricted, whereas the tricarboxylic acid cycle enzymes are widely distributed. Regardless, mitochondria are descended from the α-proteobacteria and eukaryotic ALA synthase probably has an antecedent from that group.

Bacterial ALA synthase, encoded by *hemA*, has been most studied in *Rhodobacter sphaeroides* and in the rhizobia. The interest in *R. sphaeroides* is founded primarily in the ALA requirement for bacteriochlorophyll formation, but photosynthesis and microaerobic metabolism are part of the same physiological process in that organism. *R. sphaeroides* contains two *hemA* homologues designated *hemA* and *hemT* (Neidle and Kaplan, 1993a, 1993b). Only *hemA* mRNA is expressed under aerobic or photosynthetic growth in wild-type cells, suggesting that the allele is sufficient for tetrapyrrole synthesis. However, *hemT* is expressed in a *hemA* mutant, which explains why both genes must be inactivated for ALA auxotrophy (Neidle and Kaplan, 1993b).

Interestingly, a *R. sphaeroides* ALA auxotrophic mutant derived from chemical mutagenesis was deficient only in the *hemA* gene, whereas the *hemT* coding region was unaltered (Zeilstra-Ryalls and Kaplan, 1995b). Furthermore, the auxotroph could be complemented by *hemA in trans*, but not by *hemT* even though *hemT* complements a *hemA hemT* double mutant. The basis for these observations is uncertain, but ALA synthase is a homodimer in animals (Tan and Ferreira, 1996), and thus the non-functional *hemA* gene product in *R. sphaeroides* may form mixed dimers yielding a dominant-negative phenotype.

Rhizobia live as free-living organisms or as endosymbionts of certain legumes within specialized root organs called nodules. The interest in ALA synthase, and haem synthesis in general, in the rhizobia was originally founded on the hypothesis that plant nodule haemoglobin haem is derived from the bacterium. Although this idea is unlikely to be correct (O'Brian, 1996, 2000; Santana *et al.*, 1998), the haem pathway in *Rhizobium* is highly regulated due to the metabolic diversity associated with its aerobic, microaerobic and symbiotic growth. In the latter case, the bacteria carry out the energy intensive process of nitrogen fixation in an O_2-limited environment of a nodule, which requires *de novo* synthesis of cytochromes to accommodate that demand (O'Brian, 2000). ALA synthase mutants of *Bradyrhizobium japonicum* can establish a successful symbiosis on their soybean host due to an ALA uptake activity as described below.

3.2. C_5 Pathway for ALA Synthesis

The informal reference of the C_5 pathway as the 'alternate' route for ALA synthesis has both a historic and anthropocentric basis, when in fact it is by far the predominant route in nature. The C_5 pathway was discovered in the mid 1970s in plants in the context of chlorophyll biosynthesis, and is now known to be the sole mechanism of ALA formation for all plant and algal tetrapyrroles (reviewed in Beale and Weinstein, 1990). The committed enzymes glutamyl-tRNA reductase and glutamate 1-semialdehyde (GSA) aminotransferase are nuclear-encoded and localized to plastids. The C_5 pathway is found in Bacteria exclusive of the α-proteobacteria and in the Archae.

In the first step, glutamyl-tRNA synthetase charges tRNA^{Glu} with glutamate as occurs in protein synthesis; thus this step is not committed to ALA formation (Fig. 3). Glutamyl-tRNA^{Glu} is reduced to GSA by glutamyl-tRNA reductase, and finally GSA is converted to ALA by a transamination reaction. Thus, the 5-carbon skeleton of glutamate is incorporated intact into ALA, hence the name C_5 pathway. The tRNA^{Glu} that activates glutamate has the UUC anticodon, and is also used in protein synthesis (Schon *et al.*, 1986; Schneegurt and Beale, 1988). However, a point mutation in the T-loop of chloroplast tRNA^{Glu} from *Euglena gracilis* affects chlorophyll formation, but not protein synthesis (Stange-Thomann *et al.*, 1994). As expected from that phenotype, the altered tRNA was acylated by the synthetase, but Glu-tRNA^{Glu} was a poor substrate for tRNA reductase. Thus, the tRNA has different specificity elements for the reductase than for the protein elongation factor Tu.

Glutamyl-tRNA reductase uses NADPH in the reduction of glutamyl-tRNA to GSA in the first committed step in tetrapyrrole synthesis in organisms that express the C$_5$ pathway. *E. coli* tRNA^{Glu} is a much better substrate for *E. coli* glutamyl-tRNA reductase than it is for other reductases (Avissar and Beale, 1990; Jahn *et al.*, 1991). This point may appear to belie the fact that ALA prototrophy can be restored in an *E. coli* glutamyl-tRNA reductase mutant by introduction of heterologous DNA encoding the reductase from other organisms (Ilag *et al.*, 1994; Sangwan and O'Brian, 1999), but activity recovered from extracts of the transformed cells is extremely low.

The primary structures of bacterial glutamyl-tRNA reductases are fairly well conserved, but the *E. coli* enzyme is a monomer of 45 kDa (Jahn *et al.*, 1991), whereas it may be a multimer in *B. subtilis* (Schröder *et al.*, 1992). The designation *hemA* refers to genes encoding both bacterial glutamyl-tRNA reductase and ALA synthase even though they are dissimilar proteins that catalyse different reactions. This confusing nomenclature is based on the original assumption that bacteria as a whole possess ALA synthase because of studies with animals and on work with *Rhizobium* (Avissar and Nadler, 1978; Leong *et al.*, 1982) and *Rhodobacter* (Nandi and Shemin, 1977), where the presence of ALA synthase was demonstrated. Thus, the mutated genetic loci of *E. coli* and *S. typhimurium* ALA auxotrophs presumed to encode ALA synthase were designated *hemA* (Sasarman *et al.*, 1968, 1970), and the *bona fide* cloned ALA synthase gene isolated from *Sinorhizobium meliloti* was named *hemA* as well (Leong *et al.*, 1982). The subsequent molecular characterization of the enteric bacterial genes revealed that *hemA* encodes glutamyl-tRNA reductase (Avissar and Beale, 1989; Elliott, 1989; Li *et al.*, 1989a, 1989b; Verkamp and Chelm, 1989), and made clear that the C$_5$ pathway is not unique to plants or to chlorophyll synthesis. We suggest that the gene encoding glutamyl-tRNA reductase be renamed *gtr*, a designation used for the plant enzyme (Sangwan and O'Brian, 1999), and *hemA* designate the bacterial gene encoding ALA synthase.

The *prfA* gene encoding peptide release factor 1 (RF1) lies immediately downstream of the *gtr* (*hemA*) gene in an operon in *E. coli* and *S. typhimurium*. Elliott proposed a regulatory mechanism for autoregulation of *prfA* by translational readthrough of the *gtr* gene. The *gtr* coding region terminates with a UAG (amber) codon recognized by RF1. Low level of RF1 results in readthrough of the *gtr* amber codon, which may bring a ribosome to the ribosome-binding site of *prfA* mRNA, thereby promoting translation reinitiation. This model suggests that *gtr* function is incidental to *prfA* control, and in principle any gene would work upstream of *prfA* provided that it contained an amber codon, could tolerate a C-terminal extension, and is translated at a fairly constant rate. The latter criterion would ensure that *prfA* is autoregulated rather than controlled by the upstream gene. Indeed, glutamyl-tRNA reductase is regulated, but that control is primarily at the level of protein stability rather than degradation (see below). *prfA* is immediately downstream of *ptsP* in *B. japonicum* and in other rhizobia (Kaneko *et al.*, 2000; Galibert, 2001; N.D. King and M.R. O'Brian, unpublished). PtsP controls ALA transport in *B. japonicum* (King and O'Brian, 2001), raising the possibility of a functional link between RF1 and ALA metabolism.

GSA aminotransferase activity results in the net transfer of an amino group from the 2-carbon to the 1-carbon position (Fig. 3). However, this reaction involves intermolecular nitrogen transfer between two GSA molecules (Mau and Wang, 1988; Mayer *et al.*, 1993), and therefore the enzyme is not an aminomutase as it is sometimes referred. GSA aminotransferase is a pyridoxal phosphate-dependent enzyme. The bacterial enzyme is encoded by the *hemL* gene, and like the glutamyl-tRNA reductase gene, it is prevalent in the sequenced prokaryotic genomes that make tetrapyrroles. The conversion of GSA to ALA occurs non-enzymatically *in vitro*, which may explain the leaky phenotype of *E. coli hemL* mutants (Ilag *et al.*, 1991). The X-ray crystal structure of GSA aminotransferase from *Synechococcus* has been resolved, and reveals it to be a dimer with asymmetry in pyridoxal phosphate cofactor binding and active site reactivity (Hennig *et al.*, 1997). Pyridoxal phosphate binds to Lys-273 of one of the subunits, and a region that is disordered on that subunit is ordered on the other. The structure bound to substrate analogue gabaculine, and presumably the physiological substrate GSA, reveals that the active site includes residues from the N-terminal domain.

3.3. ALA Dehydratase

ALA dehydratases from prokaryotes and eukaryotes share a high degree of sequence similarity with each other. Phylogenetic analysis shows that plant ALA dehydratase, a nuclear-encoded plastid enzyme, is more similar to prokaryotic ALA dehydratase than to the enzyme from other eukaryotes, and suggests that the plant gene is descended from the bacterial ancestor of chloroplasts (Kaczor *et al.*, 1994).

ALA dehydratase is a homooctamer that catalyses the condensation of two ALA molecules to form the pyrrole porphobilinogen (PBG) (Fig. 4). There are two ALA-binding sites, the so-called A and P sites corresponding to the ALA that contributes to the acetate or propionate portion of PBG, respectively. The

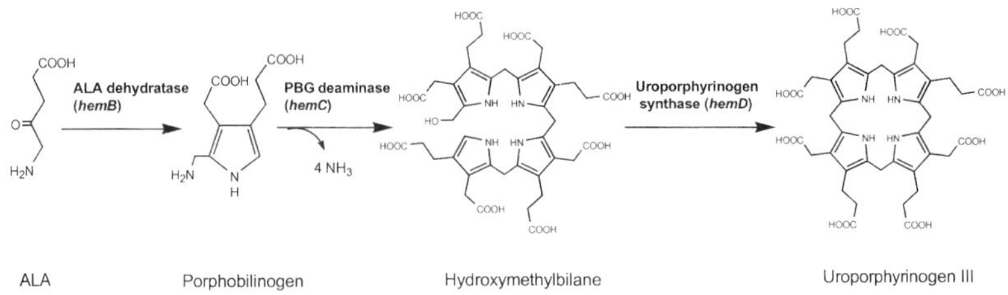

Figure 4   Synthesis of uroporphyrinogen III from ALA. This portion of the pathway is common to all tetrapyrroles.

P site ALA forms a Schiff's base with a conserved lysine residue (lys-247 in *E. coli*). Features of the A site probably differ among the different enzymes, depending on the metal requirement. Animal and yeast ALA dehydratases are zinc-binding proteins whereas the plant enzyme requires Mg^{2+} for activity. Bacterial ALA dehydratases are diverse in their metal requirement, which may be Zn^{2+}, Mg^{2+} or both, and may also bind to monovalent ions as well (Jaffe, 2000).

Studies with the bovine zinc enzyme indicated two types of zinc, a structural zinc involving three proximal cysteines, and a catalytic zinc coordinated by other ligands (Dent *et al.*, 1990). However, analysis of bacterial mutants and structural information shows that the triple cysteine region is involved in catalysis. *B. japonicum* ALA dehydratase is a Mg^{2+}-dependent enzyme (Chauhan and O'Brian, 1993). Amino acid substitutions that place three cysteines in positions normally found in the zinc enzymes of animal and yeast enzymes are sufficient to alter the metal specificity from Mg^{2+} to Zn^{2+} (Chauhan and O'Brian, 1995). Thus, it was possible to identify regions responsible for each metal in the respective enzymes. The modified enzyme also acquires sensitivity to lead as is observed for the naturally occurring Zn^{2+} ALA dehydratases (Chauhan *et al.*, 1997). In support of those observations, X-ray crystallographic analysis of the *S. cerevisiae* ALA dehydratase shows that those three cysteines bind zinc and lead directly (Erskine *et al.*, 1997). Furthermore, the modified *B. japonicum* enzyme has a pH optimum 3.5 units lower than the wild-type enzyme, and is much more thermostable (Chauhan *et al.*, 1997). These observations suggest that the cysteine-rich domain in the zinc dehydratases is involved in both structure and catalysis. Furthermore, the crystal structures of the yeast (Erskine *et al.*, 1999) and *E. coli* (Erskine *et al.*, 1999) ALA dehydratases show that the liganding cysteines found in the wild-type zinc enzymes and introduced in the mutant *B. japonicum* enzyme are part of the active site and proximal to lysine that forms a Schiff base with ALA.

E. coli ALA dehydratase is a zinc metalloenzyme that is stimulated by Mg^{2+} (Mitchell and Jaffe, 1993). It is an octomer comprised of four asymmetric dimers. The X-ray crystal structure shows that the cysteine domain coordinates with an active site zinc (Erskine *et al.*, 1999) as the mutant studies suggested, and a second zinc atom probably contributes to the stability of the octomer. Mg^{2+} may have a stabilizing effect as well.

Binding and kinetic studies indicate that *B. japonicum* ALA dehydratase contains four active site Mg^{2+} atoms per octamer, and eight allosteric Mg^{2+} ions (Petrovich *et al.*, 1996). The Mg^{2+} enzyme from *Pseudomonas aeruginosa* contains only four allosteric ions per octamer, but activity is not strictly metal-dependent (Frankenberg *et al.*, 1999b). Similarly, ALA dehydratase from *Chlorobium vibrioforme* is enhanced by Mg^{2+}, but is not essential (Rhie *et al.*, 1996). The crystal structure of the *P. aeruginosa* enzyme shows that it is composed of four asymmetric dimers. The monomers in each dimer differ by having either a closed active site pocket shielded from solvent by a 'lid' or an open active site (Frankenberg *et al.*, 1999a). The Mg^{2+} is 14 Å from the Schiff base lysine, which is too far to play a direct catalytic role. However, it is suggested that the open and closed forms of each monomer are governed by Mg^{2+} binding (Frankenberg *et al.*, 1999a). It is unclear whether plant ALA dehydratases will be similar to either *B. japonicum* or *P. aeruginosa* with respect to the role of Mg^{2+}, but the strict metal requirement for the plant enzyme is similar to that observed in *B. japonicum*.

3.4. Porphobilinogen Deaminase

PBG is polymerized to the open chain tetrapyrrole 1-hydroxymethylbilane by the enzyme porphobilinogen deaminase (Fig. 4). The *hemC* gene encoding PBG deaminase is clustered with the *hemD* in many bacterial genomes. *E. coli* PBG deaminase is a 34-kDa monomer that has been crystallized and characterized (Scott *et al.*, 1988; Warren and Jordan, 1988; Louie *et al.*, 1992). Interestingly, PBG is both a substrate and a component of the cofactor for PBG deaminase. The dipyrromethane cofactor is a PBG dimer covalently linked to Cys-242 (*E. coli* numbering) through its free amino group (Jordan *et al.*, 1988b; Rose *et al.*, 1988; Warren and Jordan, 1988). This explains the deficiency of PBG deaminase activity in ALA synthase ALA dehydratase mutants (Umanoff *et al.*, 1988; Sangwan and O'Brian, 1991). The dipyrromethane cofactor in the active site primes the sequential addition and deamination of the PBG tetrapolymerization reaction. Thus, there is a transient PBG hexamer of which the hydroxymethylbilane tetrapyrrole is hydrolysed leaving the cofactor intact to prime the subsequent tetrapolymerization reaction. Polymerization terminates after four units either because the size of the active site cavity sterically prevents additional PBG molecules, or else the chain is pulled through the active site (Louie *et al.*, 1992).

3.5. Uroporphyrinogen III Synthase

Hydroxymethylbilane is an unstable intermediate that is readily cyclized to uroporphyrinogen I nonenzymatically, which can be decarboxylated to form coproporphyrinogen I. However, these symmetric isomers are not physiologically useful. Uroporphyrinogen III synthase catalyses the cyclization of hydroxymethylbilane and inversion of the D ring to the asymmetric III isomer of uroporphyrinogen. Uroporphyrinogen III is the final common intermediate for all tetrapyrroles. Recombinant uroporphyrinogen III synthase from *E. coli* (Jordan *et al.*, 1988a; Alwan *et al.*, 1989) and *B. subtilis* (Stamford *et al.*, 1995) has been obtained and characterized to some extent. Mouse uroporphyrinogen III synthase shares only about 22% identity with the *B. subtilis* enzyme and very little homology with that from *E. coli*. Remarkably, uroporphyrinogen III synthase activity has been described in plants and is necessary for chlorophyll synthesis (Higuchi and Bogorad, 1975), but the gene has not been identified in a plant, nor has a gene homologue been implicated in the sequenced *Arabidopsis thaliana* genome.

The *hemD* gene is clustered with *hemC* in many cases in bacteria, and with other tetrapyrrole synthesis genes in some instances, such as in the genome of *B. subtilis* (Hansson *et al.*, 1991). The *hemC* and *hemD* genes are found immediately downstream of *algR* in *Pseudomonas aeruginosa* (Mohr *et al.*, 1994). AlgR regulates *algD*, a virulence factor gene involved in the mucoid phenotype that is expressed by the organism in people with cystic fibrosis. Mutations in *hemC* or *hemD* affect *algD* promoter activity under nitrate growth (Mohr *et al.*, 1994). The observations suggest a physiological link between tetrapyrrole formation and mucoidy and virulence in that organism.

An intriguing *hemD*-like gene is found in the strict anaerobe *Clostridium josui*, that is a gene fusion of the *cobA* gene with *hemD* predicted to encode a hybrid protein (Fujino *et al.*, 1995). The *cobA* gene encodes uroporphyrinogen III methyltransferase, an activity involved in cobalamin and sirohaem synthesis. The *cobA/hemD* fusion can complement an *E. coli hemD* mutant; thus it has uroporphyrinogen III synthase activity. It should be noted that the *cobA/hemD* fusion from *C. josui* has led to confusing annotation of other uncharacterized genes from sequenced genomes that might lead one to the erroneous conclusion that uroporphyrinogen III synthase and uroporphyrinogen III methyltransferase and their respective genes are equivalent to each other. In fact, the vast majority of those genes encode proteins similar to either the synthase or the methyltransferase.

The CobA (uroporphyrinogen III methyltransferase) protein is functionally and structurally homologous to the C-terminal portion of sirohaem synthase (CysG), a multifunctional protein that catalyses the methylation reaction of CobA, in addition to two additional reactions necessary for sirohaem formation from uroporphyrinogen III. Interestingly, *C. josui* contains a *hemA/cysG* gene fusion in which the 5′ portion of the *cysG* is found. Thus, the *cysG* sequence is separated in two sections that are fused to *hemA* and *hemD*. Accordingly, the gene cluster containing *hemA/cysG-hemC-cobA/hemD-hemB* can complement a *cysG* mutant of *E. coli* (Fujino *et al.*, 1995), strongly indicating that the CysG domains of each protein can collectively carry out the reactions normally carried out by an intact *cysG* gene product.

3.6. Uroporphyrinogen Decarboxylase

Uroporphyrinogen decarboxylase catalyses the sequential decarboxylation of the four acetate residues of uroporphyrinogen III to methyl groups yielding coproporphyrinogen III (Fig. 5). It is also capable of decarboxylating the other nonphysiological isomers of uroporphyrinogen as well. At high substrate concentrations, the decarboxylation sequence is random (Jones and Jordan, 1993), but occurs in a specific sequence physiologically, beginning with the acetate on ring D, followed by A, B and finally C (Luo and Lim,

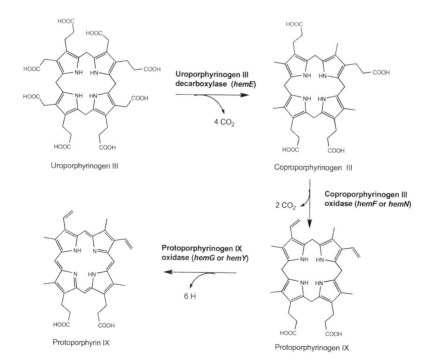

Figure 5   Synthesis of protoporphyrin IX from uroporphyrinogen III.

1993). Using site-directed mutations of yeast uroporphyrinogen decarboxylase and assaying decarboxylation intermediates, Labbe-Bois' group suggested that all four carboxylations occur at a single active site (Chelstowska *et al.*, 1992). The X-ray crystal structure of the human recombinant decarboxylase supports a single active site as well (Whitby *et al.*, 1998). The crystal structure also reveals that uroporphyrinogen decarboxylase is a dimer (Whitby *et al.*, 1998), whereas the protein from numerous sources was previously reported to be a monomer (reviewed in Elder and Roberts, 1995).

Genes or cDNAs encoding uroporphyrinogen decarboxylase have been isolated from animal (Romana *et al.*, 1987a, 1987b), yeast (Chelstowska *et al.*, 1992) and bacterial sources (Kiel *et al.*, 1990; Hansson and Hederstedt, 1992; Nishimura *et al.*, 1993), and genome sequencing projects indicate *hemE* gene homologues in prokaryotes that make haem or chlorophyll. The proteins are highly conserved, and are likely to be functionally very similar. The major interest in uroporphyrinogen decarboxylase is based on the human disease porphyria cutanea tarda, which is caused by a mutation in the corresponding gene (Elder and Roberts, 1995).

3.7. Coproporphyrinogen III Oxidase

The propionate groups at positions 2 and 4 of coproporphyrinogen III are oxidatively decarboxylated to vinyl groups to form protoporphyrinogen IX by the enzyme coproporphyrinogen III oxidase (Fig. 5). Eukaryotic coproporphyrinogen oxidases are structurally conserved and require O_2 for the oxidative reaction. Numerous bacteria also contain an O_2-dependent coproporphyrinogen oxidase, a protein approximately 300 amino acids in length that shares 60–70% similarity with the plant and animal oxidases, and about 45% similarity to the yeast enzyme.

In the yeast *Saccharyomyces cerevisiae,* control of genes for oxygen is thought to be sensed through haem biosynthesis via this O_2-dependent step of the pathway (Kastaniotis and Zitomer, 2000); thus there is a positive correlation between haem and oxygen. However, unlike eukaryotes, bacteria can carry out respiratory metabolism using nitrate, sulphate or other electron acceptors in place of O_2. Thus, hypoxia or anaerobiosis is not necessarily associated with fermentative metabolism, and the demand for haem persists and may actually increase. Similarly, hypoxia is required for induction of chlorophyll synthesis for photosynthesis in bacteria such as *Rhodobacter,* which includes coproporphyrinogen oxidase activity. This problem is resolved by the expression of an O_2-independent coproporphyrinogen oxidase in prokaryotes encoded by the *hemN* gene.

The *hemN* gene product is dissimilar from the O_2-dependent enzyme encoded by *hemF* in its primary sequence even though the porphyrinogen substrates and products are identical in both cases. The O_2-independent copro-porphyrinogen oxidase was originally identified based on a *R. sphaeroides* mutant that could not synthesize bacteriochlorophyll, but accumulated the oxidized precursor coproporphyrin (Coomber *et al.*, 1992; Gibson *et al.*, 1992). The subsequent identification of *hemN* genes in *S. typhimurium* (Xu and Elliott, 1994) and *E. coli* (Troup *et al.*, 1995) demonstrated that the O_2-independent oxidase is not dedicated to chlorophyll synthesis. In *S. typhimurium*, which contains both *hemF* and *hemN*, a mutation in *hemF* shows no growth impairment, whereas a *hemN* mutant is a haem auxotroph only under anaerobic growth (Xu *et al.*, 1992). Similarly, a *hemF hemN* double mutant can be complemented by *hemF* only for aerobic growth, but *hemN* confers normal aerobic or anaerobic growth. Thus, the O_2-independent coproporphyrinogen oxidase can function in the presence of O_2, but does not use it for catalysis.

An O_2-independent coproporphyrinogen oxidase activity has been detected in extracts of *R. sphaeroides* (Seehra *et al.*, 1983) and *Bradyrhizobium japonicum* (Keithly and Nadler, 1983) that apparently uses $NADP^+$ in the presence of ATP and S-adenosylmethionine. However, Xu and Elliott (Xu and Elliott, 1994) could not demonstrate that activity from the *S. typhimurium hemN* product. Thus, the physiological oxidant of coproporphyrinogen III used by the O_2-independent oxidase is unresolved.

Whole genome sequencing indicates that *hemN* is prevalent in bacteria that synthesize haem. Even organisms that contain a *hemF* gene also contain a *hemN* homologue. Furthermore, several bacteria have two *hemN* homologues. In *B. japonicum*, *hemN2* is required for anaerobic growth and symbiosis with soybean, whereas a *hemN1* mutant has no phenotype (Fischer *et al.*, 2001). Furthermore, the *B. japonicum hemN2*, but not *hemN1*, was able to complement a *S. typhimurium hemF hemN* strain, indicating that the *hemN1* product may not be an active protein. Similarly, a mutation in only one of two *hemN* homologues of *Bacillus subtilis* affected aerobic and anaerobic growth (Hippler *et al.*, 1997; Homuth *et al.*, 1999).

The *hemN* gene is clustered with other genes involved in microaerobic respiration in *R. sphaeroides* (Zeilstra-Ryalls and Kaplan, 1995a), *Mesorhizobium loti* (Kaneko *et al.*, 2000) and *Paracoccus denitrificans* (de Gier *et al.*, 1996). In particular, it is proximal to the *ccoNOQP* (*fixNOQP*) operon encoding the subunits of the cytochrome cbb_3 oxidase. This oxidase has an extremely high affinity for O_2, and is induced under microaerobic conditions (Preisig *et al.*, 1993, 1996). An *fnr* gene homologue is also found in this cluster. Fnr is a transcriptional regulator that mediates O_2-dependent gene expression. This gene clustering indicates that the O_2-independent coproporphyrinogen oxidase is not an anaerobic oxidase in the strict sense of the term, but allows the organism to respire and maintain oxidative metabolism under oxygen deprivation.

3.8. Protoporphyrinogen IX Oxidase

Protoporphyrinogen oxidase catalyses the six electron oxidation of protoporphyrinogen to protoporphyrin in the penultimate step of the haem pathway, and the final common step of haem and chlorophyll formation (Fig. 5). This oxidation results in conjugated double bonds in the tetrapyrrole macrocycle that makes protoporphyrin, haems and chlorophylls chromophoric. A protoporphrinogen oxidase deficiency is the genetic basis for variegate porphyria in humans (Nordmann and Deybach, 1990; Warren *et al.*, 1996). In plants, protoporphyrinogen oxidase is the target of diphenyl ether herbicides, such as acifluorfen-methyl (Matringe *et al.*, 1989; Camadro *et al.*, 1993). The phytotoxicity of these herbicides is based on the accumulation of protoporphyrinogen, which diffuses from the chloroplast to other parts of the cell, where it is then oxidized non-enzymatically to generate reactive oxygen species, thereby damaging cellular components (Camadro *et al.*, 1993; Jacobs and Jacobs, 1993). The deleterious effects of protoporphyrin deficiency in animals and plants underscores the physiological need for an enzyme to catalyse protoporphyrinogen oxidation despite the fact that the reduced porphyrinogen can be readily oxidized non-enzymatically.

In strictly aerobic organisms, molecular oxygen is the electron acceptor in protoporphyrinogen oxidation, producing H_2O_2 (Dailey and Dailey, 1996). In anaerobes or facultative aerobes, another electron acceptor is involved, and oxidation may be linked to the respiratory chain (Jacobs and Jacobs, 1976, 1977, 1981). Although unproven, this link may explain the accumulation of porphyrins and deficiency in protoporphyrinogen oxidase activity of mutants defective in cytochrome *c* biogenesis (O'Brian *et al.*, 1987; Biel and Biel, 1990; Ramseier *et al.*, 1991; Yeoman *et al.*, 1997). The purified enzyme from the anaerobic bacterium *Desulfovibrio gigas* is composed of three dissimilar subunits; it cannot use molecular oxygen, but does use the artificial electron acceptor 2,6-dichlorophenol-indophenol *in vitro* (Klemm and Barton, 1987).

The molecular characterization of bacterial protoporphyrinogen oxidase has focused primarily on *E. coli* and *Bacillus subtilis*, which express entirely different oxidases. The *E. coli hemG* gene locus was identified on analysis of a mutant that accumulates protoporphyrinogen and is defective in protoporphyrinogen oxidase activity (Sasarman *et al.*, 1979). The *hemG* gene expressed *in trans* from a plasmid complements a *hemG* mutant with respect to activity in crude extract and membrane preparations (Sasarman *et al.*, 1993). Enzyme activity of the purified 21 kDa product has not been demonstrated; therefore *hemG* may not be sufficient for protoporphyrinogen oxidation, and *hemG* could encode a subunit of a larger complex. The apparent requirement for electron transport for protoporphyrinogen oxidation in *E. coli* may make it difficult to characterize unambiguously the *hemG* product in pure form. In fact, direct evidence that the *hemG* product is a protoporphyrinogen oxidase protein is absent.

The 51-kDa protoporphyrinogen oxidase protein encoded by *hemY* in *Bacillus subtilis* is homologous to the oxidase in eukaryotes (Hansson and Hederstedt, 1992), and both the eukaryotic and bacterial proteins use O_2 as the terminal electron acceptor. However, anaerobic growth of *B. subtilis* is affected in a *hemY* mutant (Homuth *et al.*, 1999). Eukaryotic protoporphyrinogen oxidase and bacterial HemY are flavin-containing proteins (Proulx and Dailey, 1992; Camadro *et al.*, 1994; Dailey and Dailey, 1996, 1997) and belong to an FAD superfamily that also includes monoamine oxidase and phytoene desaturase (Dailey and Dailey, 1998). There is one FAD per dimer (Dailey and Dailey, 1996), indicating that the six electron oxidation occurs in three steps.

Interestingly, *B. subtilis* protoporphyrinogen oxidase is resistant to diphenyl ether herbicides, whereas the enzyme from the bacterium *Myxococcus xanthus* is sensitive (Dailey and Dailey, 1996), as are the eukaryotic oxidases. Furthermore, transgenic rice plants that express the *B. subtilis hemY* gene are resistant to the diphenyl ether herbicide oxyfluorfen, with protein targeted to plastids showing a greater resistance than those targeted to the cytoplasm (Lee *et al.*, 2000). However, the effect of the transgene may be due to a high level of enzyme activity in those plants rather than to the resistance of the protein *per se* since transgenic *Arabidopsis* plants that overexpress the tobacco protoporphyrinogen oxidase are also herbicide-resistant (Lermontova and Grimm, 2000).

The light sensitivity of an *E. coli hemH* (ferrochelatase) gene was reversed by a second mutation in a gene named *hemK* (Nakayashiki *et al.*, 1995), found in the putative *hemA-prfA-hemK* operon. The *hemK* mutant accumulated coproporphyrin and protoporphyrin when growth medium was supplemented with ALA, suggesting to the authors that *hemK* may be deficient in protoporphyrinogen oxidase activity. However, a mutation in a *hemK* homologue in yeast is not deficient in protoporphyrinogen oxidase, nor does overexpression of the *hemK* homologue confer activity (Le Guen *et al.*, 1999). Furthermore, HemK-related proteins have a consensus motif for S-adenosyl-methionine-dependent methyltransferases, and may be involved in that function. Finally, *hemK* homologues are prevalent in prokaryotes as indicated by genome sequencing, including the genomes of *Borellia burgdorferi* (Fraser *et al.*, 1997) and *Treponema pallidum* (Fraser *et al.*, 1998), organisms that do not make haem nor contain *bona fide* haem synthesis genes. From this, it is unlikely that *hemK* encodes a protoporphyrinogen oxidase, and the annotation of *hemK* homologues as such in the genome databases is not based on direct evidence.

3.9. Ferrochelatase

Ferrochelatase, encoded by the *hemH* gene in prokaryotes, catalyses the insertion of ferrous iron into the protoporphyrin IX macrocycle to yield protohaem (Fig. 6). Iron in the environment is primarily in the ferric form, and thus it must

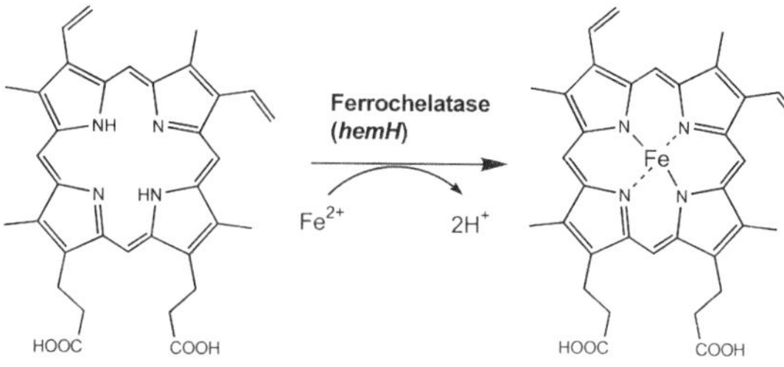

Figure 6 Chelation of ferrous iron into protoporphyrin IX via ferrochelatase to form protohaem.

be reduced before chelation. Other divalent ions such as Co^{2+} and Zn^{2+} can serve as a substrate *in vitro*, whereas others such as Mn^{2+}, Cd^{2+} and Pb^{2+} are competitive inhibitors (Dailey *et al.*, 2000). Interestingly, porphyrin metallation can be catalysed by catalytic antibodies (Cochran and Schultz, 1990), DNA (Li and Sen, 1997) and RNA (Morgan *et al.*, 1996). Ferrochelatase is functionally similar to magnesium chelatase, the enzyme that inserts Mg^{2+} into protoporphyrin in the first committed step in chlorophyll (and bacteriochlorophyll) formation (Walker and Willows, 1997). However, the enzymology of the two proteins is quite different. Ferrochelatase is a monomer or homodimer, whereas Mg-chelatase is comprised of three distinct subunits. Furthermore, ferrochelatase activity requires no ATP, in contrast to Mg-chelatase that requires ATP in a two-step activation and chelation process.

An *E. coli* ferrochelatase mutant was initially identified based on light sensitivity, which is caused by light-dependent formation of reactive oxygen species by the protoporphyrin that accumulated in those cells (Miyamoto *et al.*, 1991; Nakahigashi *et al.*, 1991). The mutated gene, originally designated *visA*, encodes a protein with similarity to the yeast ferrochelatase, and which was shown to have ferrochelatase activity (Frustaci and O'Brian, 1993b). Similarly, a *Bradyrhizobium japonicum* ferrochelatase mutant was screened based on the formation of fluorescent colonies due to the accumulation of protoporphyrin (Frustaci and O'Brian, 1992, 1993a). The *hemH* gene of *B. subtilis* was cloned as part of the *hemEHY* cluster by complementation of haem-deficient strains (Hansson and Hederstedt, 1992). *E. coli* normally does not transport haem from the medium, but haem mutants can be rendered haem auxotrophs by a poorly defined secondary mutation which allows uptake of haem (haemin). Conversely, *B. japonicum* and *B. subtilis* ferrochelatase (*hemH*) mutants are strict haem auxotrophs (Frustaci and O'Brian, 1992; Hansson and Hederstedt, 1992), showing that haem uptake is not confined to pathogenic bacteria.

Ferrochelatases are conserved among prokaryotes and eukaryotes, but few amino acid residues are strictly conserved in all of them. Animal ferrochelatases are [2Fe-2S] iron-sulphur proteins (Dailey *et al.*, 1994; Ferreira *et al.*, 1994) that possess coordinating cysteines in the C-terminus; these cysteines are absent in plant ferrochelatases, and the C-terminus is altogether absent in many bacteria. The physiological function of the iron-sulphur cluster is not known. More recently, bacteria with C-terminal extensions have been identified that contain cysteine clusters, but are distinct from that found in animals (Dailey *et al.*, 2000). The structural heterogeneity is also seen in a phylogenetic analysis of characterized and putative ferrochelatases (Dailey *et al.*, 2000). The bacterial enzymes do not all cluster together, but rather form two groups that do not coincide with the taxonomic groupings of the organisms. The significance of this is unclear.

Eukaryotic ferrochelatase is membrane-associated in mitochondria (Dailey, 1990) and in chloroplasts of plants (Matringe *et al.*, 1994), and is associated with membranes in some bacteria (Jacobs and Jacobs, 1984; Dailey *et al.*, 1986; Frustaci and O'Brian, 1993a). However, the enzyme from *B. subtilis* is soluble (Hansson and Hederstedt, 1994). This is an interesting feature that needs to be reconciled with the lipophilic nature of the substrate and product of ferrochelatase.

The X-ray crystal structure of human (Wu *et al.*, 2001) and *B. subtilis* (Alkaradaghi *et al.*, 1997; Lecerof *et al.*, 2000) ferrochelatases have been solved. The eukaryotic enzyme is a homodimer which is stabilized by interactions between the C-terminal extension that is missing in the bacterial enzyme. Accordingly, the *B. subtilis* ferrochelatase is a monomer. It remains to be determined whether bacterial ferrochelatases possessing a C-terminal extension are dimers as well. Furthermore, human ferrochelatase contains a 12 residue N-terminal region that is probably involved in membrane association and which is missing in the soluble *B. subtilis* enzyme (Gora *et al.*, 1999; Wu *et al.*, 2001). This stretch is conserved in some bacteria, but not others. Dailey *et al.* (2000) suggest that predictions about the oligomerization and solubility of bacterial ferrochelatases are plausible based on the N-terminal and C-terminal regions.

4. MODIFICATIONS OF PROTOHAEM FOR CYTOCHROME OXIDASE HAEMS

Protohaem is the functional group of *b*-type cytochromes, cytochromes P450, and haemoglobins. Protohaem is modified further to haems *a*, *o* or *d* found in terminal cytochrome oxidases, and to a substituted mesohaem, haem *c*, for the cytochrome *c* prosthetic group (see Fig. 1). Haems *o* and *d* are unique to prokaryotes, but haem o is a precursor for haem a, which is found in eukary-otes as well. Haem d is a prosthetic group in the cytochrome bd oxidase complex and is also the active group for $E.$ $coli$ catalase type HPII. Haem d is modified from protohaem by the addition of two hydroxyl groups in the cis conformation at positions 5 and 6 of the macrocycle (Fig. 1). $E.$ $coli$ catalase HPII catalyses the modification of protohaem to haem d (Loewen et $al.$, 1993). Mutant proteins have been identified that retain only residual activity and con-tain protohaem rather than haem d or mixtures of protohaem and cis and $trans$ isomers of haem d (Mate et $al.$, 1999).

Haems o and a are prosthetic groups in haem-copper oxidases. Both haems contain a 17-carbon hydroxyethylfarnesyl chain linked to the A ring vinyl group of protohaem. Haem o is further modified by oxidation of the D ring methyl group to a formyl group to yield haem a (reviewed in Mogi et $al.$, 1994). Overexpression of $E.$ $coli$ $cyoE$ gene encoding haem o synthase increases the haem o content in cytoplasmic membranes, and can be carried out in $vitro$ using farnesyl diphosphate, ferrous protohaem and Mg^{2+} (Saiki et $al.$, 1993). Haem o is likely to be an intermediate in haem a synthesis in $vivo$. Haem a is synthesized in $E.$ $coli$ cells that overexpress $B.$ $subtilis$ $ctaAB$ genes (Svensson et $al.$, 1993; Svensson and Hederstedt, 1994). The $ctaB$ gene is a $cyoE$ homologue and a $ctaA$ mutation of $B.$ $subtilis$ cannot synthesize haem a. CtaA is a haem protein that probably catalyses the monooxygenation of the methyl side group of haem in the formylation of haem o (Svensson and Hederstedt, 1994; Svensson et $al.$, 1996). Furthermore, a haem o synthase homologue, Cox10p, is present in $S.$ $cerevisiae$ even though yeasts do not use haem o as a prosthetic group (Nobrega et $al.$, 1990; Tzagoloff et $al.$, 1990, 1993). A Cox10p mutant does not synthesize haem a, and a lesion in the COX11 gene results in the accumulation of haem o. These observations pro-vide compelling supportive evidence that haem o is an intermediate in haem a synthesis in prokaryotes and eukaryotes.

5. CYTOCHROME c BIOGENESIS

Besides the cytochromes carrying various types of haem in a non-covalent arrangement (i.e. haems $a,$ $b,$ $d,$ o), the c-type cytochromes represent a distin-guished class, as they contain covalently bound haem. Most cytochrome c polypeptides possess one or several signature sequence(s) C-X-X-C-H, to which haem is attached. Two thioether bonds are formed between the vinyl side chains of the porphyrin rings of haem and the cysteine thiols of the apocytochrome in a stereo-specific reaction that is catalysed by an enzyme referred to as cytochrome c haem lyase. Such enzymatic activity was described more than 10 years ago for the maturation of yeast and $Neurospora$ $crassa$ mitochondrial cytochromes c and c_1, where *in vitro* haem ligation assays have resulted in the formation of holo-cytochromes c. Since then, many different laboratories have been hunting for a bacterial cytochrome c haem lyase (CCHL). Although these attempts have not yet resulted in the isolation of a bacterial CCHL, they have unraveled the high complexity of this, at first sight simple, process and, surprisingly different solutions to this problem which have evolved in nature.

5.1. The Paradigm of Mitochondrial Cytochrome c Maturation

The biosynthesis of cytochrome c was studied first in yeast and *N. crassa* mitochondria. In an attempt to understand the cause of the stereo-specific attachment of haem to apocytochrome c, Basile *et al.* found cytochrome c synthetase activity in a cell-free system of yeast mitochondria (Basile *et al.*, 1980). The corresponding enzyme was partially purified and characterized (Veloso *et al.*, 1981; Taniuchi *et al.*, 1983), and there were actually two enzymes, a cytochrome c haem lyase (CCHL) and a cytochrome c_1 haem lyase (CC_1HL) with differing specificities for their apocytochrome targets (Visco *et al.*, 1985). During studies of import of cytochrome c into mitochondria of *N. crassa*, it was recognized that CCHL plays an essential role, and that import and haem attachment are coupled (Hennig and Neupert, 1981; Dumont *et al.*, 1988). Haem attachment to apocytochrome c was performed *in vitro*; however, the system was not defined completely (Nicholson *et al.*, 1987, 1988). There are no reports of a reconstituted cytochrome c haem ligation reaction with purified enzyme. Thus, the mechanism of haem ligation and in particular the establishment of stereo-specificity remain to be elucidated. The only requirement that has become clear from the biochemical characterization of the $CCHL/CC_1HL$ activities is that for reductant of both the haem iron and the cysteine thiols of the haem-binding site (Nicholson and Neupert, 1989; Nicholson *et al.*, 1989). In addition, the location of haem attachment in the intermembrane space at the periphery of the inner mitochondrial membrane is well established (Dumont *et al.*, 1991; Lill *et al.*, 1992; Mayer *et al.*, 1995).

Genes for CCHL and CC_1HL have been identified in *Saccharomyces cerevisiae* (Dumont *et al.*, 1987; Dujon *et al.*, 1994). Mutants in these genes do not grow on non-fermentable carbon sources due to loss of mature c-type cytochromes, leading to deficiencies in respiration. From *N. crassa*, only the CCHL sequence is known (Drygas *et al.*, 1989). Haem lyases share amino acid sequence similarities in the range of 50% and contain at least one Cys-Pro-Val sequence motif that seems to be involved in haem binding (Steiner *et al.*, 1996). Genes encoding homologous CCHLs have also been identified in various metazoa including humans (Steiner *et al.*, 1996); the products of these nuclear genes are imported into the organelles. However, based on whole genome sequences it appears that in these organisms only one copy of a CCHL exists, and that the presence of cytochrome c- and c_1-specific haem lyases found in fungal species cannot be considered as a general characteristic for this type of mitochondrial cytochrome c maturation.

5.2. Cytochrome c-Deficient Bacterial Mutants and the Discovery of Novel Genes

Research on bacterial cytochrome c biosynthesis started with the identification of mutants defective in TMPD oxidation, nitrogen fixation and photosynthesis (for review see Thöny-Meyer et al., 1994; Kranz and Beckman, 1995; Kranz et al., 1998), which were found to lack all c-type cytochromes. Later, mutations affecting metabolic processes such as copper resistance (Yang et al., 1996), pyoverdin production (Gaballa et al., 1996), Mn^{2+} oxidation (Caspi et al., 1998; de Vrind et al., 1998), Fe^{2+} oxidation (Cabrejos et al., 1999), isomerization of unsaturated fatty acids (Holtwick et al., 1999), and gluconate oxidation (Pujol and Kado, 2000) were found to map to cytochrome c biogenesis genes as well. In all cases, the observed phenotypes were a consequence of the loss of c-type cytochromes, which are involved in electron transfer reactions coupled to the respective metabolic pathways. None of the bacterial cytochrome c-deficient mutants affected a gene whose product shared similarity to the mitochondrial CCHL. Rather, these genes were novel and specific for cytochrome c biogenesis. Because many of these genes were identified in the context of diverse physiological processes, the nomenclature has become very complicated. In this review, the general gene names ccm for cytochrome c maturation and ccs/ccd for cytochrome c synthesis will be used preferentially; alternative gene names can be found in Thöny-Meyer (2000). To date, 14 bacterial genes have been shown to be required directly for cytochrome c biogenesis, i.e. $ccmA$, $ccmB$, $ccmC$, $ccmD$, $ccmE$, $ccmF$, $ccmG$, $ccmH$, $ccmI$, $ccsA$, $ccsB$, $ccsX$, $ccdA$, $dsbD$. The genes $dsbA$ and $dsbB$ were also reported to be essential for cytochrome c maturation; however, this requirement appears to be conditional (L. Thöny-Meyer, manuscript in preparation). Cytochrome c-synthesizing bacteria do not possess all of these genes, but rather subsets of them and, depending on which subset is present, cytochrome c maturation is classified as system I or II (see below).

5.3. The Complexity of Cytochrome c Biogenesis

Maturation of c-type cytochromes culminates in the stereo-specific covalent attachment of a haem molecule to the apocytochrome polypeptide. The binding site in the protein is in most cases a C-X-X-C-H sequence motif. The cysteines must be reduced, i.e. in the thiolate form, in order to ligate to the vinyl groups of the haem moiety. The histidine is an axial ligand of the haem iron. Ligation of haem and polypeptide is an irreversible reaction and results in two thioether bonds. Bacteria often contain multi-haem c-type cytochromes. For example, a high-molecular weight c-type cytochrome of *Desulfovibrio vulgaris* has been described which carries 16 haem-binding motifs of the C-X-X-C-H type (Pollock *et al.*, 1991). Hence, a mechanism must exist that guarantees the correct ligation of the various cysteines with the respective haem vinyl groups.

All c-type cytochromes are extracytoplasmic proteins. In bacteria, they reside in the periplasm, where they can be either soluble or attached to the cytoplasmic membrane, but there have also been reports of outer membrane c-type cytochromes. In eukaryotes, type c cytochromes are found in organelles such as mitochondria or chloroplasts, where they are exposed to the intermembrane space or the thylakoid lumen, respectively. It has been well-established that haem attachment occurs after the translocation of the apocytochrome across the membrane (Page and Ferguson, 1989, 1990; Howe and Merchant, 1994; Howe *et al.*, 1995; Thöny-Meyer *et al.*, 1996; Thöny-Meyer and Künzler, 1997). Thus, not only the apocytochrome and haem, but also the catalytic (domains of) proteins involved in haem ligation should be targeted to the appropriate extracytoplasmic location. While the mitochondrial haem lyases are intermembrane space enzymes, their bacterial counterparts are expected to be periplasmic enzymes. Translocation of polypeptide, haem and reducing power are the crucial events during cytochrome c maturation that have to take place in a spatially and temporally coordinated manner.

5.4. Insights from Whole Genome Analyses: Three Systems of Cytochrome c Maturation

Since the discovery of the mitochondrial haem lyases, comparative sequence searches for the bacterial and chloroplast haem lyases have not been successful in identifying homologous proteins. Instead, other unrelated genes and their products were discovered that were clearly required for cytochrome c biogenesis, among them the *Bradyrhizobium japonicum cyc* genes (Ramseier *et al.*, 1991), the *Rhodobacter capsulatus hel* and *ccl* genes (Beckman *et al.*, 1992; Beckman and Kranz, 1993), the *Bacillus subtilis* gene *ccdA* (Schiött *et al.*, 1997b) and the *Chlamydomonas rheinhardtii* nuclear and chloroplast genes *ccs1* (Inoue *et al.*, 1997) and *ccsA* (Huang and Barrett, 1991; Xie and Merchant, 1996), respectively. Homologues of these genes were found in many different organisms producing c-type cytochromes; however, only subsets of them were found in each organism. When the first complete bacterial genome sequences became available, in particular those of *E. coli* (Blattner *et al.*, 1997) and *B. subtilis* (Kunst *et al.*, 1997), a new picture emerged, which predicted the existence of two different systems for bacterial cytochrome *c* maturation (Kranz *et al.*, 1998). System I is represented by the genes *ccmABCDEFGHI* of γ-proteobacteria such as *E. coli* and *H. influenzae*, or by the homologous *cyc/hel/ccl* genes of the α-proteobacteria *B. japonicum* and *R. capsulatus*. The requirement of these genes in cytochrome *c* maturation of *E. coli* has been shown (Thöny-Meyer *et al.*, 1995; Grove *et al.*, 1996). In addition, the specific Ccm proteins are linked to a general system of periplasmic redox control by the protein DsbD/DipZ, which, among other functions, also has an essential role in cytochrome *c* formation (Crooke and Cole, 1995). No homology to any of the previously described mitochondrial haem lyases, which have been assigned to system III maturation (Kranz *et al.*, 1998), was found.

The comparison of the genome of the Gram-positive *B. subtilis* with that of *E. coli* revealed unambiguously that the Ccm system is absent in the former. However, *Bacillus* contained, downstream of the *resA* gene encoding a thioredoxin-like protein, the genes *resB* and *resC* encoding cytochrome *c* maturation factors homologous to Ccs1 and CcsA, respectively, of chloroplasts. These genes are also present in the cyanobacterium *Synechocystis*, where they are called *ccsB* and *ccsA* (Tichy and Vermaas, 1999), but are absent in organisms using system I. Mutations in these genes and their homologues lead to deficiency of *c*-type cytochromes (Cabrejos *et al.*, 1999; Tichy and Vermaas, 1999; Le Brun *et al.*, 2000). In addition to ResB and ResC, system II also comprises factors with predicted activities in redox control. The genetically defined *ccdA* gene codes for a protein with some similarity to DsbD sequence motifs (Schiött *et al.*, 1997a, 1997b), and the *ccsX* gene recently identified in *Bordetella pertussis* (Beckett *et al.*, 2000) encodes a periplasmic thioredoxin similar to CcmG/DsbE of system I.

Although systems I and II operate with different cytochrome *c* maturation factors, some common features can be found: (1) the involvement of a DsbD/CcdA-like protein that presumably connects a cytoplasmic with a periplasmic thioredoxin; and (2) the presence of at least one integral membrane protein with a periplasmic domain that contains the W-rich sequence motif WGXXWXWD. To date there are only four genes known to be involved in system II cytochrome *c* maturation, but the collection may not be complete. In *C. rheinhardtii*, one chloroplast and at least four nuclear loci involved in cytochrome *c* formation have been identified genetically (Xie *et al.*, 1998). By contrast, the *E. coli ccm* genes, when provided in multiple copies on a plasmid, are sufficient to stimulate maturation of overexpressed *c*-type cytochromes. This indicates that the set of cytochrome *c*-biogenesis-specific genes in this organism is complete.

Interestingly, it can be derived from whole genome sequences that organelle cytochrome *c* maturation involves system I in plant and protozoal mitochondria, system II in chloroplasts and system III with the classical CCHL (see Section 5.1) in fungal, vertebrate and non-vertebrate mitochondria.

5.5. Redox Requirements

Redox control is an important prerequisite for cytochrome *c* synthesis, as haem and the haem-binding site cysteines of the apocytochrome must be reduced before haem ligation can take place. In particular, when these residues reside in an oxidative environment like the bacterial periplasm, there is a demand for a mechanism by which the formation of disulphide bonds is prevented or antagonized. Extracytoplasmic protein thiol disulphide oxidoreductases with thioredoxin activity are ideal candidates to fulfil such function (Fabianek *et al.*, 2000). CcmG and CcmH of system I have thioredoxin-like active sites containing a C-X-X-C motif, and their cysteines have been shown to be essential for cytochrome *c* maturation (Fabianek *et al.*, 1998, 1999). The soluble periplasmic domain of the CcmG homologue of *B. japonicum* (CycY) has been crystallized (Edeling *et al.*, 2001), and the structure at 1.14 Å resolution matched very nicely with that of thioredoxin, except for an additional α-helix close to the active side (M. Edeling *et al.*, manuscript in preparation). The involvement of this protein in redox control is also confirmed by its independent discovery in *E. coli*, where a *dsbE* (*ccmG*) mutant was found to be sensitive to oxidized DTT (Missiakas and Raina, 1997). There is evidence for a transfer of reducing equivalents from CcmG via CcmH to cytochrome *c* from the determination of redox potentials of the *R. capsulatus* homologues HelX, Ccl2 and a cytochrome *c* model peptide (Stetterdahl *et al.*, 2000). The electron donor for CcmG has been proposed to be DsbD due to the detection *in vivo* of mixed disulphides between these proteins (Katzen and Beckwith, 2000). DsbD is an integral membrane protein with various conserved cysteines, whose central domain is believed to transport electrons from cytoplasmic thioredoxin across the membrane to a periplasmic thioredoxin-like domain, and from there to periplasmic thioredoxin-like proteins such as DsbC or CcmG (Stewart *et al.*, 1999; Chung *et al.*, 2000; Gordon *et al.*, 2000; Krupp *et al.*, 2001). This is in agreement with the finding that *dsbD* and *trxA* (encoding thioredoxin) mutants are defective in cytochrome *c* maturation (Sambongi and Ferguson, 1994; Crooke and Cole, 1995; Metheringham *et al.*, 1996; Reid *et al.*, 1998). Recently, a role for CcmG after CcmH in the cytochrome *c* assembly pathway has been proposed based on the finding that oxidized CcmG accumulates in a *ccmH* mutant (Reid *et al.*, 2001). The precise flow of reducing equivalents through the periplasmic thioredoxins remains to be elucidated. There are some organisms using system I cytochrome *c* maturation, in which the function of DsbD is replaced by CcdA (Deshmukh *et al.*, 2000).

System II cytochrome *c* maturation apparently has at least two components involved in redox control. (1) CcdA shares homology with the central domain of DsbD that is believed to transport electrons across the membrane. (2) CcsX (Beckett *et al.*, 2000) is homologous to the gene product of *resA* that is encoded upstream of *resBC* in *Bacillus subtilis*. However, ResA lacks a signal sequence and thus is believed to be cytoplasmic, whereas *Bordetella pertussis* CcsX was shown to be periplasmic, but anchored to the membrane. In this regard, CcsX more closely resembles the CcmG subfamily of thioredoxins. Also, in *B. pertussis*, CcdA seems to be replaced by a DsbD homologue. The current speculation is that in system II, electrons are also passed through the membrane by a membrane-integral protein of either the CcdA- or the DsbD-type. A cytoplasmic thioredoxin such as ResA might be the electron donor, and a periplasmic thioredoxin such as CcsX the electron acceptor, which finally reduces the haem-binding site of apocytochromes *c*. So far, a complete description of all thioredoxin-type proteins involved in a redox pathway for type II cytochrome *c* maturation is not available.

It should be noted that some variability in the use of either CcdA or DsbD was also seen for type I cytochrome *c* maturation. In *R. capsulatus* and *Paracoccus pantotrophus*, CcdA functions instead of DsbD (Deshmukh *et al.*, 2000; Bardischewsky and Friedrich, 2001). An integrated view on the similarities and differences of the flux of reducing power in system I versus system II is depicted in Fig. 7.

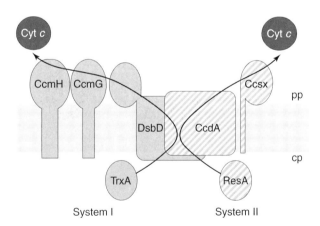

Figure 7   Redox pathway for cytochrome *c* maturation. The components in the diagram are described in the text.

5.6. Haem Delivery

The key step of cytochrome c maturation is the covalent ligation of haem to the apocytochrome. This requires that haem is translocated across the membrane and placed in a stereospecifically correct orientation close to the haem-binding site of the cytochrome c polypeptide. This should allow the formation of thioether bonds between the haem vinyl groups and the cysteines of the apoprotein. This pathway is referred to as haem delivery. It is known best from system I maturation (Thöny-Meyer, 2000) (Fig. 8). In this system, the CcmE protein (Reid *et al.*, 1998; Schulz *et al.*, 1998) represents an intermediate of the haem delivery pathway in that it binds haem transiently in the periplasm and then donates it to apocytochrome c, thus acting as a periplasmic haem chaperone. Most strikingly, the binding of haem to CcmE is covalent and occurs at a strictly conserved histidine (Schulz *et al.*, 1998). This histidine has been shown to be essential for haem delivery not only in *E. coli* CcmE (Schulz *et al.*, 1998), but also in the CcmE homologues of *R. capsulatus*, *B. japonicum* and *Arabidopsis thaliana* (Deshmukh *et al.*, 2000; Schulz and Thöny-Meyer, 2000; Spielewoy *et al.*, 2001). In *A. thaliana*, *AtCCME* is a nuclear gene whose translation product is targeted and attached to the mitochondrial inner membrane (Spielewoy *et al.*, 2001). The reason for a covalent haem binding to CcmE may be a stereospecific selection of one of the two haem vinyl groups for the subsequent haem ligation (Thöny-Meyer, 2000).

When cytochrome c biosynthesis is blocked at a stage after haem transfer to CcmE, the haem-binding form of CcmE can accumulate to detectable levels. This is the case in *ccmF*, *ccmG* and *ccmH* mutants (Fabianek *et al.*, 1999; Schulz *et al.*, 1999). While CcmGH are believed to participate in reduction of the apocytochrome c haem binding site, CcmF remains the best candidate for a system I haem lyase. In *E. coli*, a paralogue of CcmF, NrfE was shown to be

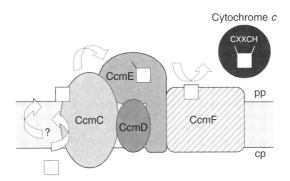

Figure 8 Haem delivery during cytochrome c maturation. The components in the diagram are described in the text.

specifically involved in attachment of haem to the unusual haem-binding site C-X-X-C-K in the *c*-type cytochrome NrfA, a formate-dependent nitrite reductase (Eaves *et al.*, 1998). Apparently, CcmF and NrfE can discriminate between the C-X-X-C-H and the C-X-X-C-K haem-binding sites. This makes it likely that these proteins interact with the apocytochrome *c*. An interesting question with regard to haem ligation is whether the formation of two thioether bonds occurs in one or in two consecutive steps. At present no cell-free or *in vitro* haem ligation experiment has been successful, and we are far from understanding the mechanism of the covalent haem attachment to *c*-type cytochromes.

More details on components involved in haem delivery to CcmE have been reported recently. The integral membrane protein CcmC is both essential and sufficient for haem attachment to CcmE (Schulz *et al.*, 1999). Haem attachment to CcmE is more efficient in the presence of the small membrane protein CcmD that influences the levels of CcmE protein in the membrane. CcmC and CcmE interact directly, as shown by co-immunoprecipitation experiments (Ren and Thöny-Meyer, 2001). CcmC has six transmembrane helices (Gaballa *et al.*, 1998; Goldman *et al.*, 1998) and the striking W-rich sequence motif in the second periplasmic loop domain. This motif, together with conserved histidines in adjacent periplasmic loops, was proposed to function as a haem-binding site. This motif was also found in CcmF and its paralogue NrfE, and in the CcsA family of type II maturation factors (Xie and Merchant, 1998). Site-directed mutagenesis of amino acids within this motif and of the histidines affected cytochrome *c* synthesis in many cases (Thöny-Meyer, 2000). Its direct involvement in holo-CcmE formation can be modulated in the presence or absence of CcmD (Schulz *et al.*, 2000), and it seems to participate in CcmC–CcmE interaction (Ren and Thöny-Meyer, 2001). Although it was possible to assess haem binding to CcmC experimentally, mutational alterations in the W-rich motif and in the histidines did not abolish the observed CcmC–haem interactions (Ren and Thöny-Meyer, 2001). From all these experiments it is clear that a complex between CcmC, CcmD, CcmE and haem is formed at an early stage of cytochrome *c* maturation, but the precise mechanism of haem transfer to CcmE has not yet been unraveled.

Does CcmC catalyse the covalent attachment of haem to CcmE, or does it function as a haem exporter, or both? The question of haem export is still unsolved. Kranz and collaborators have suggested, based on co-immunoprecipitation experiments, that the *R. capsulatus* CcmC homologue HelC is part of an ABC transporter of the subunit composition HelABCD (CcmABCD) used for haem export (Goldman *et al.*, 1997; Goldman and Kranz, 2001). By contrast, the finding that CcmE can be loaded with haem in the absence of CcmAB indicates a function of CcmC separate from that of the ABC transporter with a CcmAB composition (Schulz *et al.*, 1999). Assays for haem export using a periplasmic haem reporter system (Goldman *et al.*, 1996; Throne-Holst *et al.*, 1997), complementation of *ccm*-defective mutants with external haem (Page *et al.*, 1997; Ren and Thöny-Meyer, 2001) and direct measurements of haem transport into everted membrane vesicles of wild-type and *ccm* mutants (Cook and Poole, 2000) were attempted to solve the question of haem transport, but no conclusive answers have been obtained. In-frame deletion mutations in *ccmA* and *ccmB* prevent haem binding to CcmE unless CcmC is provided in large excess. Under such conditions, CcmE could bind haem, but cytochrome *c* maturation was still arrested, suggesting that the function of the ABC transporter is in part related to haem attachment to CcmE, and in another part at a later stage of maturation (Schulz *et al.*, 1999). In summary, we cannot rule out the possibility that CcmAB or CcmC participates in haem export.

For system II maturation, CcsA and CcsB are believed to be involved in haem delivery by analogy to system I, because CcsA carries a W-rich motif. CcsA is absent in a *ccs1* mutant (Xie *et al.*, 1998). The formation of a complex between CcsA and CcsB(Ccs1) is supported by the finding that their homologues in *Helicobacter pylori* are encoded by a single gene.

5.7. Conclusions

To date, we can distinguish between three systems for cytochrome *c* maturation. The simplest system is probably system III and has evolved more recently than the other systems, as it is present only in mitochondria of fungi and metazoa. The only known component is a CCHL enzyme with, in some cases, specificity for either cytochrome *c* or c_1. Recently, yeast cytochrome *c* was expressed in the cytoplasm of *E. coli* and shown to bind haem when the mitochondrial CCHL was co-expressed (Pollock *et al.*, 1998; Sanders and Lill, 2000). This suggests that the single subunit CCHL is sufficient to attach haem covalently to the apoprotein. Systems I and II do not share any similarity with system III, but have a few common features. It seems that system II is simpler than system I in having only one haem lyase protein with a W-rich motif that may transfer haem directly to cytochrome *c*. System I uses an intermediate, the haem chaperone CcmE, and thus needs two haem lyase-like proteins with W-rich motifs, CcmC specific for CcmE and CcmF specific for cytochrome *c*. The advantage of these additional components is not obvious, but they may facilitate the stereo-specificity of haem insertion. The presence of the CcmE haem chaperone is a hallmark of system I cytochrome *c* maturation. It has been proposed that Archaea also use system I maturation due to the presence of a gene encoding a CcmC-like protein (Kranz *et al.*, 1998). However, *ccmE* is missing in archaeal genomes. This indicates that Archaea may use a maturation pathway more similar to system II.

Besides the mechanism of haem attachment, one of the most intriguing questions is that of haem transport. The last step of haem biosynthesis, i.e. insertion of iron into protoporphyrin IX occurs on the cytoplasmic side of the bacterial membrane, and on the matrix/stromal side of the organelles. Yet, haem insertion takes place at the opposite side of the membrane, making translocation of haem across the lipid bilayer essential for cytochrome c maturation. As haem transport across the membrane is not understood in any of the cytochrome c maturation systems, it will be the challenge of future research to define how haem is targeted to the subcellular locations of its use.

6. ALA TRANSPORT

The fact that ALA synthesis mutants can be maintained by supplementation of growth medium with ALA shows that many, if not all, bacteria can take up the precursor. However, evidence for a physiological role for ALA transport has been presented in *Bradyrhizobium japonicum* (Sangwan and O'Brian, 1991). ALA synthase mutants of *Sinorhizobium meliloti* (Leong *et al.*, 1982), *Azorhizobium caulinodans* (Pawlowski *et al.*, 1993) and *Rhizobium* sp. NGR234 (Stanley *et al.*, 1988) are all ALA auxotrophs in minimal media cultures. They are also unable to establish a symbiosis on their respective plant hosts as expected, since haem is required for viability and nitrogen fixation. However, a *hemA* strain of *Bradyrhizobium japonicum* is also an auxotroph but it elicits functional nodules on soybeans (Guerinot and Chelm, 1986) and contains haem (Sangwan and O'Brian, 1991). The *B. japonicum hemA* mutant can also establish a symbiosis on cowpea and mungbean, hosts which the *hemA* mutant of *Rhizobium* sp. NGR234 cannot successfully infect (McGinnis and O'Brian, 1995). *B. japonicum* has a vigorous ALA uptake activity that is deficient in the other rhizobia, and evidence indicates that the *hemA* mutant can be rescued by acquisition of ALA from the plant host and plant ALA synthesis is induced in nodules (Sangwan and O'Brian, 1991; McGinnis and O'Brian, 1995).

ALA uptake in *Salmonella typhimurium* (Elliott, 1993) and *E. coli* (Verkamp *et al.*, 1993) is catalysed by a dipeptide permease (Dpp) system. ALA is structurally similar to glycyl-glycine, and Dpp can transport any dipeptide containing L-amino acids. A mutation in the regulatory gene *lrp* confers oligopeptide and ALA uptake in a *dpp* strain, suggesting an additional mechanism for ALA uptake (King and O'Brian, 1997). The *lrp* gene encodes the leucine-responsive regulatory protein (Lrp), a global transcriptional regulator found in numerous enteric bacteria, which activates or represses many genes and operons including *oppABCDF* in *E. coli* (Newman *et al.*, 1992; Calvo and Matthews, 1994).

Overexpression of the *E. coli lrp* gene inhibits ALA uptake in an *E. coli dpp+* strain, indicating that Lrp can repress *dpp* expression as well. Similarly, overexpression of *B. japonicum lrp* had a negative effect on ALA uptake in *B. japonicum*. Competition experiments suggest that *B. japonicum* transports ALA via a mechanism that transports oligopeptides (King and O'Brian, 1997). A screen for uptake-defective mutants yielded *lysC* and *ptsP*, proximal genes in the *B. japonicum* genome (King and O'Brian, 2001). Aspartokinase, encoded by *lysC*, catalyses the phosphorylation of aspartate for synthesis of three amino acids, but the *lycC* strain was not an amino acid auxotroph. The *ptsP* gene encodes Enzyme I^{Ntr} (EI^{Ntr}), a paralogue of Enzyme I of the phosphoenolpyruvate:sugar phosphotransferase (PTS) system. Aspartokinase appears to affect EI^{Ntr} activity by regulating its phosphorylation by a mechanism that involves direct interaction of the two proteins (King and O'Brian, 2001). The findings are unusual for a PTS protein because it is involved in the transport of non-sugar solutes, and positively affects transport in a dephosphorylated state rather than when phosphorylated.

7. REGULATION OF HAEM BIOSYNTHESIS

The cellular demand for haem and other tetrapyrroles can vary considerably depending on growth conditions and other environmental variables. In principle, haem levels can be regulated at the level of synthesis or degradation. However, haem oxygenase, a haem degradation enzyme, has been described only in relatively few organisms that use haem as an iron source (Schmitt, 1997a, 1997b; Zhu *et al.*, 2000), or degrade haem to an open chain tetrapyrrole for photosynthetic function (Willows *et al.*, 2000). A strong case for degradation as a general feature in prokaryotes remains to be made. With this in mind, it is striking that supplementation of bacterial cultures with ALA or overexpression of ALA synthesis genes greatly increases the level of porphyrin intermediates in cells and in the medium, but does not in itself increase overall haem levels. Laboratory manipulations to increase cellular haem have been accomplished only by overexpression of a haem protein such as haemoglobin (Hart *et al.*, 1994; Verderber *et al.*, 1997). Collectively, these observations suggest several things. Firstly, control of haem synthesis is probably coupled with that of haem apoprotein formation in some way that is not currently understood. Second, it suggests that accumulation of free haem cannot be tolerated, which is consistent with its known toxicity. Finally, in the absence of haem degradation or excretion, control of haem concentration in cells must be primarily at the synthesis level. Subsequent identification of haem degradation as a general feature in prokaryotes would obviously warrant a revision of this latter point.

7.1. Regulation by Iron

Ferrous iron is inserted into protoporphyrin IX in the final step of haem biosynthesis catalysed by ferrochelatase. The cytotoxicity of porphyrins has been readily demonstrated in animals (Kappas *et al*., 1995), plants (Mock and Grimm, 1997) and bacteria (Nakahigashi *et al*., 1991), and is caused by their ability to catalyse light-dependent formation of reactive oxygen species. Iron must be acquired exogenously and may be a limiting nutrient, thus a prima facie argument can be made for the coordination of the haem pathway with the cellular iron level to prevent protoporphyrin synthesis from exceeding iron availability. This problem has been partially addressed in mammalian erythroid cells, where the iron-regulatory protein, IRP (reviewed in Hentze and Kuhn (1996) Rouault and Klausner (1996)) inhibits translation of mRNA encoding the haem synthesis enzyme ALA synthase under low iron conditions (Bhasker *et al*., 1993; Melefors *et al*., 1993). However, it has not been established that this mechanism is sufficient to regulate the haem pathway as a whole or that it prevents excess protoporphyrin synthesis under iron limitation.

Expression of *hemA* and *hemB*, the genes encoding ALA synthase and ALA dehydratase respectively, are positively affected by iron in *Bradyrhizobium japonicum* (Page *et al*., 1994; Chauhan *et al*., 1997). *hemB* mRNA and protein levels varying over 80-fold as a function of the iron concentration in which the cells are grown (Chauhan *et al*., 1997). The iron response regulator (Irr) protein from the bacterium *B. japonicum* coordinates the haem biosynthetic pathway to prevent the accumulation of toxic porphyrin precursors under iron limitation (Hamza *et al*., 1998). Loss of function of the *irr* gene is sufficient to uncouple the pathway from iron-dependent control as discerned by the accumulation of protoporphyrin under iron limitation. Irr belongs to the Fur family of transcriptional regulators involved in metal-dependent control of gene expression. The Irr protein accumulates in cells under iron limitation, where it negatively regulates the pathway at *hemB* (Hamza *et al*., 1998). Thus, *hemB* expression is constitutively high in an *irr* mutant strain.

Irr is a conditionally stable protein that degrades rapidly when cells are exposed to iron, allowing derepression of synthesis (Qi *et al*., 1999). This iron-dependent degradation is mediated by haem (Qi *et al*., 1999). Haem binds directly to Irr at a haem regulatory motif (HRM), which is necessary for its degradation. Accordingly, Irr persists in haem synthesis mutant strains in the presence of iron. Furthermore, mutation of an invariant cysteine in the HRM stabilizes Irr in the presence of iron (Qi *et al*., 1999). Thus, haem is an effector molecule in Irr degradation that reflects the availability of iron for haem synthesis. Irr interacts with ferrochelatase, and responds to haem and its precursors at the site of haem synthesis (Qi and O'Brian, 2002).

Although the *hemA* and *hemB* genes encode enzymes belonging to common pathways, they are not regulated by iron in the same way in *B. japonicum*.

Whereas *hemB* is regulated by Irr, *hemA* is controlled by Fur (Hamza *et al.*, 2000), a global regulator of iron metabolism and homeostasis found throughout the eubacterial kingdom. Like *hemB*, *hemA* expression is induced in iron replete cells, but only *hemA* is altered in a *fur* mutant strain, where mRNA levels remain high even under iron limitation (Hamza *et al.*, 2000). A phenotype for the *fur* strain under iron limitation shows that *B. japonicum* Fur is functional under those conditions to directly or indirectly regulate *hemA*, which differs from the generally held view that Fur is active as a regulator only in the presence of iron (Escolar *et al.*, 1999). It is unclear why *hemA* and *hemB* expressions are coordinated with iron differently. ALA synthase uses glycine and succinyl-CoA as substrate, and Fur has been implicated in succinate metabolism, at least in *E. coli* (Escolar *et al.*, 1999). Thus, perhaps *hemA* needs to be coordinated with its substrate or with respiratory metabolism more generally, whilst ALA dehydratase, the *hemB* product, uses a committed tetrapyrrole precursor as substrate.

7.2. Regulation by O_2

Many prokaryotes are able to support oxidative respiration using nitrate, sulphate or some oxidant other than oxygen as a terminal electron acceptor. Furthermore, many bacteria can respire under very low oxygen tensions due to the expression of terminal oxidases with high affinity for O_2. Thus, anaerobiosis or O_2 limitation does not necessarily elicit a switch to fermentative metabolism, and may actually require an increase in tetrapyrrole synthesis. Oxygen can also be a substrate in the oxidative steps of haem biosynthesis, and hypoxia may result in a change in expression from an O_2-dependent coproporphyrinogen oxidase gene to an O_2-independent one in organisms that contain genes for both. Glutamyl-tRNA reductase activity may be the rate-limiting step in haem synthesis in *E. coli*, but O_2 control, which is transcriptional, is modest (Darie and Gunsalus, 1994) and the predominant regulation may be by haem by a post-transcriptional mechanism (see below). However, in other organisms, the O_2 status is a major regulatory factor in control of haem biosynthesis.

Regulation of tetrapyrrole synthesis genes by oxygen in *Rhodobacter sphaeroides* is under numerous regulatory circuits for photosynthetic and anaerobic metabolism. The *fnrL* locus was identified in a mutant screen for elevated *hemA* (ALA synthase) gene expression under aerobic conditions, and is required for photosynthetic growth and anaerobic growth with dimethylsulfoxide (Zeilstra-Ryalls and Kaplan, 1995a, 1998). Fnr is a transcriptional regulator involved in anaerobic or microaerobic gene expression, and FnrL probably functions both to activate *hemA* under anaerobic conditions and repress it aerobically. Interestingly, *hemA* is also upregulated in a mutant

defective in the *ccoNOQP* operon encoding the *cbb₃* terminal oxidase
(Zeilstra-Ryalls and Kaplan, 1996). This oxidase has a very high affinity for O_2
(Preisig *et al.*, 1993, 1996; Thöny-Meyer *et al.*, 1994), allowing microaerobic
respiration. This mutation also results in oxygen-insensitive development of
photosynthetic membranes, indicating that the lesion elicits a global response.
The microaerobic/anaerobic phenotype of the *ccoNOQP* mutant in the presence of oxygen revealed that normal regulation is not due to O_2 per se, but
rather an alteration of the cellular redox state that O_2 limitation creates. This
redox signalling involves intramolecular electron transfer within the *cbb₃* oxidase resulting in a direct relationship between repression of photosynthesis
genes, probably including *hemA*, and electron flow (Oh and Kaplan, 2000).
The signal transduced by the oxidase mutant under aerobic conditions is not
mediated by FnrL as judged by studies using an *fnrL ccoP* double mutant,
where the effect of each mutation on *hemA* promoter activity was roughly
additive (Oh *et al.*, 2000). A mutation in the *hbdA* gene involved in polyhy-
droxybutyrate metabolism also results in elevated aerobic *hemA* expression,
which might affect the redox state by altering the cellular NAD⁺/NADH ratio
(Fales *et al.*, 2001).

R. *sphaeroides* contains two *hemN* gene homologues encoding O_2-inde-
pendent coproporphyrinogen oxidase designated *hemN* and *hemZ*
(Zeilstra-Ryalls and Kaplan, 1995a). Both genes are repressed by FnrL under
aerobic conditions and derepressed under microaerobic or anaerobic conditions
(Yeliseev and Kaplan, 1999; Oh *et al.*, 2000). The *hemZ* gene is clustered
near *fnrL* and *ccoNOQP*, and is more strictly regulated by FnrL than is *hemN*
(Zeilstra-Ryalls and Kaplan, 1995a; Oh *et al.*, 2000). *hemN* is negatively
regulated by the outer membrane protein TspO apparently by a post-transcriptional control mechanism (Yeliseev and Kaplan, 1999). TspO negatively
regulates carotenoid and bacteriochlorophyll production aerobically.
Introduction of multiple copies of *hemN in trans* induced bacteriochlorophyll
formation, a phenotype similar to a *tspO* mutant (Yeliseev and Kaplan, 1999).
The two coproporphyrinogen oxidase isozymes must have different functions
because expression of *hemZ in trans* did not alter gene expression. It was proposed that TspO attenuates HemN activity resulting in the accumulation of a
tetrapyrrole precursor, which in turn acts as a corepressor of bacteriochlorophyll and carotenoid gene expression (Yeliseev and Kaplan, 1999).

Oxygen is a key regulator of symbiotic differentiation in the rhizobia. *De
novo* haem biosynthesis for synthesis of cytochrome *cbb₃* oxidase and other
cytochromes is required for microaerobic metabolism, hence a need to control
haem biosynthesis. The signal transduction pathway leading to a global
response to O_2 originates with the FixL/FixJ two-component regulatory
system. FixL is a haem protein that binds O_2 directly and becomes autophos-
phorylated in the deoxy form to phosphorylate FixJ. $FixK_2$, an Fnr family
protein, is part of this regulatory cascade and is controlled directly or indirectly

by FixJ, depending on the rhizobial species (Fischer, 1994). Accordingly, the induction of a *hemA-lacZ* fusion by O_2 deprivation is abolished in a *fixL* or *fixJ* background, and mutation of a FixK-like element in *hemA* renders the fusion unresponsive to O_2 (Page and Guerinot, 1995). Direct examination of *hemB* RNA accumulation and synthesis reveals strong regulation of the *hemB* gene in response to the oxygen status (Chauhan and O'Brian, 1997). *hemB* expression is unaffected in a *fixJ* strain when grown in air, but the rapid induction in response to hypoxia observed in wild-type cells does not occur in the mutant. The cytochrome cbb_3 oxidase (David *et al.*, 1989; Fischer, 1994) encoded by the *fixNOQP* operon (*ccoNOQP* in other organisms) is controlled by FixJ, and thus a role for the regulator in coordination of haem and apoprotein synthesis in O_2-limited cells was deduced.

Changes in the cytochrome composition during differentiation for microaerobic metabolism in *Rhizobium* involve repression, as well as activation. Cytochrome aa_3 is a terminal oxidase necessary for aerobic respiration; its expression is diminished or absent in microaerobically cultured cells and in symbiotic bacteria (bacteroids), and the message level of at least one subunit-encoding gene is under O_2 control (Gabel and Maier, 1993). Interestingly, cytochrome aa_3 persists in a *fixJ* strain of *B. japonicum* cells grown under O_2 restriction (Chauhan and O'Brian, 1997). Thus, FixJ appears to be involved in repression of the oxidase and therefore positive and negative regulation of cytochrome expression share a common regulatory feature.

Like *R. sphaeroides*, *B. japonicum* contains two genes for O_2-independent coproporphyrinogen oxidase, designated *hemN*$_1$ and *hemN*$_2$ (Fischer *et al.*, 2001). Neither gene is expressed aerobically, but both are strongly induced under microaerobic or anaerobic conditions. Induction of both genes is mediated by FixK$_2$. Ironically, the *hemN*$_1$ gene was identified in a study of the 400 kb 'symbiotic gene region' of the *B. japonicum* chromosome, but it is not required for symbiosis on soybean or for anaerobic growth on nitrate (Nienaber *et al.*, 2000; Fischer *et al.*, 2001). On the other hand, a *hemN*$_2$ mutant is unable to grow anaerobically on nitrate and does not elicit functional nodules on its plant host (Fischer *et al.*, 2001). Furthermore, *hemN*$_2$, but not *hemN*$_1$, was capable of complementing a *S. typhimurium hemF hemN* double mutant *in trans*, indicating that *hemN*$_1$ does not encode a functional protein (Fischer *et al.*, 2001).

The presence of multiple *hemN* homologues and control by Fnr-like proteins is not confined to the α-proteobacteria *Rhodobacter* and *Rhizobium*. A *hemN* gene from *Pseudomonas aeruginosa* is under control of two Fnr-type regulators Dnr and Anr, both of which are needed for anaerobic induction (Rompf *et al.*, 1998). The *hemN* gene in that organism is also expressed aerobically, which is dependent only on Anr. A *hemN hemF* double mutant of *P. aeruginosa* grows both aerobically and anaerobically, indicating a second O_2-independent coproporphyrinogen oxidase gene (Rompf *et al.*, 1998).

Surprisingly, the *hemF* gene encoding the O_2-dependent oxidase is also strongly induced by anaerobiosis which is mediated by both Anr and Dnr, which if fully expressed suggests an alternative function for that enzyme.

7.3. Regulation by Haem

Haem serves as a signalling molecule in Irr degradation for iron control of haem synthesis in *B. japonicum* (Qi *et al.*, 1999) (see above). However, haem probably reflects the availability of iron for haem synthesis in that case, and is not an end-product feedback control mechanism. In *S. typhimurium*, there is excellent evidence that haem regulates its own synthesis at the level of the first committed step catalysed by glutamyl-tRNA reductase (GTR). In experiments where haem content was controlled by exogenous supplementation of haem synthesis mutants, it was demonstrated that GTR levels were increased up to 25-fold in haem-starved cells, with only a two-fold increase in *gtr* (*hemA*) promoter activity (Wang *et al.*, 1997). The reductase is a conditionally stable protein with a half-life of about 20 min in the presence of haem, which is increased to greater than 5 h under haem starvation (Wang *et al.*, 1999a). Degradation was energy-dependent and was blocked in mutants defective in the proteolytic enzymes Lon or ClpAP. A GTR–LacZ fusion shows similar haem-dependent degradation as native protein, and the N-terminal 18 amino acids of GTR is sufficient to destabilize LacZ as well. However, in the latter fusion, degradation is no longer haem-dependent. The authors suggest that the 18-amino acid degradation tag is accessible to protease digestion when GTR is bound to haem, but is always accessible in a GTR(1–18)–LacZ fusion (Wang *et al.*, 1999a). Substitution of two adjacent leucine residues for lysine in the degradation tag of GTR abrogates haem-dependent turnover (Wang *et al.*, 1999b). Although direct haem binding has not been shown to be involved in GTR degradation, recombinant barley GTR fused with glutathione purifies with the haem bound, and its activity is inhibited by haem (Vothknecht *et al.*, 1996).

8. A GENOMIC PERSPECTIVE OF HAEM BIOSYNTHESIS

Haemophilus influenzae, the organism from which the first complete genome sequence was learned, derives its genus name from its reliance on exogenous haem for growth ('haem loving'). Thus, it seems particularly fitting to take a genomic view of prokaryotic haem biosynthesis. *H. influenzae* requires either protoporphyrin or haem for growth and accordingly it contains only the ferrochelatase-encoding gene. There are many other examples where the known status of tetrapyrrole synthesis in prokaryotes is in accordance with the corresponding genes. However, this is not always the case, and the exceptions should provide the impetus for taking new experimental directions. Furthermore, the sequences of several genomes are now available from prokaryotes such as *Aeropyrum pernix* and *Halobacterium* sp. NRC-1 in which, at the time of this writing, there are fewer than 30 published papers on each organism in the Current Contents database. Thus, unlike model organisms, the genome represents a starting point for the characterization of these unusual prokaryotes.

8.1. Haem synthesis genes in bacteria with reduced genomes

Numerous bacteria that form obligate associations with eukaryotes, either pathogenic or symbiotic, have very small genomes. Genome degeneracy is probably caused by the lack of selection pressure to maintain genes that the host renders non-essential. Thus, the bacterial genome reflects, at least to some extent, its metabolic relationship with its eukaryotic host.

Members of the genus *Rickettsia* are obligate intracellular bacteria responsible for epidemic typhus in humans. They belong to the α-proteobacteria, the closest known relatives of mitochondria. The genome of *R. prowazekii* (Andersson *et al.*, 1998) indicates a haem biosynthetic pathway, as well as respiratory haem proteins, superoxide dismutase, TCA cycle enzymes, which are all consistent with aerobic metabolism and similar to the other known α-proteobacterial genomes. However, some other aspects of iron metabolism are absent, most notably genes encoding proteins for iron transport and regulation.

The Fur protein is a global transcriptional regulator of iron metabolism prevalent in eubacteria and the Archaea. It controls many facets of iron metabolism, including the *hemA* gene in *Bradyrhizobium japonicum* (Hamza *et al.*, 2000). Irr and PerR are structurally similar to Fur, but are functionally distinct. Nevertheless, they also mediate iron control of haem synthesis in *B. japonicum* (Hamza *et al.*, 1998) and *B. subtilis* (Bsat *et al.*, 1998), respectively. Interestingly, there are no *fur* gene homologues in the *R. prowazekii* genome, indicating the abandonment of bacteria-type iron regulation.

Most eubacteria have iron transport systems, either a permease system that can transport the metal directly, or siderophore-mediated transport. Siderophores are iron chelators which bind to a cognate cellular receptor for uptake. The α-proteobacteria *Caulobacter crescentus*, *Mesorhizobium loti*, and *Sinorhizobium meliloti* have ferric iron ABC transport system proteins, and the last has siderophore-mediated iron transport as well. The *R. prowazekii* genome appears to lack genes encoding these iron transport proteins despite the plethora of iron-containing proteins in that organism. However, *R. prowazekii* does contain a highly conserved gene homologue of *atm1*, a mitochondrial iron transport protein prevalent in eukaryotes, and which is associated with X-linked sideroblastic anaemia and ataxia in humans (Allikmets *et al.*, 1999). Atm1 is also found in *M. loti*, and is therefore probably a retained gene, rather than one acquired from its eukaryotic host by horizontal gene transfer. The concomitant retention of *atm1* and loss of the bacteria-type iron transporter genes in both *R. prowazekii* and mitochondria may be evolutionary consequences of an obligatory intracellular milieu.

For the most part, genome reduction is not an adaptive response, but rather it reflects a lack of selective pressure to maintain genes rendered non-essential by the obligate association with a eukaryotic host (Ochman and Moran, 2001). An exception to this is observed in *Borrelia burgdorferi*, the causative agent of Lyme disease. Most microbial pathogens overcome iron limitation, or so-called nutritional immunity, created by the host by sequestering iron in one or more ways. *B. burgdorferi* has apparently circumvented this problem by eliminating the need for iron altogether (Posey and Gherardini, 2000), which is reflected by a genome that does not encode most iron proteins. Accordingly, the haem pathway genes, as well as those for haem proteins, are absent. A *fur* gene homologue was identified (Posey and Gherardini, 2000), but its product has greater similarity to PerR, which responds directly to manganese in *B. subtilis*, and the influences of iron are indirect (Bsat *et al.*, 1998). It is possible that *B. burgdorferi* represents the extreme case of a strategy employed by other pathogens with reduced genomes. *Mycoplasma pneumoniae*, a respiratory pathogen and *Treponema pallidum*, the syphilis spirochete, lack the haem pathway enzymes, cytochromes, iron transport systems and most iron proteins. Exceptions are genes for rubredoxin, an iron-sulphur protein, found in *T. pallidum*, and NifS, a protein required in iron-sulphur cofactor synthesis. Both *M. pneumonia* and *T. pallidum* have genes encoding dinucleotide reductase, an iron protein involved in nucleotide synthesis, which is missing in *B. burgdorferi*.

Buchnera is an obligate endosymbiont of aphids (Baumann *et al.*, 1995) with a genome of only 590 genes. The *Buchnera* sp. APS genome encodes the four subunits for the cytochrome *o* complex, as well as protohaem farnesyltransferase, the enzyme that places a farnesylethyl group on protohaem to form the cytochrome *o* prosthetic group. In addition, the genome encodes sulphite reductase, which has sirohaem as the prosthetic group, and also the enzymes that synthesize sirohaem from uroporphyrinogen. However, *Buchnera* contains only a homologue of *hemC*, the gene encoding porphobilinogen deaminase, and a truncated *hemN* homologue encoding O_2-independent coproporphyrinogen oxidase. The exchange of nutrients between *Buchnera* and its aphid host is documented (Baumann *et al.*, 1995), and the bacterial genome raises the possibility that the aphid may provide *Buchnera* with haem or haem intermediates. By loose analogy, the haem pathway in non-photosynthetic eukaryotes is spatially separated between mitochondria and the cytosol, and exchange of haem precursors between the two compartments occurs.

8.2. Correlation of Haem Pathway Genes with Optimal Growth Temperature in Thermophilic Archaebacteria?

Kawashima *et al.* (2000) determined the genome sequence of the thermophilic archaeon *Thermoplasma volcanium*, and correlated the optimal growth temperature of numerous thermophiles in which the genome sequence is known with several genomic and metabolic parameters. They observe that *T. volacanium*, which has the lowest optimal growth temperature of 60°C, contains all of the genes of the haem pathway (except for protoporphyrinogen oxidase, but see Section 3.8 above). At the other extreme, *Pyrococcus furiosus*, P. OT3 and *P. abyssi* have no haem genes, and are the most thermophilic, growing in the 96–100°C range. *Methanococcus jannaschii*, *Methanobacterium thermoautotrophicum* and *Archaeoglobus fulgidus*, which have intermediate optimal growth temperature of 65–85°C, have a partial pathway containing the early steps of the pathway to uroporphyrinogen formation.

The authors attribute the correlation of haem pathway enzymes (genes) with optimal growth temperature to the thermal instability of haem and its precursors, and suggest that haem proteins are replaced by proteins with more thermostable prosthetic groups. However, this explanation assumes that the haem intermediates in the pathway leading to uroporphyrinogen formation are more thermostable than the latter intermediates and haem itself. If temperature were the driving factor, then the correlation should apply to taxonomically distant organisms. However, the thermophilic aerobe *Aquifex aeolicus*, which belongs to the Eubacterial kingdom rather than the Archaea, grows at 95°C and contains haem proteins (Schütz *et al.*, 2000) and haem biosynthesis gene homologues, showing that haem and its precursors can withstand high temperature. Also, haem proteins are found in numerous thermophilic Archaea as well (Gartner, 1991; Lübben and Morand, 1994; Schütz *et al.*, 2000). *T. volcanium* can live aerobically as well as anaerobically and contains genes encoding haem proteins, superoxide dismutase and tricarboxylic acid enzymes, all characteristic of an organism adapted to aerobic metabolism. On the other hand, *Pyrococcus* species are anaerobic, do not require haem and are missing respiratory enzymes. It does contain a gene encoding a protein with similarity to a cytochrome c_3 hydrogenase from *Desulfovibrio* species, but whether the protein is functional is unknown. However, a functional cytochrome in *Pyrococcus* would suggest an alternative mechanism for synthesis or the ability to acquire haem exogenously, and not an alternative cofactor that is functionally equivalent to haem.

The methanogenic thermophiles *M. jannaschi* and *M. thermoautotrophicum* require coenzyme F430 as the prosthetic group of methyl coenzyme M reductase, the enzyme that catalyses the final step in methane formation. Coenzyme F430 is a nickel tetrapyrrole synthesized from uroporphyrinogen III (Friedmann *et al.*, 1991), hence the presence of a partial haem pathway in those organisms. Thus, the distribution of haem biosynthesis genes in the thermophiles seems to correlate directly with metabolism, not with the thermostability of haem or its intermediates.

8.3. The Absence of Identifiable Uroporphyrinogen III Synthase Gene Homologues in Numerous Prokaryotic Genomes

Recall that uroporphyrinogen III synthase, product of the *hemD* gene in bacteria, synthesizes uroporphyrinogen III from the unstable linear tetrapyrrole hydroxymethylbilane. All of the enzymes necessary for uroporphyrinogen synthesis, including uroporphyrinogen synthase, should be found in organisms that make tetrapyrrole, since uroporphyrinogen is the branch point for all of them. Thus, *hemB*, *hemC* and *hemD* genes should be present in all tetrapyrrole-synthesizing prokaryotes (there are two ALA synthesis routes, thus the genes will not be universal). Initial perusal of 63 complete genomes suggested that some organisms did not contain a *hemD* gene that are predicted to synthesize tetrapyrroles. We searched these genomes for those that contained a *hemB* and *hemC* gene, but not a *hemD* gene. This was done by a protein BLAST search of each genome using bona fide uroporphyrinogen III synthases from *E. coli*, *B. subtilis* and mouse (protein id No. 41676, 399875 and 1363253, respectively). It is important to note that numerous open reading frames annotated as *hemD* are actually *cobA* or *cysG* homologues because *Clostridium josui* contains a bifunctional *cobA–hemD* gene and protein fusion (Fujino *et al.*, 1995) (see Section 3.5 above).

Nevertheless, we identified 12 genomes representing eight genera and 10 species that contained no identifiable uroporphyrinogen synthase gene. *Agrobacterium tumefaciens*, *Mesorhizobium loti*, *Rickettsia prowazekii*, *R. conorii* and *Caulobacter crescentus* are α-proteobacteria. *Aeropyrum pernix* and *Thermoplasma acidophilum* are Archaea. *Chlamydophila pneumoniae* (strains AR39, J138, CWL029), *Chlamydia muridarum* and *C. trachomatis* are chlamydiaceae. Furthermore *Aquifex aeolicus* contains only a very weak uroporphyrinogen synthase homologue (protein id No. 2984071). The gene annotated as *hemD* in *Sinorhizobium meliloti* is immediately downstream of *hemC* and is found in numerous other bacteria, but is only weakly similar to the uroporphyrinogen III synthase gene from *S. cerevisiae* in the C-terminal region. Interestingly, *Synechocystis* sp. PCC6803, *Mycobacterium leprae* and *M. tuberculosis* have the *cobA–hemD* fusion found in *C. josui*.

It should be noted that bacterial *hemD* gene products have been shown to be bona fide uroporphyrinogen synthases only in *E. coli* and *B. subtilis*, which share only 25% identity to each other. In the absence of uroporphyrinogen III synthase activity, hydroxymethylbilane will spontaneously cyclize to the I isomer of uroporphyrinogen, which cannot be used to make functional tetrapyrroles. Thus, if we assume that this activity is essential, then uroporphyrinogen III synthases must be diverse in their primary sequence in prokaryotes. Moreover, a uroporphyrinogen III synthase gene has not been identified in a plant, nor is a recognizable homologue found in the *Arabidopsis thaliana* genome. This is surprising considering the large chlorophyll requirement in plants and the successful purification of the protein from wheat (Higuchi and Bogorad, 1975).

8.4. The Absence of Identifiable Protoporphyrinogen Oxidase Gene Homologues in Many Prokaryotes

Protoporphyrinogen oxidase catalyses the oxidation of protoporphyrinogen to protoporphyrin, and this activity is required for the synthesis of haem, chlorophyll and bilins. The two types of prokaryotic protoporphyrinogen oxidases are typified by those expressed in *E. coli* and *B. subtilis* encoded by *hemG* and *hemY*, respectively (see above). As stated previously, the *hemK* gene product is not a protoporphyrinogen oxidase, and the annotation of *hemK* homologues as such can be confusing. Furthermore, an *E. coli* ORF has been designated as *hemY*, but is unrelated to the *Bacillus*-type *hemY* and has not been demonstrated to be a haem biosynthesis gene.

We searched for protoporphrinogen oxidase gene homologues in the genomes of organisms that contained the *hemE*, *hemN* and *hemH* genes, as would be expected in organisms that synthesize haem. Thirty-nine genomes representing 25 genera and 31 species contain the three genes (Table 1). Of these, 11 genera contained no recognizable protoporphyrinogen oxidase gene, defined as those that encode homologues of the *hemG* or *hemY* products (Table 1).

In those organisms that do contain a protoporphyrinogen oxidase gene, *hemY* is more prevalent than *hemG*, and additional *hemY* genes have been identified in organisms whose genomes have not been fully sequenced. Furthermore, the bacterial *hemY* product is homologous to protoporphyrinogen oxidases from eukaryotes, including mammals, plants and yeast. By contrast, a *hemG* gene is more restricted, and has been found only in *E. coli*, *Salmonella typhimurium*, *Vibrio cholerae*, *Pasteurella multicoda*, *M. loti* and *Yersinia pestis*. All but *M. loti* are in the γ-subdivision of the proteobacteria, but that group also includes *P. aeruginosa* and *X. fastidiosa*, which have no identifiable protoporphyrinogen oxidase gene. In addition, *M. loti* does not

Table 1   Status of protoporphyrinogen oxidase gene homologues in genomes of prokaryotes that contain other haem synthesis genes.

Organisms with a *hemY* gene homologue
Aquifex aeolicus
Bacillus halodurans C-125
Bacillus subtilis
Chlamydophila pneumoniae strains CWL029, AR39 and J138
Chlamydia trachomatis
Chlamydia muridarum
Deinococcus radiodurans
Listeria innocua
Listeria monocytogenes
Mycobacterium tuberculosis strains H37Rv and CDC1551
Mycobacterium leprae
Staphylococcus aureus strains N315 and Mu50

Organisms with a *hemG* gene homologue
Escherichia coli, strains K12, O157:H7 and O157:H7 EDL933
Mesorhizobium loti
Pasteurella multocida
Salmonella typhi
Salmonella typhimurium
Vibrio cholerae
Yersinia pestis

Organisms with no identifiable protoporphyrinogen oxidase gene homologue
Agrobacterium tumefaciens
Campylobacter jejuni
Caulobacter crescentus
Helicobacter pylori strains 26695 and J99
Neisseria meningitidis strains MC58 and Z2491
Pseudomonas aeruginosa
Rickettsia conorii
Rickettsia prowazekii
Sinorhizobium meliloti
Synechocystis PCC6803
Thermoplasma volcanium
Xylella fastidiosa

The 63 complete genomes listed at the National Center for Biotechnology Information (NCBI) web site (www.ncbi.nlm.nih.gov:80/PMGifs/Genomes/micr.html) were analysed by BLAST protein homology searches, using the *hemG* gene product from *E. coli* K12, the *hemY* gene product from *B. subtilis*, and the protoporphyrinogen oxidase from mouse. The criteria used for organisms expected to contain a protoporphyrinogen oxidase gene in the table is that they contain the other three genes committed to haem synthesis, namely *hemE*, *hemN* and *hemH*. Thus organisms lacking a protoporphyrinogen oxidase gene, but that are also lacking other haem synthesis genes are not included in the list. Furthermore, several organisms described in section 9 contain haem proteins, but are lacking the final four enzymes of the haem pathway, including protoporphyrinogen oxidase. They are not included in the table, because they also lack *hemE*, *hemN* and *hemH*.

The genomes analysed include those listed in the table, as well as the following complete genomes: *Aeropyrum pernix* K1, *Archaeoglobus fulgidus*, *Borrelia burgdorferi*, *Buchnera* sp. *APS*, *Clostridium acetobutylicum*, *Halobacterium* sp. *NRC-1*, *Haemophilus influenzae*, *Lactococcus lactis* subsp. *lactis*, *Methanobacterium thermoautotrophicum*, *Methanococcus jannaschii*, *Mycoplasma genitalium*, *Mycoplasma pneumoniae*, *Mycoplasma pulmonis*, *Pyrococcus abyssi*, *Pyrococcus horikoshii*, *Streptococcus pneumoniae TIGR4* and *R6*, *Streptococcus pyogenes*, *Sulfolobus solfataricus*, *Sulfolobus tokodaii*, *Thermoplasma acidophilum*, *Treponema pallidum*, *Thermotoga maritima*, *Ureaplasma urealyticum*.

typify the α-proteobacteria since *S. meliloti*, *C. crescentus* or *Rickettsia* do not contain *hemG* or any recognizable protoporphyrinogen oxidase gene.

Unlike uroporphyrinogen III synthase, the reaction catalysed by protoporphyrinogen oxidase can proceed nonenzymatically *in vitro*, formally raising the possibility that this enzyme is not required in organisms where the gene has not been found. However, protoporphyrinogen oxidase mutants of *E. coli* (Sasarman *et al.*, 1993) and *B. subtilis* (Homuth *et al.*, 1999) are haem-defective. It seems more likely that a previously unidentified protoporphyrinogen oxidase exists in prokaryotes, or else a known protein has an additional activity. In support of the latter idea, the oxygen-dependent coproporphyrinogen oxidase of *E. coli* (*hemF*) has protoporphyrinogen oxidase activity, and a secondary site suppressor mutant of a *hemG* strain has elevated *hemF* promoter activity (Narita *et al.*, 1999). Furthermore, overexpression of *hemF* rescues a *hemG* mutant. It remains to be established whether the *hemF* product normally participates in protoporphyrinogen oxidation in wild-type cells.

9. ARE THERE ALTERNATIVE HAEM PATHWAYS?

9.1. An Alternative Route of Coproporphyrinogen III Synthesis from Uroporphyrinogen III in *Desulfovibrio vulgaris*

The intermediates in haem biosynthesis are well conserved and, with the exceptions noted above, the enzymes that metabolize those intermediates are homologous amongst prokaryotes and eukaryotes. However, alternative haem synthesis has been proposed in the anaerobic bacterium *Desulfovibrio vulgaris* whereby two methyl groups of haem c_3 are derived from the S-methyl group of methionine rather than from ALA (Akutsu *et al.*, 1993; Ishida *et al.*, 1998). In the conventional pathway, the four methyl groups of the porphyrin ring are derived from decarboxylation of the acetate groups of uroporphyrinogen III to yield coproporphyrinogen III via uroporphyrinogen III decarboxylase. Thus, the methyl groups and all carbon atoms are derived from ALA. In the alternative scheme (Fig. 9), it is proposed that uroporphyrinogen III is first methylated at positions C-1 and C-3 (Fischer numbering system) to yield precorrin 2, a precursor for sirohaem and vitamin B_{12}. The methyl donor is S-adenosylmethionine (SAM). The acetate groups at those carbon positions are removed and the remaining acetates are decarboxylated to methyl groups yielding coproporphyrinogen III. Thus, the methyl groups at positions 5 and 8 are derived from ALA, but the 1 and 3 methyl groups originate from SAM.

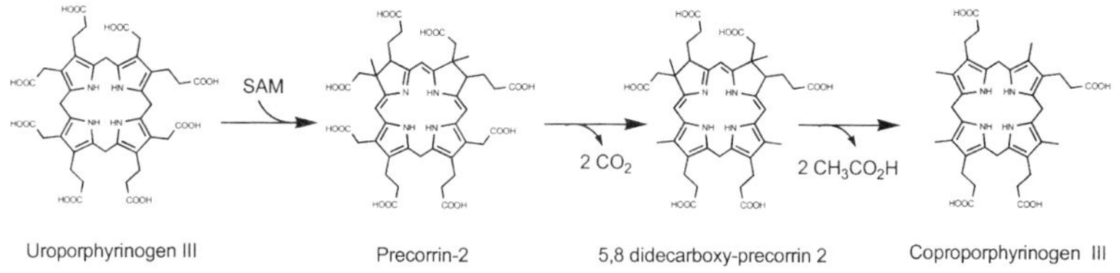

Uroporphyrinogen III   Precorrin-2   5,8 didecarboxy-precorrin 2   Coproporphyrinogen III

Figure 9 Proposed alternative route of coproporphyrinogen III synthesis from uro-porphyrinogen III in *Desulfovibrio vulgaris* as descrbed in Ishida *et al.* (1998). Note that in this scheme the methyl groups at positions 1 and 3 are derived from S-adenosylmethionine (SAM), rather than from ALA. Numbering of carbons is according to the Fischer system.

The pathway described in *D. vulgaris* is not generally applicable to anaer-obic tetrapyrrole synthesis. Bollivar *et al.* (1995) showed that the methyl groups of protoporphyrin synthesized anaerobically in the strict anaerobe *Chlorobium vibrioforme* or in the facultative anaerobe *R. sphaeroides* are derived from ALA and not from L-methionine. Furthermore, a *S. typhimurium cysG* mutant that cannot make precorrin 2 is capable of syn-thesizing haem anaerobically (Bollivar *et al.*, 1995). Finally, the alternative haem pathway described by Sato's group (Ishida *et al.*, 1998) can only sub-stitute for the step catalysed by uroporphyrinogen III synthase in the conventional pathway, a reaction that does not require O_2. Thus, the alterna-tive pathway does not circumvent the O_2 problem associated with the oxidation steps catalysed by coproporphyrinogen III oxidase and protopor-phyrinogen IX oxidase.

9.2. The Apparent Lack of the Pathway from Uroporphyrinogen III to Protohaem in Prokaryotes that Contain Haem Proteins

The absence of identifiable haem synthesis genes in an organism known to contain haem can suggest the possibility of an alternative synthetic pathway. In the cases where a single enzyme cannot be identified by homology (see above), it is likely that a similar reaction is carried out by a structurally disparate enzyme since both the substrate and product of the missing step can be accounted for. However, genome sequencing projects reveal cases where numerous enzymes catalysing successive steps of haem formation cannot be found in organisms that do express haem enzymes.

The Archaea *Sulfolobus acidocaldarius* and *S. solfataricus* respire aerobically, contain haem proteins (Becker and Schäfer, 1991; Lübben and Morand, 1994; Lübben *et al.*, 1994; Iwasaki *et al.*, 1995; Wright *et al.*, 1996; Yano *et al.*, 2000), and the genome of *S. solfataricus* indicates a plethora of haem protein-encoding genes (www-archbac.u-psud.fr/projects/sulfolobus/). However, our own homology searches, as well as the annotations of other investigators, find genes corresponding to the haem pathway only to uroporphyrinogen III formation, but the final four genes cannot be found by those criteria. Similarly, the last four enzymes of the pathway are unaccounted for in *Thermoplasma acidophilum* despite the presence of haem proteins (Gartner, 1991) and numerous haem protein gene homologues (Ruepp *et al.*, 2000). Although we could find no citations that identify haem proteins from *Aeropyrum pernix*, it is an obligately aerobic thermophile with many haem protein-encoding genes, but it too lacks identifiable genes that metabolize uroporphyrinogen III to haem (Kawarabayasi *et al.*, 1999). Finally, the same phenomenon is observed in the extreme halophilic archaeon *Halobacterium* sp. NRC-1 (Ng *et al.*, 2000), which contains many haem protein genes, and *H. halobium* haem proteins have been characterized extensively (Denda *et al.*, 1991, 1995; Fujiwara *et al.*, 1993; Hallberg Gradin and Colmsjo, 1989).

The discrepancy between a haem protein requirement and the apparent lack of enzymes that metabolize uroporphyrinogen III to haem is thus far confined to aerobic extremophilic Archaea. However, these omissions are not observed in the thermophilic proteobacterium *Aquifex aeolicus* (Deckert *et al.*, 1998) or in the thermophilic archaeon *Thermoplasma volcanium* (Kawashima *et al.*, 2000) (except for protoporphyrinogen oxidase, but see above) and thus neither thermophily nor taxonomy place an intrinsic restriction upon the expression of a conventional haem biosynthetic pathway.

The ecological niches of these extremophiles do not provide an obvious source of exogenous haem for uptake, and therefore it is very probable that haem is synthesized *de novo*. Even if they possess the alternative route for coproporphyrinogen III synthesis from uroporphyrinogen III as described above for *D. vulgaris*, the subsequent oxidations and iron chelation cannot be deduced based on homology. One of the most astonishing features of archaeal genomes revealed by sequencing projects is the vast numbers of genes with no known homologues. Understanding the novel enzymes or intermediates in haem biosynthesis implicated by at least some aerobic archaeal genomes represents an important frontier in prokaryotic metabolism.

ACKNOWLEDGEMENTS

Work from M.R.O'B.'s laboratory was supported by National Science Foundation grants MCB-0089928 and MCB-0077628 and US Department of Agriculture grant 99-35305-8062. Work from L.T.-M.'s laboratory was supported by a Grant of the Swiss National Foundation for Scientific Research.

REFERENCES

Akutsu, H., Park, J.-S. and Sano, S. (1993) L-methionine methyl is specifically incorporated into the C-2 and C-7 positions of the porphyrin of cytochrome c_3 in a strictly anaerobic bacterium, *Desulfovibrio vulgaris*. *J. Am. Chem. Soc.* **115**, 12185–12186.

Alkaradaghi, S., Hansson, M., Nikonov, S., Jonsson, B. and Hederstedt, L. (1997) Crystal structure of ferrochelatase: the terminal enzyme in heme biosynthesis. *Structure* **5**, 1501–1510.

Allikmets, R., Raskind, W.H., Hutchinson, A., Schueck, N.D., Dean, M. and Koeller, D.M. (1999) Mutation of a putative mitochondrial iron transporter gene (ABC7) in X-linked sideroblastic anemia and ataxia (XLSA/A). *Hum. Mol. Genet.* **8**, 743–749.

Alwan, A.F., Mgbeje, B.I. and Jordan, P.M. (1989) Purification and properties of uroporphyrinogen III synthase (co-synthase) from an overproducing recombinant strain of *Escherichia coli* K-12. *Biochem. J.* **264**, 397–402.

Andersson, S.G.E., Zomorodipour, A., Andersson, J.O., Sicheritz-Ponten, T., Alsmark, U.C.M., Podowski, R.M., Naslund, A.K., Eriksson, A.S., Winkler, H.H. and Kurland, C.G. (1998) The genome sequence of *Rickettsia prowazekii* and the origin of mitochondria. *Nature* **396**, 133–140.

Avissar, Y.J. and Beale, S.I. (1989) Identification of the enzymatic basis for δ-aminolevulinic acid auxotrophy in a *hemA* mutant of *Escherichia coli*. *J. Bacteriol.* **171**, 2919–2924.

Avissar, Y.J. and Beale, S.I. (1990) Cloning and expression of a structural gene from *Chlorobium vibrioforme* that complements the *hemA* mutation in *Escherichia coli*. *J. Bacteriol.* **172**, 1656–1659.

Avissar, Y.J. and Nadler, K.D. (1978) Stimulation of tetrapyrrole formation in *Rhizobium japonicum* by restricted aeration. *J. Bacteriol.* **135**, 782–789.

Bardischewsky, F. and Friedrich, C. (2001) Identification of *ccdA* in *Paracoccus pantotrophus* GB17: disruption of *ccdA* causes complete deficiency in *c*-type cytochromes. *J. Bacteriol.* **183**, 257–263.

Basile, G., Di Bello, C. and Taniuchi, H. (1980) Formation of an Iso-1-cytochrome *c*-like species containing a covalently bonded heme group from the apoprotein by a yeast cell-free system in the presence of hemin. *J. Biol. Chem.* **255**, 7181–7191.

Baumann, P., Baumann, L., Lai, C.-Y. and Rouhbakhsh, D. (1995) Genetics, physiology and evolutionary relationships of the genus *Buchnera*: intracellular symbionts of aphids. *Annu. Rev. Microbiol.* **49**, 55–94.

Beale, S.I. and Weinstein, J.D. (1990) Tetrapyrrole metabolism in photosynthetic organisms. In: *Biosynthesis of Heme and Chlorophylls* (H.A. Dailey, ed.) pp. 287–391, McGraw-Hill Publishing, New York.

Becker, M. and Schäfer, G. (1991) Purification and spectral characterization of a *b*-type cytochrome from the plasma membrane of the archaebacterium *Sulfolobus acidocaldarius*. *FEBS Lett.* **291**, 331–335.

Beckett, C.S., Loughman, J.A., Karberg, K.A., Donato, G.M., Goldman, W.E. and Kranz, R.G. (2000) Four genes are required for the system II cytochrome *c* biogenesis pathway in *Bordetella pertussis*, a unique bacterial model. *Mol. Microbiol.* **38**, 465–481.

Beckman, D.L. and Kranz, R.G. (1993) Cytochromes *c* biogenesis in a photosynthetic bacterium requires a periplasmic thioredoxin-like protein. *Proc. Natl Acad. Sci. USA* **90**, 2179–2183.

Beckman, D.L., Trawick, D.R. and Kranz, R.G. (1992) Bacterial cytochromes *c* biogenesis. *Genes Dev.* **6**, 268–283.

Bhasker, C.R., Burgiel, G., Neupert, B., Emery-Goodman, A., Kuhn, L.C. and May, B.K. (1993) The putative iron-responsive element in the human erythroid 5-aminolevulinate synthase mRNA mediates translational control. *J. Biol. Chem.* **268**, 12699–12705.

Biel, S.W. and Biel, A.J. (1990) Isolation of a *Rhodobacter capsulatus* mutant that lacks *c*-type cytochromes and excretes porphyrins. *J. Bacteriol.* **172**, 1321–1326.

Blattner, F.R., Plunkett III, G., Bloch, C.A., Perna, N.T., Burland, V., Riley, M., Collado-Vides, J., Glasner, J.D., Rode, C.K., Mayhew, G.F., *et al.* (1997) The complete genome sequence of *Escherichia coli* K-12. *Science* **277**, 1453–1462.

Bollivar, D.W., Elliott, T. and Beale, S.I. (1995) Anaerobic protoporphyrin biosynthesis does not require incorporation of methyl groups from methionine. *J. Bacteriol.* **177**, 5778–5783.

Bsat, N., Herbig, A., Casillas-Martinez, L., Setlow, P. and Helmann, J.D. (1998) *Bacillus subtilis* contains multiple Fur homologs: identification of the iron uptake (Fur) and peroxide regulon (PerR) repressors. *Mol. Microbiol.* **29**, 189–198.

Cabrejos, M.-E., Zhao, H.-L., Guacucano, M., Bueno, S., Levican, G., Garcia, E., Jedlicki, E. and Holmes, D.S. (1999) IST1 insertional inactivation of the *resB* gene: implication for phenotype switching in *Thiobacillus ferrooxidans*. *FEMS Microbiol. Lett.* **175**, 223–229.

Calvo, J.M. and Matthews, R.G. (1994) The leucine-responsive regulatory protein, a global regulator of metabolism in *Escherichia coli*. *Microbiol. Mol. Biol. Rev.* **58**, 466–490.

Camadro, J.-M., Matringe, M., Brouillet, N., Thome, F. and Labbe, P. (1993). Characterization of plant and yeast protoporphyrinogen oxidase: Molecular target of diphenyl ether type herbicides. In: *Porphyric Pesticides. Chemistry, Toxicology and Pharmaceutical Applications* (S.O. Duke and C.A. Rebeiz, eds), American Chemical Society, Washington, DC.

Camadro, J.M., Thome, F., Brouillet, N. and Labbe, P. (1994) Purification and properties of protoporphyrinogen oxidase from the yeast *Saccharomyces cerevisiae*. Mitochondrial location and evidence for a precursor form of the protein. *J. Biol. Chem.* **269**, 32085–32091.

Caspi, R., Tebo, B.M. and Haygood, M.G. (1998) *c*-type cytochromes and manganese oxidation in *Pseudomonas putida* MnB1. *Appl. Environ. Microbiol.* **64**, 3549–3555.

Castresana, J. and Moreira, D. (1999) Respiratory chains in the last common ancestor of living organisms. *J. Mol. Evol.* **49**, 453–460.

Chauhan, S. and O'Brian, M.R. (1993) *Bradyrhizobium japonicum* δ-aminolevulinic acid dehydratase is essential for symbiosis with soybean and contains a novel metal-binding domain. *J. Bacteriol.* **175**, 7222–7227.

Chauhan, S. and O'Brian, M.R. (1995) A mutant *Bradyrhizobium japonicum* δ-aminolevulinic acid dehydratase with an altered metal requirement functions *in situ* for tetrapyrrole synthesis in soybean root nodules. *J. Biol. Chem.* **270**, 19823–19827.

Chauhan, S. and O'Brian, M.R. (1997) Transcriptional regulation of δ-aminolevulinic acid dehydratase synthesis by oxygen in *Bradyrhizobium japonicum* and evidence for developmental control of the *hemB* gene. *J. Bacteriol.* **179**, 3706–3710.

Chauhan, S., Titus, D.E. and O'Brian, M.R. (1997) Metals control activity and expression of the heme biosynthesis enzyme δ-aminolevulinic acid dehydratase in *Bradyrhizobium japonicum*. *J. Bacteriol.* **179**, 5516–5520.

Chelstowska, A., Zoladek, T., Garey, J., Kushner, J., Rytka, J. and Labbe-Bois, R. (1992) Identification of amino acid changes affecting yeast uroporphyrinogen decarboxylase activity by sequencing analysis of *hem12* mutant alleles. *Biochem. J.* **288**, 753–757.

Chung, J., Chen, T. and Missiakas, D. (2000) Transfer of electrons across the cytoplasmic membrane by DsbD, a membrane protein involved in thiol-disulphide exchange and protein folding in the bacterial periplasm. *Mol. Microbiol.* **35**, 1099–1109.

Cochran, A.G. and Schultz, P.G. (1990) Antibody-catalyzed porphyrin metallation. *Science* **249**, 781–783.

Cook, G.M. and Poole, R.K. (2000) Oxidase and periplasmic cytochrome assembly in *Escherichia coli* K-12: CydDC and CcmAB are not required for haem-membrane association. *Microbiology* **146**, 527–536.

Coomber, S.A., Jones, R.M., Jordan, P.M. and Hunter, C.N. (1992) A putative anaerobic coproporphyrinogen III oxidase in *Rhodobacter sphaeroides*. I. Molecular cloning, transposon mutagenesis and sequence analysis of the gene. *Mol. Microbiol.* **6**, 3159–3169.

Crooke, H. and Cole, J. (1995) The biogenesis of *c*-type cytochromes in *Escherichia coli* requires a membrane-bound protein, DipZ, with a protein disulphide isomerase-like domain. *Mol. Microbiol.* **15**, 1139–1150.

Dailey, H.A. (1990). *Biosynthesis of Hemes and Chlorophylls*, McGraw-Hill Publishing, New York.

Dailey, H.A. and Dailey, T.A. (1996) Protoporphyrinogen oxidase of *Myxococcus xanthus* – expression, purification and characterization of the cloned enzyme. *J. Biol. Chem.* **271**, 8714–8718.

Dailey, H.A. and Dailey, T.A. (1997) Characteristics of human protoporphyrinogen oxidase in controls and variegate porphyrias. *Cell. Mol. Biol.* **43**, 67–73.

Dailey, H.A., Fleming, J.E. and Harbin, B.M. (1986) Ferrochelatase from *Rhodopseudomonas sphaeroides*: substrate specificity and role of sulfhydryl and arginyl residues. *J. Bacteriol.* **165**, 1–5.

Dailey, H.A., Finnegan, M.G. and Johnson, M.K. (1994) Human ferrochelatase is an iron-sulfur protein. *Biochemistry* **33**, 403–407.

Dailey, H.A., Dailey, T.A., Wu, C.K., Medlock, A.E., Wang, K.F., Rose, J.P. and Wang, B.C. (2000) Ferrochelatase at the millennium: structures, mechanisms and [2Fe-2S] clusters. *Cell. Mol. Life Sci.* **57**, 1909–1926.

Dailey, T.A. and Dailey, H.A. (1998) Identification of an FAD superfamily containing protoporphyrinogen oxidases, monoamine oxidases and phytoene desaturase. Expression and characterization of phytoene desaturase of *Myxococcus xanthus*. *J. Biol. Chem.* **273**, 13658–13662.

Darie, S. and Gunsalus, R.P. (1994) Effect of heme and oxygen availability on *hemA* gene expression in *Escherichia coli*: role of the *fnr*, *arcA*, and *himA* gene products. *J. Bacteriol.* **176**, 5270–5276.

David, M., Daveran, M.L., Batut, J., Dedieu, A., Domergue, O., Ghai, J., Hertwig, C., Boistard, P. and Kahn, D. (1989) Cascade regulation of *nif* gene expression in *Rhizobium meliloti*. *Cell* **54**, 671–683.

de Gier, J.W., Schepper, M., Reijnders, W.N., van Dyck, S.J., Slotboom, D.J., Warne, A., Saraste, M., Krab, K., Finel, M., Stouthamer, A.H. *et al.* (1996) Structural and functional analysis of aa_3-type and cbb_3-type cytochrome *c* oxidases of *Paracoccus denitrificans* reveals significant differences in proton-pump design. *Mol. Microbiol.* **20**, 1247–1260.

de Vrind, J.P.M., Brouwers, G.J., Corstjens, P.L.A.M., den Dulk, J. and deVrind-deJong, E.W. (1998) The cytochrome *c* maturation operon is involved in manganese oxidation in *Pseudomonas putida* GB-1. *Appl. Environ. Microbiol.* **64**, 3556–3562.

Deckert, G., Warren, P.V., Gaasterland, T., Young, W.G., Lenox, A.L., Graham, D.E., Overbeek, R., Snead, M.A., Keller, M., Aujay, M., Huber, R., Feldman, R.A., Short, J.M., Olsen, G.J. and Swanson, R.V. (1998) The complete genome of the hyperthermophilic bacterium *Aquifex aeolicus*. *Nature* **392**, 353–358.

Denda, K., Fujiwara, T., Seki, M., Yoshida, M., Fukumori, Y. and Yamanaka, T. (1991) Molecular cloning of the cytochrome aa_3 gene from the archaeon (Archaebacterium) *Halobacterium halobium*. *Biochem. Biophys. Res. Commun.* **181**, 316–322.

Denda, K., Mogi, T., Anraku, Y., Yamanaka, T. and Fukumori, Y. (1995) Characterization of chimeric heme-copper respiratory oxidases using subunits I of *Escherichia coli* cytochrome *bo* and *Halobacterium salinarium* cytochrome aa_3. *Biochem. Biophys. Res. Commun.* **217**, 428–436.

Dent, A.J., Beyersmann, D., Block, C. and Hasnain, S.S. (1990) Two different zinc sites in bovine 5-aminolevulinate dehydratase distinguished by extended X-ray absorption fine structure. *Biochemistry* **29**, 7822–7828.

Deshmukh, M., Brasseur, G. and Daldal, F. (2000) Novel *Rhodobacter capsulatus* genes required for the biogenesis of various *c*-type cytochromes. *Mol. Microbiol.* **35**, 123–138.

Drygas, M.E., Lambowitz, A.M. and Nargang, F.E. (1989) Cloning and analysis of the *Neurospora crassa* cytochrome *c* heme lyase. *J. Biol. Chem.* **264**, 17897–17906.

Dujon, B., Alexandraki, D., Andre, B., Ansorge, W., Baladron, V., Ballesta, J.P., Banrevi, A., Bolle, P.A., Bolotin-Fukuhara, M. and Bossier, P. *et al.* (1994) Complete DNA sequence of yeast chromosome XI. *Nature* **369**, 371–378.

Dumont, M.E., Ernst, J.F., Hampsey, D.M. and Sherman, F. (1987) Identification and sequence of the gene encoding cytochrome *c* heme lyase in the yeast *Saccharomyces cerevisiae*. *EMBO J.* **6**, 235–241.

Dumont, M.E., Ernst, J.F. and Sherman, F. (1988) Coupling of heme attachment to import of cytochrome *c* into yeast mitochondria. Studies with heme lyase-deficient mitochondria and altered apocytochromes *c*. *J. Biol. Chem.* **263**, 15928–15937.

Dumont, M.E., Cardillo, T.S., Hayes, M.K. and Sherman, F. (1991) Role of cytochrome *c* heme lyase in mitochondrial import and accumulation of cytochrome *c* in *Saccharomyces cerevisiae*. *Mol. Cell. Biol.* **11**, 5487–5496.

Eaves, D.J., Grove, J., Staudenmann, W., James, P., Poole, R.K., White, S.A., Griffiths, I. and Cole, J.A. (1998) Involvement of products of the *nrfEFG* genes in the covalent attachment of haem *c* to a novel cysteine-lysine motif in the cytochrome c_{552} nitrite reductase from *Escherichia coli*. *Mol. Microbiol.* **28**, 205–216.

Edeling, M.A., Guddat, L.W., Fabianek, R.A., Halliday, J.A., Jones, A., Thöny-Meyer, L. and Martin, J.L. (2001) Crystallization and preliminary diffraction studies of native and selenomethionine CcmG (CycY, DsbE). *Acta Cryst.* **57**, 1293–1295.

Elder, G.H. and Roberts, A.G. (1995) Uroporphyrinogen decarboxylase. *J. Bioenerg. Biomemb.* **27**, 207–214.

Elliott, T. (1989) Cloning, genetic characterization and nucleotide sequence of the *hemA-prfA* operon of *Salmonella typhimurium*. *J. Bacteriol.* **171**, 3948–3960.

Elliott, T. (1993) Transport of 5-aminolevulinic acid by the dipeptide permease in *Salmonella typhimurium*. *J. Bacteriol.* **175**, 325–331.

Erskine, P.T., Senior, N., Awan, S., Lambert, R., Lewis, G., Tickle, L.J., Sarwar, M., Spencer, P., Thomas, P., Warren, M.J. *et al.* (1997) X-ray structure of 5-aminolaevulinate dehydratase, a hydrid aldolase. *Nature Struct. Biol.* **4**, 1025–1031.

Erskine, P.T., Norton, E., Cooper, J.B., Lambert, R., Coker, A., Lewis, G., Spencer, P., Sarwar, M., Wood, S.P., Warren, M.J. and Shoolingin-Jordan, P.M. (1999) X-ray structure of 5-aminolevulinic acid dehydratase from *Escherichia coli* complexed with the inhibitor levulinic acid at 2.0 Å resolution. *Biochemistry* **38**, 4266–4276.

Escolar, L., Perez-Martin, J. and de Lorenzo, V. (1999) Opening the iron box: transcriptional metalloregulation by the Fur protein. *J. Bacteriol.* **181**, 6223–6229.

Fabianek, R.A., Hennecke, H. and Thöny-Meyer, L. (1998) The active-site cysteines of the periplasmic thioredoxin-like protein CcmG of *Escherichia coli* are important but not essential for cytochrome *c* maturation. *J. Bacteriol.* **180**, 1947–1950.

Fabianek, R.A., Hofer, T. and Thöny-Meyer, L. (1999) Characterization of the *Escherichia coli* CcmH protein reveals new insights into the redox pathway required for cytochrome *c* maturation. *Arch. Microbiol.* **171**, 92–100.

Fabianek, R.A., Hennecke, H. and Thöny-Meyer, L. (2000) Periplasmic protein thiol:disulphide oxidoreductases of *Escherichia coli*. *FEMS Microbiol. Rev.* **24**, 303–316.

Fales, L., Kryszak, L. and Zeilstra-Ryalls, J. (2001) Control of *hemA* expression in

Rhodobacter sphaeroides 2.4.1: effect of a transposon insertion in the *hbdA* gene. *J. Bacteriol.* **183**, 1568–1576.

Ferreira, G.C., Franco, R., Lloyd, S.G., Pereira, A.S., Moura, I., Moura, J.J.G. and Huynh, B.H. (1994) Mammalian ferrochelatase, a new addition to the metalloenzyme family. *J. Biol. Chem.* **269**, 7062–7065.

Fischer, H. (1994) Genetic regulation of nitrogen fixation in rhizobia. *Microbiol. Rev.* **58**, 352–386.

Fischer, H.M., Velasco, L., Delgado, M.J., Bedmar, E.J., Schären, S., Zingg, D., Göttfert, M. and Hennecke, H. (2001) One of two *hemN* genes in *Bradyrhizobium japonicum* is functional during anaerobic growth and in symbiosis. *J. Bacteriol.* **183**, 1300–1311.

Frankenberg, N., Erskine, P.T., Cooper, J.B., Shoolingin-Jordan, P.M., Jahn, D. and Heinz, D.W. (1999a) High resolution crystal structure of a Mg^{2+}-dependent porphobilinogen synthase. *J. Mol. Biol.* **289**, 591–602.

Frankenberg, N., Jahn, D. and Jaffe, E.K. (1999b) *Pseudomonas aeruginosa* contains a novel type V porphobilinogen synthase with no required catalytic metal ions. *Biochemistry* **38**, 13976–13982.

Fraser, C.M., Casjens, S., Huang, W.M., Sutton, G.G., Clayton, R., Lathigra, R., White, O., Ketchum, K.A., Dodson, R., Hickey, E.K., Gwinn, M., Dougherty, B., Tomb, J.F., Fleischmann, R.D., Richardson, D., Peterson, J., Kerlavage, A.R., Quackenbush, J., Salzberg, S., Hanson, M., van Vugt, R., Palmer, N., Adams, M.D., Gocayne, J., Venter, J.C. *et al.* (1997) Genomic sequence of a Lyme disease spirochaete, *Borrelia burgdorferi*. *Nature* **390**, 580–586.

Fraser, C.M., Norris, S.J., Weinstock, G.M., White, O., Sutton, G.G., Dodson, R., Gwinn, M., Hickey, E.K., Clayton, R., Ketchum, K.A. *et al.* (1998) Complete genome sequence of *Treponema pallidum*, the syphilis spirochete. *Science* **281**, 375–388.

Friedmann, H.C., Klein, A. and Thauer, R.K. (1991) Biochemistry of coenzyme F430, a nickel porphinoid involved in methanogenesis. In: *Biosynthesis of Tetrapyrroles*, (P.M. Jordan, ed.), pp. 139–154, Elsevier Scientific, New York.

Frustaci, J.M. and O'Brian, M.R. (1992) Characterization of a *Bradyrhizobium japonicum* ferrochelatase mutant and isolation of the *hemH* gene. *J. Bacteriol.* **174**, 4223–4229.

Frustaci, J.M. and O'Brian, M.R. (1993a) Analysis of the *Bradyrhizobium japonicum hemH* gene and its expression in *Escherichia coli*. *Appl. Environ. Microbiol.* **59**, 2347–2351.

Frustaci, J.M. and O'Brian, M.R. (1993b) The *Escherichia coli visA* gene encodes ferrochelatase, the final enzyme of the heme biosynthetic pathway. *J. Bacteriol.* **175**, 2154–2156.

Fujino, E., Fujino, T., Karita, S., Sakka, K. and Ohmiya, K. (1995) Cloning and sequencing of some genes responsible for porphyrin biosynthesis from the anaerobic bacterium *Clostridium josui*. *J. Bacteriol.* **177**, 5169–5175.

Fujiwara, T., Fukumori, Y. and Yamanaka, T. (1993) *Halobacterium halobium* cytochrome *b*-558 and cytochrome *b*-562: purification and some properties. *J. Biochem. (Tokyo)* **113**, 48–54.

Gaballa, A., Koedam, N. and Cornelis, P. (1996) A cytochrome *c* biogenesis gene involved in pyoverdine production in *Pseudomonas fluorescens* ATCC 17400. *Mol. Microbiol.* **21**, 777–785.

Gaballa, A., Baysse, C., Koedam, N., Muyldermans, S. and Cornelis, P. (1998) Different residues in periplasmic domains of the CcmC inner membrane protein of *Pseudomonas fluorescens* ATCC 17400 are critical for cytochrome *c* biogenesis and pyoverdine-mediated iron uptake. *Mol. Microbiol.* **30**, 547–555.

Gabel, C. and Maier, R.J. (1993) Oxygen-dependent transcriptional regulation of cytochrome aa_3 in *Bradyrhizobium japonicum*. *J. Bacteriol.* **175**, 128–132.

Galibert, F., Finan, T.M., Long, S.R., Puhler, A., Abola, P., Ampe, F., Barloy-Hubler, F., Barnett, M.J., Becker, A., Boistard, P. *et al.* (2001) The composite genome of the legume symbiont *Sinorhizobium meliloti. Science* **293**, 668–672.

Gartner, P. (1991) Characterization of a quinol-oxidase activity in crude extracts of *Thermoplasma acidophilum* and isolation of an 18-kDa cytochrome. *Eur. J. Biochem.* **200**, 215–222.

Gibson, L.C., McGlynn, P., Chaudhri, M. and Hunter, C.N. (1992) A putative anaerobic coproporphyrinogen III oxidase in *Rhodobacter sphaeroides*. II. Analysis of a region of the genome encoding *hemF* and the *puc* operon. *Mol. Microbiol.* **6**, 3171–3186.

Goldman, B.S. and Kranz, R.G. (2001) ABC transporters associated with cytochrome *c* biogenesis. *Res. Microbiol.* **152**, 323–329.

Goldman, B.S., Gabbert, K.K. and Kranz, R.G. (1996) Use of heme reporters for studies of cytochrome biosynthesis and heme transport. *J. Bacteriol.* **178**, 6338–6347.

Goldman, B.S., Beckman, D.L., Bali, A., Monika, E.M., Gabbert, K.K. and Kranz, R.G. (1997) Molecular and immunological analysis of an ABC transporter complex required for cytochrome *c* biogenesis. *J. Mol. Biol.* **268**, 724–738.

Goldman, B.S., Beck, D.L., Monika, E.M. and Kranz, R.G. (1998) Transmembrane heme delivery systems. *Proc. Natl Acad. Sci. USA* **95**, 5003–5008.

Gora, M., Rytka, J. and Labbe-Bois, R. (1999) Activity and cellular location in *Saccharomyces cerevisiae* of chimeric mouse/yeast and *Bacillus subtilis*/yeast ferrochelatases. *Arch. Biochem. Biophys.* **361**, 231–240.

Gordon, E.H.J., Page, M.D., Willis, A.C. and Ferguson, S.J. (2000) *Escherichia coli* DipZ: anatomy of a transmembrane protein disulphide reductase in which three pairs of cysteine residues, one in each of three domains, contribute differentially to function. *Mol. Microbiol.* **35**, 1360–1374.

Grove, J., Tanapongpipat, S., Thomas, G., Griffiths, L., Crooke, H. and Cole, J. (1996) *Escherichia coli* K-12 genes essential for the synthesis of *c*-type cytochromes and a third nitrate reductase located in the periplasm. *Mol. Microbiol.* **19**, 467–481.

Guerinot, M.L. and Chelm, B.K. (1986) Bacterial δ-aminolevulinic acid synthase is not essential for leghemoglobin formation in the soybean/*Bradyrhizobium japonicum* symbiosis. *Proc. Natl Acad. Sci. USA* **83**, 1837–1841.

Hallberg Gradin, C. and Colmsjo, A. (1989) Four different *b*-type cytochromes in the halophilic archaebacterium, *Halobacterium halobium. Arch. Biochem. Biophys.* **272**, 130–136.

Hamza, I., Chauhan, S., Hassett, R. and O'Brian, M.R. (1998) The bacterial Irr protein is required for coordination of heme biosynthesis with iron availability. *J. Biol. Chem.* **273**, 21669–21674.

Hamza, I., Qi, Z., King, N.D. and O'Brian, M.R. (2000) Fur-independent regulation of iron metabolism by Irr in *Bradyrhizobium japonicum. Microbiology* **146**, 669–676.

Hansson, M. and Hederstedt, L. (1992) Cloning and characterization of the *Bacillus subtilis hemEHY* gene cluster, which encodes protoheme IX biosynthetic enzymes. *J. Bacteriol.* **174**, 8081–8093.

Hansson, M. and Hederstedt, L. (1994) Purification and characterization of a water-soluble ferrochelatase from *Bacillus subtilis. Eur. J. Biochem.* **220**, 201–208.

Hansson, M., Rutberg, L., Schröder, I. and Hederstedt, L. (1991) The *Bacillus subtilis hemAXCDBL* gene cluster, which encodes enzymes of the biosynthetic pathway from glutamate to uroporphyrinogen III. *J. Bacteriol.* **173**, 2590–2599.

Hart, R.A., Kallio, P.T. and Bailey, J.E. (1994) Effect of biosynthetic manipulation of heme on insolubility of *Vitreoscilla* hemoglobin in *Escherichia coli. Appl. Environ. Microbiol.* **60**, 2431–2437.

Hennig, B. and Neupert, W. (1981) Assembly of cytochrome *c*. Apocytochrome *c* is bound to specific sites on mitochondria before its conversion to holocytochrome c. *Eur. J. Biochem.* **121**, 203–212.

Hennig, M., Grimm, B., Contestabile, R., John, R.A. and Jansonius, J.N. (1997) Crystal structure of glutamate-1-semialdehyde aminomutase: an alpha2- dimeric vitamin B6-dependent enzyme with asymmetry in structure and active site reactivity. *Proc. Natl Acad. Sci. USA* **94**, 4866–4871.

Hentze, M.W. and Kuhn, L.C. (1996) Molecular control of vertebrate iron metabolism: mRNA-based regulatory circuits operated by iron, nitric oxide and oxidative stress. *Proc. Natl Acad. Sci. USA* **93**, 8175–8182.

Higuchi, M. and Bogorad, L. (1975) The purification and properties of uroporphyrinogen I synthases and uroporphyrinogen III cosynthase. *Ann. NY Acad. Sci.* **244**, 401–418.

Hippler, B., Homuth, G., Hoffmann, T., Hungerer, C., Schumann, W. and Jahn, D. (1997) Characterization of *Bacillus subtilis hemN. J. Bacteriol.* **179**, 7181–7185.

Holtwick, R., Keweloh, H. and Meinhardt, F. (1999) *cis/trans* isomerase of unsaturated fatty acids of *Pseudomonas putida* P8: evidence for a heme protein of the cytochrome c type. *Appl. Environ. Microbiol.* **65**, 2644–2649.

Homuth, G., Rompf, A., Schumann, W. and Jahn, D. (1999) Transcriptional control of *Bacillus subtilis hemN* and *hemZ. J. Bacteriol.* **181**, 5922–5929.

Howe, G. and Merchant, S. (1994) Role of heme in the biosynthesis of cytochrome c_6. *J. Biol. Chem.* **269**, 5824–5832.

Howe, G., Mets, L. and Merchant, S. (1995) Biosynthesis of cytochrome f in *Chlamydomonas reinhardtii*: analysis of the pathway in gabaculine-treated cells and in the heme attachment mutant B6. *Mol. Gen. Genet.* **246**, 156–165.

Huang, C.J. and Barrett, E.L. (1991) Sequence analysis and expression of the *Salmonella typhimurium asr* operon encoding production of hydrogen sulfide from sulfite. *J. Bacteriol.* **173**, 1544–1553.

Ilag, L.L., Jahn, D., Eggertsson, G. and Söll, D. (1991) The *Escherichia coli hemL* gene encodes glutamate 1-semialdehyde aminotransferase. *J. Bacteriol.* **173**, 3408–3413.

Ilag, L.L., Kumar, A.M. and Söll, D. (1994) Light regulation of chlorophyll biosynthesis at the level of 5-aminolevulinate formation in Arabidopsis. *Plant Cell.* **6**, 265–275.

Inoue, K., Dreyfuss, B.W., Kindle, K.L., Stern, D.B., Merchant, S. and Sodeinde, O.A. (1997) Ccs1, a nuclear gene required for the post-translational assembly of chloroplast c-type cytochromes. *J. Biol. Chem.* **272**, 31747–31754.

Ishida, T., Yu, L., Akutsu, H., Ozawa, K., Kawanishi, S., Seto, A., Inubushi, T. and Sano, S. (1998) A primitive pathway of porphyrin biosynthesis and enzymology in *Desulfovibrio vulgaris. Proc. Natl Acad. Sci. USA* **95**, 4853–4858.

Iwasaki, T., Wakagi, T. and Oshima, T. (1995) Resolution of the aerobic respiratory system of the thermoacidophilic archaeon, *Sulfolobus* sp. strain 7. III. The archaeal novel respiratory complex II (succinate:caldariellaquinone oxidoreductase complex) inherently lacks heme groups. *J. Biol. Chem.* **270**, 30902–30908.

Jacobs, N.J. and Jacobs, J.M. (1976) Nitrate, fumarate and oxygen as electron acceptors for a late step in microbial heme synthesis. *Biochim. Biophys. Acta* **449**, 1–9.

Jacobs, N.J. and Jacobs, J.M. (1977) Evidence for involvement of the electron transport system at a late step of anaerobic microbial heme synthesis. *Biochim. Biophys. Acta* **459**, 141–144.

Jacobs, N.J. and Jacobs, J.M. (1981) Protoporphyrinogen oxidation in *Rhodopseudomonas spheroides*, a step in heme and bacteriochlorophyll synthesis. *Arch. Biochem. Biophys.* **211**, 305–311.

Jacobs, J.M. and Jacobs, N.J. (1984) Protoporphyrinogen oxidation, an enzymatic step in heme and chlorophyll synthesis: partial characterization of the reaction in plant organelles and comparison with mammalian and bacterial systems. *Arch. Biochem. Biophys.* **229**, 312–319.

Jacobs, J.M. and Jacobs, N.J. (1993) Porphyrin accumulation and export by isolated barley (*Hordeum vulgare*) plastids. *Plant Physiol.* **101**, 1181–1187.

Jaffe, E.K. (2000) The porphobilinogen synthase family of metalloenzymes. *Acta Crystallogr. D. Biol. Crystallogr.* **56**, 115–128.

Jahn, D., Michelsen, U. and Söll, D. (1991) Two glutamyl-tRNA reductase activities in *Escherichia coli. J. Biol. Chem.* **266**, 2542–2548.

Jones, R.M. and Jordan, P.M. (1993) Purification and properties of the uroporphyrinogen decarboxylase from *Rhodobacter sphaeroides. Biochem. J.* **293**, 703–712.

Jordan, P.M., Mgbeje, B.I., Thomas, S.D. and Alwan, A.F. (1988a) Nucleotide sequence for the *hemD* gene of *Escherichia coli* encoding uroporphyrinogen III synthase and initial evidence for a *hem* operon. *Biochem. J.* **249**, 613–616.

Jordan, P.M., Warren, M.J., Williams, H.J., Stolowich, N.J., Roessner, C.A., Grant, S.K. and Scott, A.I. (1988b) Identification of a cysteine residue as the binding site for the dipyrromethane cofactor at the active site of *Escherichia coli* porphobilinogen deaminase. *FEBS Lett.* **235**, 189–193.

Kaczor, C.M., Smith, M.W., Sangwan, I. and O'Brian, M.R. (1994) Plant δ-aminolevulinic acid dehydratase. Expression in soybean root nodules and evidence for a bacterial lineage of the *Alad* gene. *Plant Physiol.* **104**, 1411–1417.

Kaneko, T., Nakamura, Y., Sato, S., Asamizu, E., Kato, T., Sasamoto, S., Watanabe, A., Idesawa, K., Ishikawa, A., Kawashima, K. *et al.* (2000) Complete genome structure of the nitrogen-fixing symbiotic bacterium *Mesorhizobium loti. DNA Res.* **7** (Suppl.), 381–406.

Kappas, A., Sassa, S., Galbraith, R.A. and Nordman, Y. (1995) The porphyrias. In: *The Metabolic and Molecular Bases of Inherited Disease*, (C.R. Scriver, A.L. Beaudet, W.S. Sly, D. Valle, J.B. Stanbury, J.B. Wyngaarden and D.S. Frederickson, eds) pp. 2103–2159, McGraw-Hill, New York.

Kastaniotis, A.J. and Zitomer, R.S. (2000) Rox1 mediated repression. Oxygen dependent repression in yeast. *Adv. Exp. Med. Biol.* **475**, 185–195.

Katzen, F. and Beckwith, J. (2000) Transmembrane electron transfer by the membrane protein DsbD occurs via a disulfide bond cascade. *Cell* **103**, 769–779.

Kawarabayasi, Y., Hino, Y., Horikawa, H., Yamazaki, S., Haikawa, Y., Jin-no, K., Takahashi, M., Sekine, M., Baba, S., Ankai, A. *et al.* (1999) Complete genome sequence of an aerobic hyper-thermophilic crenarchaeon, *Aeropyrum pernix* K1. *DNA Res.* **6**, 83–101.

Kawashima, T., Amano, N., Koike, H., Makino, S., Higuchi, S., Kawashima-Ohya, Y., Watanabe, K., Yamazaki, M., Kanehori, K., Kawamoto, T. *et al.* (2000) Archaeal adaptation to higher temperatures revealed by genomic sequence of *Thermoplasma volcanium. Proc. Natl Acad. Sci. USA* **97**, 14257–14262.

Keithly, J.H. and Nadler, K.D. (1983) Protoporphyrin formation in *Rhizobium japonicum. J. Bacteriol.* **154**, 838–845.

Kiel, J.A., Boels, J.M., Beldman, G. and Venema, G. (1990) Nucleotide sequence of the *Synechococcus* sp. PCC7942 branching enzyme gene (*glgB*): expression in *Bacillus subtilis. Gene* **89**, 77–84.

King, N.D. and O'Brian, M.R. (1997) Identification of the *lrp* gene in *Bradyrhizobium japonicum* and its role in regulation of δ-aminolevulinic acid uptake. *J. Bacteriol.* **179**, 1828–1831.

King, N.D. and O'Brian, M.R. (2001) Evidence for direct interaction between Enzyme I[Ntr] and aspartokinase to regulate bacterial oligopeptide transport. *J. Biol. Chem.* **276**, 21311–21316.

Klemm, D.J. and Barton, L.L. (1987) Purification and properties of protoporphyrinogen oxidase from an anaerobic bacterium, *Desulfovibrio gigas. J. Bacteriol.* **169**, 5209–5215.

Kranz, R.G. and Beckman, D.L. (1995) Cytochrome biogenesis. In: *Anoxygenic Photosynthetic Bacteria* (R.E. Blankenship, M.T. Madigan and C.E. Bauer, eds), pp. 709–723. Kluwer Academic, Dordrecht.

Kranz, R., Lill, R., Goldman, B., Bonnard, G. and Merchant, S. (1998) Molecular mechanisms of cytochrome *c* biogenesis: three distinct systems. *Mol. Microbiol.* **29**, 383–396.

Krupp, R., Chan, C. and Missiakas, D. (2001) DsbD-catalyzed transport of electrons across the membrane of *Escherichia coli. J. Biol. Chem.* **276**, 3696–3701.

Kunst, F., Ogasawara, N., Moszer, I., Albertini, A.M., Alloni, G., Azevedo, V., Bertero, M.G., Bessieres, P., Bolotin, A., Borchert, S. *et al.* (1997) The complete genome sequence of the Gram-positive bacterium *Bacillus subtilis. Nature* **390**, 249–256.

Le Brun, N.E., Bengtsson, J. and Hederstedt, L. (2000) Genes required for cytochrome *c* synthesis in *Bacillus subtilis. Mol. Microbiol.* **36**, 638–650.

Le Guen, L., Santos, R. and Camadro, J.M. (1999) Functional analysis of the *hemK* gene product involvement in protoporphyrinogen oxidase activity in yeast. *FEMS Microbiol. Lett.* **173**, 175–182.

Lecerof, D., Fodje, M., Hansson, A., Hansson, M. and Al-Karadaghi, S. (2000) Structural and mechanistic basis of porphyrin metallation by ferrochelatase. *J. Mol. Biol.* **297**, 221–232.

Lee, H.J., Lee, S.B., Chung, J.S., Han, S.U., Han, O., Guh, J.O., Jeon, J.S., An, G. and Back, K. (2000) Transgenic rice plants expressing a *Bacillus subtilis* protoporphyrinogen oxidase gene are resistant to diphenyl ether herbicide oxyfluorfen. *Plant Cell. Physiol.* **41**, 743–749.

Leong, S.A., Ditta, D.S. and Helinski, D.R. (1982) Heme synthesis in *Rhizobium*. Identification of a cloned gene coding for δ-aminolevulinic acid synthetase from *Rhizobium meliloti. J. Biol. Chem.* **257**, 8724–8730.

Lermontova, I. and Grimm, B. (2000) Overexpression of plastidic protoporphyrinogen IX oxidase leads to resistance to the diphenyl-ether herbicide acifluorfen. *Plant Physiol.* **122**, 75–84.

Li, J.M., Brathwaite, O., Cosloy, S.D. and Russell, C.S. (1989a) 5-Aminolevulinic acid synthesis in *Escherichia coli. J. Bacteriol.* **171**, 2547–2552.

Li, J.M., Russell, C.S. and Cosloy, S.D. (1989b) Cloning and structure of the *hemA* gene of *Escherichia coli* K-12. *Gene* **82**, 209–217.

Li, Y. and Sen, D. (1997) Toward an efficient DNAzyme. *Biochemistry* **36**, 5589–5599.

Lill, R., Stuart, R.A., Drygas, M.E., Nargang, F.E. and Neupert, W. (1992) Import of cytochrome *c* heme lyase into mitochondria: a novel pathway into the intermembrane space. *EMBO J.* **11**, 449–456.

Loewen, P.C., Switala, J., von Ossowski, I., Hillar, A., Christie, A., Tattrie, B. and Nicholls, P. (1993) Catalase HPII of *Escherichia coli* catalyzes the conversion of protoheme to cis-heme d. *Biochemistry* **32**, 10159–10164.

Louie, G.V., Brownlie, P.D., Lambert, R., Cooper, J.B., Blundell, T.L., Wood, S.P., Warren, M.J., Woodcock, S.C. and Jordan, P.M. (1992) Structure of porphobilinogen deaminase reveals a flexible multidomain polymerase with a single catalytic site. *Nature* **359**, 33–39.

Lübben, M. and Morand, K. (1994) Novel prenylated hemes as cofactors of cytochrome oxidases. Archaea have modified hemes A and O. *J. Biol. Chem.* **269**, 21473–21479.

Lübben, M., Warne, A., Albracht, S.P. and Saraste, M. (1994) The purified SoxABCD quinol oxidase complex of *Sulfolobus acidocaldarius* contains a novel haem. *Mol. Microbiol.* **13**, 327–335.

Luo, J. and Lim, C.K. (1993) Order of uroporphyrinogen III decarboxylation on incubation of porphobilinogen and uroporphyrinogen III with erythrocyte uroporphyrinogen decarboxylase. *Biochem. J.* **289**, 529–532.

Mate, M.J., Sevinc, M.S., Hu, B., Bujons, J., Bravo, J., Switala, J., Ens, W., Loewen, P.C. and Fita, I. (1999) Mutants that alter the covalent structure of catalase hydroperoxidase II from *Escherichia coli*. *J. Biol. Chem.* **274**, 27717–27725.

Matringe, M., Camadro, J.-M., Labbe, P. and Scalla, R. (1989) Protoporphyrinogen oxidase as a molecular target for diphenyl ether herbicides. *Biochem. J.* **260**, 231–235.

Matringe, M., Camadro, J.M., Joyard, J. and Douce, R. (1994) Location of ferrochelatase activity within mature pea chloroplasts. *J. Biol. Chem.* **269**, 15010–15015.

Mau, Y.-H.L. and Wang, W.-Y. (1988) Biosynthesis of δ-aminolevulinic acid in *Chlamydomonas reinhardtii*. Study of the transamination mechanism using specifically labeled glutamate. *Plant Physiol.* **86**, 793–797.

Mayer, A., Neupert, W. and Lill, R. (1995) Translocation of apocytochrome *c* across the outer membrane of mitochondria. *J. Biol. Chem.* **270**, 12390–12397.

Mayer, S.M., Gawlita, E., Avissar, Y.J., Anderson, V.E. and Beale, S.I. (1993) Intermolecular nitrogen transfer in the enzymic conversion of glutamate to delta-aminolevulinic acid by extracts of *Chlorella vulgaris*. *Plant Physiol.* **101**, 1029–1038.

McGinnis, S.D. and O'Brian, M.R. (1995) The rhizobial *hemA* gene is required for symbiosis in species with deficient δ-aminolevulinic acid uptake activity. *Plant Physiol.* **108**, 1547–1552.

Melefors, O., Goossen, B., Johansson, H.E., Stripecke, R., Grary, N.K. and Hentze, M.W. (1993) Translational control of 5-aminolevulinate synthase by mRNA by iron-responsive elements in erythroid cells. *J. Biol. Chem.* **268**, 5974–5978.

Metheringham, R., Tyson, K.L., Crooke, H., Missiakas, D., Raina, S. and Cole, J.A. (1996) Effects of mutations in genes for proteins involved in disulphide bond formation in the periplasm on the activities of anaerobically induced electron transfer chains in *Escherichia coli* K12. *Mol. Gen. Genet.* **253**, 95–102.

Missiakas, D. and Raina, S. (1997) Protein folding in the bacterial periplasm. *J. Bacteriol.* **179**, 2465–2471.

Mitchell, L.W. and Jaffe, E.K. (1993) Porphobilinogen synthase from *Escherichia coli* is a Zn(II) metalloenzyme stimulated by Mg(II). *Arch. Biochem. Biophys.* **300**, 169–177.

Miyamoto, K., Nakahigashi, K., Nishimura, K. and Inokuchi, H. (1991) Isolation and characterization of visible light-sensitive mutants of *Escherichia coli* K12. *J. Mol. Biol.* **219**, 393–398.

Mock, H.-P. and Grimm, B. (1997) Reduction of uroporphyrinogen decarboxylase by antisense RNA expression affects activities of other enzymes involved in tetrapyrrole biosynthesis and leads to light-dependent necrosis. *Plant Physiol.* **113**, 1101–1112.

Mogi, T., Saiki, K. and Anraku, Y. (1994) Biosynthesis and functional role of haem O and haem A. *Mol. Microbiol.* **14**, 391–398.

Mohr, C.D., Sonsteby, S.K. and Deretic, V. (1994) The *Pseudomonas aeruginosa* homologs of *hemC* and *hemD* are linked to the gene encoding the regulator of mucoidy AlgR. *Mol. Gen. Genet.* **242**, 177–184.

Morgan, M., Prudent, J.R. and Schultz, P.G. (1996) Porphyrin metallation catalyzed by a small RNA molecule. *J. Am. Chem. Soc.* **118**, 7012–7013.

Nakahigashi, K., Nishimura, K., Miyamoto, K. and Inokuchi, H. (1991) Photosensitivity of a protoporphyrin-accumulating, light-sensitive mutant (*visA*) of *Escherichia coli* K-12. *Proc. Natl Acad. Sci. USA* **88**, 10520–10524.

Nakayashiki, T., Nishimura, K. and Inokuchi, H. (1995) Cloning and sequencing of a previously unidentified gene that is involved in the biosynthesis of heme in *Escherichia coli*. *Gene* **153**, 67–70.

Nandi, D.L. and Shemin, D. (1977) Quaternary structure of δ-aminolevulinic acid synthase from *Rhodopseudomonas spaeroides*. *J. Biol. Chem.* **252**, 2278–2280.

Narita, S., Taketani, S. and Inokuchi, H. (1999) Oxidation of protoporphyrinogen IX in

Escherichia coli is mediated by the aerobic coproporphyrinogen oxidase. *Mol. Gen. Genet.* **261**, 1012–1020.

Neidle, E.L. and Kaplan, S. (1993a) 5-Aminolevulinic acid availability and control of spectral complex formation in *hemA* and *hemT* mutants of *Rhodobacter sphaeroides. J. Bacteriol.* **175**, 2304–2313.

Neidle, E.L. and Kaplan, S. (1993b) Expression of the *Rhodobacter sphaeroides hemA* and *hemT* genes, encoding two 5-aminolevulinic acid synthase isozymes. *J. Bacteriol.* **175**, 2292–2303.

Newman, E.B., D'Ari, R. and Lin, R.T. (1992) The leucine-Lrp regulon in *E. coli*: a global response in search of a raison d'etre. *Cell* **68**, 617–619.

Ng, W.V., Kennedy, S.P., Mahairas, G.G., Berquist, B., Pan, M., Shukla, H.D., Lasky, S.R., Baliga, N.S., Thorsson, V., Sbrogna, J. *et al.* (2000) Genome sequence of *Halobacterium* species NRC-1. *Proc. Natl Acad. Sci. USA* **97**, 12176–12181.

Nicholson, D.W. and Neupert, W. (1989) Import of cytochrome *c* into mitochondria: reduction of heme, mediated by NADH and flavin nucleotides, is obligatory for its covalent linkage to apocytochrome *c. Proc. Natl Acad. Sci. USA* **86**, 4340–4344.

Nicholson, D.W., Köhler, H. and Neupert, W. (1987) Import of cytochrome *c* into mitochondria. Cytochrome *c* heme lyase. *Eur. J. Biochem.* **164**, 147–157.

Nicholson, D.W., Hergersberg, C. and Neupert, W. (1988) Role of cytochrome *c* heme lyase in the import of cytochrome *c* into mitochondria. *J. Biol. Chem.* **263**, 19034–19042.

Nicholson, D.W., Stuart, R.A. and Neupert, W. (1989) Role of cytochrome c_1 heme lyase and of the two proteolytic processing steps during import into mitochondria. *J. Biol. Chem.* **264**, 10156–10168.

Nienaber, A., Huber, A., Gottfert, M., Hennecke, H. and Fischer, H.M. (2000) Three new NifA-regulated genes in the *Bradyrhizobium japonicum* symbiotic gene region discovered by competitive DNA-RNA hybridization. *J. Bacteriol.* **182**, 1472–1480.

Nishimura, K., Nakayashiki, T. and Inokuchi, H. (1993) Cloning and sequencing of the *hemE* gene encoding uroporphyrinogen III decarboxylase (UPD) from *Escherichia coli* K-12. *Gene* **133**, 109–113.

Nobrega, M.P., Nobrega, F.G. and Tzagoloff, A. (1990) COX10 codes for a protein homologous to the ORF1 product of *Paracoccus denitrificans* and is required for the synthesis of yeast cytochrome oxidase. *J. Biol. Chem.* **265**, 14220–14226.

Nordmann, Y. and Deybach, J.C. (1990) Human hereditary porphyrias. In: *Biosynthesis of Heme and Chlorophylls* (H.A. Dailey, ed.) pp. 491–542, McGraw-Hill, New York.

O'Brian, M.R. (1996) Heme synthesis in the rhizobium-legume symbiosis: a palette for bacterial and eukaryotic pigments. *J. Bacteriol.* **178**, 2471–2478.

O'Brian, M.R. (2000). Heme biosynthesis and function in the *Rhizobium*-legume symbiosis. In: *Prokaryotic Nitrogen Fixation. Model System for the Analysis of a Biological Process* (E.W. Triplett, ed.), pp. 509–528, Horizon Scientific, Norfolk.

O'Brian, M.R., Kirshbom, P.M. and Maier, R.J. (1987) Bacterial heme synthesis is required for expression of the leghemoglobin holoprotein, but not the apoprotein in soybean root nodules. *Proc. Natl Acad. Sci. USA* **84**, 8390–8393.

Ochman, H. and Moran, N.A. (2001) Genes lost and genes found: evolution of bacterial pathogenesis and symbiosis. *Science* **292**, 1096–1098.

Oh, J.I. and Kaplan, S. (2000) Redox signaling: globalization of gene expression. *EMBO J.* **19**, 4237–4247.

Oh, J.I., Eraso, J.M. and Kaplan, S. (2000) Interacting regulatory circuits involved in orderly control of photosynthesis gene expression in *Rhodobacter sphaeroides* 2.4.1. *J. Bacteriol.* **182**, 3081–3087.

Page, M.D. and Ferguson, S.J. (1989) A bacterial *c*-type cytochrome can be translocated to the periplasm as an apo form; the biosynthesis of cytochrome cd_1 (nitrite reductase) from *Paracoccus denitrificans*. *Mol. Microbiol.* **3**, 653–661.

Page, M.D. and Ferguson, S.J. (1990) Apo forms of cytochrome c_{550} and cytochrome cd_1 are translocated to the periplasm of *Paracoccus denitrificans* in the absence of haem incorporation caused by either mutation or inhibition of haem synthesis. *Mol. Microbiol.* **4**, 1181–1192.

Page, K.M.and Guerinot, M.L. (1995) Oxygen control of the *Bradyrhizobium japonicum hemA* gene. *J. Bacteriol.* **177**, 3979–3984.

Page, K.M., Connolly, E.L. and Guerinot, M.L. (1994) Effect of iron availability on expression of the *Bradyrhizobium japonicum hemA* gene. *J. Bacteriol.* **176**, 1535–1538.

Page, M.D., Pearce, D.A., Norris, H.A. and Ferguson, S.J. (1997) The *Paracoccus denitrificans ccmA, B* and *C* genes: cloning and sequencing and analysis of the potential of their products to form a haem or apo-*c*-type cytochrome transporter. *Microbiology* **143**, 563–576.

Pawlowski, K., Gough, S.P., Kannangara, C.G. and de Bruijn, F.J. (1993) Characterization of a 5-aminolevulinic acid synthase mutant of *Azorhizobium caulinodans*. *Mol. Plant Microbe Interact.* **6**, 35–44.

Petrovich, R.M., Litwin, S. and Jaffe, E.K. (1996) *Bradyrhizobium japonicum* porphobilinogen synthase uses two Mg (II) and monovalent cations. *J. Biol. Chem.* **271**, 8692–8699.

Pollock, W.B.R., Loutfi, M., Bruschi, M., Rapp-Giles, B.J., Wall, J.D. and Voordouw, G. (1991) Cloning, sequencing and expression of the gene encoding the high-molecular-weight cytochrome *c* from *Desulfovibrio vulgaris* Hildenborough. *J. Bacteriol.* **173**, 220–228.

Pollock, W.B.R., Rosell, F.I., Twitchett, M.B., Dumont, M.E. and Mauk, A.G. (1998) Bacterial expression of a mitochondrial cytochrome *c*. Trimethylation of Lys72 in yeast *iso*-1-cytochrome *c* and the alkaline conformational transition. *Biochemistry* **37**, 6124–6131.

Posey, J.E. and Gherardini, F.C. (2000) Lack of a role for iron in the Lyme disease pathogen. *Science* **288**, 1651–1653.

Preisig, O., Anthamatten, D. and Hennecke, H. (1993) Genes for a microaerobically induced oxidase complex in *Bradyrhizobium japonicum* are essential for a nitrogen-fixing endosymbiosis. *Proc. Natl Acad. Sci. USA* **90**, 3309–3313.

Preisig, O., Zufferey, R., Thöny-Meyer, L., Appleby, C.A. and Hennecke, H. (1996) A high-affinity cbb_3-type cytochrome oxidase terminates the symbiosis-specific respiratory chain of *Bradyrhizobium japonicum*. *J. Bacteriol.* **178**, 1532–1538.

Proulx, K.L. and Dailey, H.A. (1992) Characteristics of murine protoporphyrinogen oxidase. *Protein Sci.* **1**, 801–809.

Pujol, C.J. and Kado, C.I. (2000) Genetic and biochemical characterization of the pathway in *Pantoea citrea* leading to pink disease of pineapple. *J. Bacteriol.* **182**, 2230–2237.

Qi, Z. and O'Brian, M.R. (2002) Interaction between the bacterial iron response regulator and ferrochelatase mediates genetic control of heme biosynthesis. *Mol. Cell* **9**, 155–162.

Qi, Z., Hamza, I. and O'Brian, M.R. (1999) Heme is an effector molecule for iron-dependent degradation of the bacterial iron response regulator (Irr) protein. *Proc. Natl Acad. Sci. USA* **96**, 13056–13061.

Ramseier, T.M., Winteler, H.V. and Hennecke, H. (1991) Discovery and sequence analysis of bacterial genes involved in the biogenesis of *c*-type cytochromes. *J. Biol. Chem.* **266**, 7793–7803.

Reid, E., Eaves, D.J. and Cole, J.A. (1998) The CcmE protein from *Escherichia coli* is a haem-binding protein. *FEMS Microbiol. Lett.* **166**, 369–375.

Reid, E., Cole, J. and Eaves, D.J. (2001) The *Escherichia coli* CcmG protein fulfils a specific role in cytochrome *c* assembly. *Biochem. J.* **355**, 51–58.

Ren, Q. and Thöny-Meyer, L. (2001) Physical interaction of CcmC with heme and the heme chaperone CcmE during cytochrome *c* maturation. *J. Biol. Chem.* **276,** 32591–32596.

Rhie, G., Avissar, Y.J. and Beale, S.I. (1996) Structure and expression of the *Chlorobium vibrioforme hemB* gene and characterization of its encoded enzyme, porphobilinogen synthase. *J. Biol. Chem.* **271,** 8176–8182.

Rodgers, K.R. (1999) Heme-based sensors in biological systems. *Curr. Opin. Chem. Biol.* **3,** 158–167.

Romana, M., Dubart, A., Beaupain, D., Chabret, C., Goossens, M. and Romeo, P.H. (1987a) Structure of the gene for human uroporphyrinogen decarboxylase. *Nucleic Acids Res.* **15,** 7343–7356.

Romana, M., Le Boulch, P. and Romeo, P.H. (1987b) Rat uroporphyrinogen decarboxylase cDNA: nucleotide sequence and comparison to human uroporphyrinogen decarboxylase. *Nucleic Acids Res.* **15,** 7211.

Rompf, A., Hungerer, C., Hoffmann, T., Lindenmeyer, M., Romling, U., Gross, U., Doss, M.O., Arai, H., Igarashi, Y. and Jahn, D. (1998) Regulation of *Pseudomonas aeruginosa hemF* and *hemN* by the dual action of the redox response regulators Anr and Dnr. *Mol. Microbiol.* **29,** 985–997.

Rose, S., Frydman, R.B., de los Santos, C., Sburlati, A., Valasinas, A. and Frydman, B. (1988) Spectroscopic evidence for a porphobilinogen deaminase-tetrapyrrole complex that is an intermediate in the biosynthesis of uroporphyrinogen III. *Biochemistry* **27,** 4871–4879.

Rouault, T.A. and Klausner, R.D. (1996) Iron-sulfur clusters as biosensors of oxidants and iron. *Trends Biochem. Sci.* **21,** 174–177.

Ruepp, A., Graml, W., Santos-Martinez, M.L., Koretke, K.K., Volker, C., Mewes, H.W., Frishman, D., Stocker, S., Lupas, A.N. and Baumeister, W. (2000) The genome sequence of the thermoacidophilic scavenger *Thermoplasma acidophilum*. *Nature* **407,** 508–513.

Saiki, K., Mogi, T., Ogura, K. and Anraku, Y. (1993) *In vitro* heme O synthesis by the *cyoE* gene product from *Escherichia coli*. *J. Biol. Chem.* **268,** 26041–26044.

Sambongi, Y. and Ferguson, S.J. (1994) Specific thiol compounds complement deficiency in *c*-type cytochrome biogenesis in *Escherichia coli* carrying a mutation in a membrane-bound disulphide isomerase-like protein. *FEBS Lett.* **353,** 235–238.

Sanders, C. and Lill, H. (2000) Expression of prokaryotic and eukaryotic cytochromes *c* in *Escherichia coli*. *Biochim. Biophys. Acta* **1459,** 131–138.

Sangwan, I. and O'Brian, M.R. (1991) Evidence for an inter-organismic heme biosynthetic pathway in symbiotic soybean root nodules. *Science* **251,** 1220–1222.

Sangwan, I. and O'Brian, M.R. (1999) Expression of a soybean gene encoding the tetrapyrrole synthesis enzyme glutamyl-tRNA reductase in symbiotic root nodules. *Plant Physiol.* **119,** 593–598.

Santana, M.A., Pihakaski-Maunsbach, K., Sandal, N., Marcker, K.A. and Smith, A.G. (1998) Evidence that the plant host synthesizes the heme moiety of leghemoglobin in root nodules. *Plant Physiol.* **116,** 1259–1269.

Sasarman, A., Surdeanu, M. and Horodniceanu, T. (1968) Locus determining the synthesis of δ-aminolevulinic acid in *Escherichia coli* K-12. *J. Bacteriol.* **96,** 1882–1884.

Sasarman, A., Sanderson, K.E., Surdeanu, M. and Sonea, S. (1970) Hemin-deficient mutants of *Salmonella typhimurium*. *J. Bacteriol.* **102,** 531–536.

Sasarman, A., Chartrand, P., Lavoie, M., Tardif, D., Proschek, R. and Lapointe, C. (1979) Mapping of a new *hem* gene in *Escherichia coli* K12. *J. Gen. Microbiol.* **113,** 297–303.

Sasarman, A., Letowski, J., Czaika, G., Ramirez, V., Nead, M.A., Jacobs, J.M. and Morais, R. (1993) Nucleotide sequence of the *hemG* gene involved in the protoporphyrinogen oxidase activity of *Escherichia coli* K12. *Can. J. Microbiol.* **39,** 1155–1161.

Schiött, T., Throne-Holst, M. and Hederstedt, L. (1997a) *Bacillus subtilis* CcdA defective mutants are blocked in a late step of cytochrome *c* biogenesis. *J. Bacteriol.* **179**, 4523–4529.

Schiött, T., von Wachenfeldt, C. and Hederstedt, L. (1997b) Identification and characterization of the *ccdA* gene, required for cytochrome *c* synthesis in *Bacillus subtilis*. *J. Bacteriol.* **179**, 1962–1973.

Schmitt, M.P. (1997a) Transcription of the *Corynebacterium diphtheriae hmuO* gene is regulated by iron and heme. *Infect. Immun.* **65**, 4634–4641.

Schmitt, M.P. (1997b) Utilization of host iron sources by *Corynebacterium diphtheriae*: identification of a gene whose product is homologous to eukaryotic heme oxygenases and is required for acquisition of iron from heme and hemoglobin. *J. Bacteriol.* **179**, 838–845.

Schneegurt, M.A. and Beale, S.I. (1988) Characterization of the RNA required for biosynthesis of δ-aminolevulinic acid from glutamate. Purification by anticodon-based affinity chromatography and determination that the UUC glutamate anticodon is a general requirement for function in ALA synthesis. *Plant Physiol.* **86**, 497–504.

Schon, A., Krupp, G., Gough, S.P., Berry-Lowe, S., Kannangara, C.G. and Söll, D. (1986) The RNA required in the first step of chlorophyll biosynthesis is a chloroplast tRNA. *Nature* **322**, 281–284.

Schröder, I., Hederstedt, L., Kannangara, C.G. and Gough, S.P. (1992) Glutamyl-tRNA reductase activity in *Bacillus subtilis* is dependent on the *hemA* gene product. *Biochem. J.* **281**, 843–850.

Schulz, H. and Thöny-Meyer, L. (2000) Interspecies complementation of *Escherichia coli ccm* mutants: CcmE (CycJ) from *Bradyrhizobium japonicum* acts as a heme chaperone during cytochrome *c* maturation. *J. Bacteriol.* **182**, 6831–6833.

Schulz, H., Hennecke, H. and Thöny-Meyer, L. (1998) Prototype of a heme chaperone essential for cytochrome *c* maturation. *Science* **281**, 1197–1200.

Schulz, H., Fabianek, R.A., Pellicioli, E.C., Hennecke, H. and Thöny-Meyer, L. (1999) Heme transfer to the heme chaperone CcmE during cytochrome *c* maturation requires the CcmC protein, which may function independently of the ABC-transporter CcmAB. *Proc. Natl Acad. Sci. USA* **96**, 6462–6467.

Schulz, H., Pellicioli, E. and Thöny-Meyer, L. (2000) New insights into the role of CcmC, CcmD, and CcmE in the haem delivery pathway during cytochrome *c* maturation by a complete mutational analysis of the conserved tryptophan-rich motif of CcmC. *Mol. Microbiol.* **37**, 1379–1388.

Schütz, M., Brugna, M., Lebrun, E., Baymann, F., Huber, R., Stetter, K.O., Hauska, G., Toci, R., Lemesle-Meunier, D., Tron, P., Schmidt, C. and Nitschke, W. (2000) Early evolution of cytochrome *bc* complexes. *J. Mol. Biol.* **300**, 663–675.

Scott, A.I., Roessner, C.A., Stolowich, N.J., Karuso, P., Williams, H.J., Grant, S.K., Gonzalez, M.D. and Hoshino, T. (1988) Site-directed mutagenesis and high-resolution NMR spectroscopy of the active site of porphobilinogen deaminase. *Biochemistry* **27**, 7984–7990.

Seehra, J.S., Jordan, P.M. and Akhtar, M. (1983) Anaerobic and aerobic coproporphyrinogen III oxidases of *Rhodopseudomonas spheroides*. Mechanism and stereochemistry of vinyl group formation. *Biochem. J.* **209**, 709–718.

Spielewoy, N., Schulz, H., Grienenberger, J.M., Thöny-Meyer, L. and Bonnard, G. (2001) CCME, a nuclear-encoded heme-binding protein involved in cytochrome *c* maturation in plant mitochondria. *J. Biol. Chem.* **276**, 5491–5497.

Stamford, N.P.J., Capretta, A. and Battersby, A.R. (1995) Expression, purification and characterization of the product from the *Bacillus subtilis hemD* gene, uroporphyrinogen III synthase. *Eur. J. Biochem.* **231**, 236–241.

Stange-Thomann, N., Thomann, H.-U., Lloyd, A.J., Lyman, H. and Söll, D. (1994) A point mutation in *Euglena gracilis* chloroplast tRNA^{Glu} uncouples protein and chlorophyll biosynthesis. *Proc. Natl Acad. Sci. USA* **91**, 7947–7951.

Stanley, J., Dowling, D.N. and Broughton, W.J. (1988) Cloning of *hemA* from *Rhizobium* sp. NGR234 and symbiotic phenotype of a gene-directed mutant in diverse legume genera. *Mol. Gen. Genet.* **215**, 32–37.

Steiner, H., Kispal, G., Zollner, A., Neupert, W. and Lill, R. (1996) Heme binding to a conserved Cys-Pro-Val motif is crucial for the catalytic function of mitochondrial heme lyases. *J. Biol. Chem.* **271**, 32605–32611.

Stetterdahl, A.T., Goldman, B.S., Hirasawa, M., Jacquot, P., Smith, A.J., Kranz, R.G. and Knaff, D.B. (2000) Oxidation-reduction properties of disulfide-containing proteins of the *Rhodobacter capsulatus* cytochrome *c* biogenesis system. *Biochemistry* **39**, 10172–10176.

Stewart, E.J., Katzen, F. and Beckwith, J. (1999) Six conserved cysteines of the membrane protein DsbD required for the transfer of electrons from the cytoplasm to the periplasm of *Escherichia coli. EMBO J.* **18**, 5963–5971.

Svensson, B. and Hederstedt, L. (1994) *Bacillus subtilis* CtaA is a heme-containing membrane protein involved in heme A biosynthesis. *J. Bacteriol.* **176**, 6663–6671.

Svensson, B., Lubben, M. and Hederstedt, L. (1993) *Bacillus subtilis* CtaA and CtaB function in haem A biosynthesis. *Mol. Microbiol.* **10**, 193–201.

Svensson, B., Andersson, K.K. and Hederstedt, L. (1996) Low-spin heme A in the heme A biosynthetic protein CtaA from *Bacillus subtilis. Eur. J. Biochem.* **238**, 287–295.

Tan, D. and Ferreira, G.C. (1996) Active site of 5-aminolevulinate synthase resides at the subunit interface. Evidence from *in vivo* heterodimer formation. *Biochemistry* **35**, 8934–8941.

Taniuchi, H., Basile, G., Taniuchi, M. and Veloso, D. (1983) Evidence for formation of two thioether bonds to link heme to apocytochrome *c* by partially purified cytochrome *c* synthetase. *J. Biol. Chem.* **258**, 10963–10966.

Thöny-Meyer, L. (2000) Haem–polypeptide interactions during cytochrome *c* maturation. *Biochim. Biophys. Acta* **1459**, 316–324.

Thöny-Meyer, L. and Künzler, P. (1997) Translocation to the periplasm and signal sequence cleavage of preapocytochrome *c* depend on *sec* and *lep*, but not on the *ccm* gene products. *Eur. J. Biochem.* **246**, 794–799.

Thöny-Meyer, L., Beck, C., Preisig, O. and Hennecke, H. (1994) The *ccoNOQP* gene cluster codes for a *cb*-type cytochrome oxidase that functions in aerobic respiration of *Rhodobacter capsulatus. Mol. Microbiol.* **14**, 705–716.

Thöny-Meyer, L., Ritz, D. and Hennecke, H. (1994) Cytochrome *c* biogenesis in bacteria: a possible pathway begins to emerge. *Mol. Microbiol.* **12**, 1–9.

Thöny-Meyer, L., Fischer, F., Künzler, P., Ritz, D. and Hennecke, H. (1995) *Escherichia coli* genes required for cytochrome *c* maturation. *J. Bacteriol.* **177**, 4321–4326.

Thöny-Meyer, L., Künzler, P. and Hennecke, H. (1996) Requirements for maturation of *Bradyrhizobium japonicum* cytochrome c_{550} in *Escherichia coli. Eur. J. Biochem.* **235**, 754–761.

Throne-Holst, M., Thöny-Meyer, L. and Hederstedt, L. (1997) *Escherichia coli* in-frame deletion mutants can produce periplasmic cytochrome *b*, but not cytochrome *c. FEBS Lett.* **410**, 351–355.

Tichy, M. and Vermaas, W. (1999) Accumulation of pre-apocytochrome *f* in a *Synechocystis* sp. PCC 6803 mutant impaired in cytochrome *c* maturation. *J. Biol. Chem.* **274**, 32396–32401.

Troup, B., Hungerer, C. and Jahn, D. (1995) Cloning and characterization of the *Escherichia coli hemN* gene encoding the oxygen-independent coproporphyrinogen III oxidase. *J. Bacteriol.* **177**, 3326–3331.

Tzagoloff, A., Capitanio, N., Nobrega, M.P. and Gatti, D. (1990) Cytochrome oxidase assembly in yeast requires the product of COX11, a homolog of the *P. denitrificans* protein encoded by ORF3. *EMBO J.* **9**, 2759–2764.

Tzagoloff, A., Nobrega, M., Gorman, N. and Sinclair, P. (1993) On the functions of the yeast COX10 and COX11 gene products. *Biochem. Mol. Biol. Int.* **31**, 593–598.

Umanoff, H., Russell, C.S. and Cosloy, S.D. (1988) Availability of porphobilinogen controls appearance of porphobilinogen deaminase activity in *Escherichia coli* K-12. *J. Bacteriol.* **170**, 4969–4971.

Veloso, D., Basile, G. and Taniuchi, H. (1981) Formation of a cytochrome *c*-like species from horse apoprotein and hemin catalyzed by yeast mitochondrial cytochrome *c* synthetase. *J. Biol. Chem.* **256**, 8646–8651.

Verderber, E., Lucast, L.J., Van Dehy, A., Cozart, P., Etter, J.B. and Best, E.A. (1997) Role of the *hemA* gene product and δ-aminolevulinic acid in regulation of *Escherichia coli* heme synthesis. *J. Bacteriol.* **179**, 4583–4590.

Verkamp, E. and Chelm, B.K. (1989) Isolation, nucleotide sequence and preliminary characterization of the *Escherichia coli* K-12 *hemA* gene. *J. Bacteriol.* **171**, 4728–4735.

Verkamp, E., Bachman, V.M., Bjornsson, J.M., Söll, D. and Eggertsson, G. (1993) The periplasmic dipeptide permease system transports 5-aminolevulinic acid in *Escherichia coli*. *J. Bacteriol.* **175**, 1452–1456.

Visco, C., Taniuchi, H. and Berlett, B.S. (1985) On the specificity of cytochrome *c* synthetase in recognition of the amino acid sequence of apocytochrome *c*. *J. Biol. Chem.* **260**, 6133–6138.

Vothknecht, U.C., Kannangara, C.G. and von Wettstein, D. (1996) Expression of catalytically active barley glutamyl-tRNA^Glu reductase in *Escherichia coli* as a fusion protein with glutathione S-transferase. *Proc. Natl Acad. Sci. USA* **93**, 9287–9291.

Walker, C.J. and Willows, R.D. (1997) Mechanism and regulation of Mg-chelatase. *Biochem. J.* **327**, 321–333.

Wang, L., Elliott, M. and Elliott, T. (1999a) Conditional stability of the HemA protein (glutamyl-tRNA reductase) regulates heme biosynthesis in *Salmonella typhimurium*. *J. Bacteriol.* **181**, 1211–1219.

Wang, L., Wilson, S. and Elliott, T. (1999b) A mutant HemA protein with positive charge close to the N terminus is stabilized against heme-regulated proteolysis in *Salmonella typhimurium*. *J. Bacteriol.* **181**, 6033–6041.

Wang, L.Y., Brown, L., Elliott, M. and Elliott, T. (1997) Regulation of heme biosynthesis in *Salmonella typhimurium*: activity of glutamyl-tRNA reductase (HemA) is greatly elevated during heme limitation by a mechanism which increases abundance of the protein. *J. Bacteriol.* **179**, 2907–2914.

Warren, M.J. and Jordan, P.M. (1988) Investigation into the nature of substrate binding to the dipyrromethane cofactor of *Escherichia coli* porphobilinogen deaminase. *Biochemistry* **27**, 9020–9030.

Warren, M.J., Jay, M., Hunt, D.M., Elder, G.H. and Rohl, J.C.G. (1996) The maddening business of King George III and porphyria. *Trends Biochem. Sci.* **21**, 229–234.

Weinstein, J.D. and Beale, S.I. (1983) Separate physiological roles and subcellular compartments for two tetrapyrrole biosynthetic pathways in *Euglena gracilis*. *J. Biol. Chem.* **258**, 6799–6807.

Whitby, F.G., Phillips, J.D., Kushner, J.P. and Hill, C.P. (1998) Crystal structure of human uroporphyrinogen decarboxylase. *EMBO J.* **17**, 2463–2471.

Willows, R.D., Mayer, S.M., Foulk, M.S., DeLong, A., Hanson, K., Chory, J. and Beale, S.I. (2000) Phytobilin biosynthesis: the *Synechocystis* sp. PCC 6803 heme oxygenase-encoding HO1 gene complements a phytochrome-deficient *Arabidopsis thaliana* HY1 mutant. *Plant Mol. Biol.* **43**, 113–120.

Wright, R.L., Harris, K., Solow, B., White, R.H. and Kennelly, P.J. (1996) Cloning of a potential cytochrome P450 from the archaeon *Sulfolobus solfataricus*. *FEBS Lett.* **384**, 235–239.

Wu, C.K., Dailey, H.A., Rose, J.P., Burden, A., Sellers, V.M. and Wang, B.C. (2001) The 2.0 Å structure of human ferrochelatase, the terminal enzyme of heme biosynthesis. *Nat. Struct. Biol.* **8**, 156–160.

Xie, Z. and Merchant, S. (1996) The plastid-encoded *ccaA* gene is required for heme attachment to chloroplast *c*-type cytochromes. *J. Biochem.* **271**, 4632–4639.

Xie, Z. and Merchant, S. (1998) A novel pathway for cytochromes *c* biogenesis in chloroplasts. *Biochim. Biophys. Acta.* **1365**, 309–318.

Xie, Z., Culler, D., Greyfuss, B.W., Kuras, R., Wollman, F.-A., Girard-Bascou, J. and Merchant, S. (1998) Genetic analysis of chloroplast *c*-type cytochrome assembly in *Chlamydomonas reinhardtii*: one chloroplast locus and at least four nuclear loci are required for heme attachment. *Genetics* **148**, 681–692.

Xu, K. and Elliott, T. (1994) Cloning, DNA sequence and complementation analysis of the *Salmonella typhimurium hemN* gene encoding a putative oxygen-independent coproporphyrinogen III oxidase. *J. Bacteriol.* **176**, 3196–3203.

Xu, K., Delling, J. and Elliott, T. (1992) The genes required for heme synthesis in *Salmonella typhimurium* include those encoding alternative functions for aerobic and anaerobic coproporphyrinogen oxidation. *J. Bacteriol.* **174**, 3953–3963.

Yang, C.-H., Azad, H.R. and Cooksey, D.A. (1996) A chromosomal locus required for copper resistance, competitive fitness and cytochrome *c* biogenesis in *Pseudomonas fluorescens*. *Proc. Natl Acad. Sci. USA* **93**, 7315–7320.

Yano, J.K., Koo, L.S., Schuller, D.J., Li, H., Ortiz de Montellano, P.R. and Poulos, T.L. (2000) Crystal structure of a thermophilic cytochrome P450 from the archaeon *Sulfolobus solfataricus*. *J. Biol. Chem.* **275**, 31086–31092.

Yeliseev, A.A. and Kaplan, S. (1999) A novel mechanism for the regulation of photosynthesis gene expression by the TspO outer membrane protein of *Rhodobacter sphaeroides* 2.4.1. *J. Biol. Chem.* **274**, 21234–21243.

Yeoman, K.H., Delgado, M.J., Wexler, M., Downie, J.A. and Johnston, A.W. (1997) High affinity iron acquisition in *Rhizobium leguminosarum* requires the *cycHJKL* operon and the *feuPQ* gene products, which belong to the family of two-component transcriptional regulators. *Microbiology* **143**, 127–134.

Zeilstra-Ryalls, J.H. and Kaplan, S. (1995a) Aerobic and anaerobic regulation in *Rhodobacter sphaeroides* 2.4.1: the role of the *fnrL* gene. *J. Bacteriol.* **177**, 6422–6431.

Zeilstra-Ryalls, J.H. and Kaplan, S. (1995b) Regulation of 5-aminolevulinic acid synthesis in *Rhodobacter sphaeroides* 2.4.1: the genetic basis of mutant H-5 auxotrophy. *J. Bacteriol.* **177**, 2760–2768.

Zeilstra-Ryalls, J.H. and Kaplan, S. (1996) Control of *hemA* expression in *Rhodobacter sphaeroides* 2.4.1: regulation through alterations in the cellular redox state. *J. Bacteriol.* **178**, 985–993.

Zeilstra-Ryalls, J.H. and Kaplan, S. (1998) Role of the *fnrL* gene in photosystem gene expression and photosynthetic growth of *Rhodobacter sphaeroides* 2.4.1. *J. Bacteriol.* **180**, 1496–1503.

Zhu, W., Wilks, A. and Stojiljkovic, I. (2000) Degradation of heme in gram-negative bacteria: the product of the *hemO* gene of *Neisseriae* is a heme oxygenase. *J. Bacteriol.* **182**, 6783–6790.

Global Adjustment of Microbial Physiology During Free Radical Stress

Pablo J. Pomposiello[1,2] and Bruce Demple[1]

[1]Department of Cancer Cell Biology, Harvard School of Public Health,
Boston, MA 02115, USA
[2]New address: Department of Microbiology, Morrill Science Center IV North,
University of Massachusetts, Amherst, MA 01003, USA

ABSTRACT

Oxidation can damage all biological macromolecules, and the survival of a cell therefore depends on its ability to control the level of oxidants. Microbes possess an astonishing variety of antioxidant defences, ranging from small, oxidant-scavenging molecules to self-regulating, homeostatic gene networks. Most often these antioxidant defences are activated by exposure to specific classes of oxidants. Interestingly, the isolation of pleiotropic mutations that impair or exacerbate the expression of subsets of oxidant-responsive genes led to the identification of global regulators. In a few, well-characterized cases, these regulators can transduce oxidative damage into gene regulation. Recently, the application of genomic tools to study the antioxidant responses of *E. coli* has both confirmed previous observations and provided evidence for a wealth of putative new anti-oxidant functions. Here, we review the remarkable diversity of antioxidant defence mechanisms, with emphasis on signal transduction by global regulator proteins and the corresponding genetic networks that protect the microbial cell against oxidative stress.

ADVANCES IN MICROBIAL PHYSIOLOGY VOL 46
ISBN 0-12-027746-8

ABBREVIATIONS

GSH   Glutathione (reduced)
PQ   Paraquat
SOD   Superoxide dismutase

1. THE NATURE OF FREE RADICAL STRESS

Aerobic metabolism requires exposure of cells to oxygen, which sometimes reacts non-enzymatically with cellular components to generate free radicals and other reactive molecules (Chance *et al.*, 1979; Gilbert and Colton, 1999). These reactive by-products can damage all biological macromolecules and thus interrupt growth and cause mutations (Gilbert and Colton, 1999; Halliwell and Gutteridge, 1999). This oxidative damage is limited by small molecules such as glutathione, which neutralizes some oxidants, and by enzymes such as superoxide dismutase (SOD) and catalase, which eliminate specific reactive species (Gilbert and Colton, 1999; Halliwell and Gutteridge, 1999). Under some circumstances, the rate of free radical production increases or cellular defence activities are diminished, which results in oxidative stress (Sies, 1991). Oxidative stress can arise in many ways: through metabolic changes (e.g. inactivation of some components of the electron transport chain) (Chance *et al.*, 1979); exposure to environmental agents that divert electron flow (e.g., the herbicide paraquat (PQ)) (Kappus and Sies, 1981); immune responses to bacterial infection (Babior, 1992; MacMicking *et al.*, 1997).

Aerobic organisms have evolved inducible defence mechanisms against various types of oxidative stress (Hidalgo and Demple, 1996; Demple, 1999; Storz and Imlay, 1999; Grant, 2001). From single proteins to complex, self-regulating genetic networks, these defences scavenge reactive oxygen species and mediate the repair of cellular damage. The knowledge of these antioxidant mechanisms is more advanced in enteric bacteria than in any other group of organisms (Hidalgo and Demple, 1996; Storz and Imlay, 1999), although research progress is being made in Gram-positive (Bsat *et al.*, 1998; Hecker and Volker, 1998) bacteria and yeast (Jamieson, 1998; Grant, 2001). It is clear that genes that originally aided organisms in colonizing an oxidizing atmosphere were in many cases recruited into different regulatory networks, helping the cell to fine-tune its metabolism according to a wide variety of metabolic and environmental conditions.

2. SOURCES OF FREE RADICAL STRESS

2.1. Aerobic Metabolism

Metabolic pathways involving electron exchange between biochemical intermediates (redox reactions) have the potential to generate damaging oxidants via side reactions. In this fashion, aerobic metabolism, photosynthesis and denitrification are metabolic functions whose oxidant by-products have to be kept at concentrations that are compatible with cellular integrity. In a sense, the reactive by-products may be considered as unavoidable 'leaks' in these natural processes (Gilbert and Colton, 1999; Halliwell and Gutteridge, 1999), and therefore oxidative damage to cellular components is the evolutionary cost of these energy production pathways. Oxygen competes with respiratory components to oxidize some elements of the electron transport chain and thus yields superoxide, which is readily converted to H_2O_2 by the enzyme superoxide dismutase (SOD) (Fridovich, 1995). H_2O_2 can be safely disposed of by catalases, or it can react further, especially with reduced metals such as Fe^{2+}, to generate still more unstable products, notably the highly reactive hydroxyl radical (•OH). DNA is a critical target for oxidative damage, and the DNA lesions caused by oxidants can disrupt replication and lead to mutations (Demple and Harrison, 1994). The metal centres of enzymes constitute another group of targets for oxidative damage, with critical sensitive activities such as aconitase (essential component of the citric acid cycle) (Gardner and Fridovich, 1991). Unsaturated lipid components of the membrane react to form lipid peroxides in chain reactions with many products derived from a single free radical (Gilbert and Colton, 1999; Halliwell and Gutteridge, 1999). The ultimate breakdown of lipid peroxides yields still another reactive compound, malondialdehyde, which can form mutagenic DNA damage (Chaudhary *et al.*, 1994), and some DNA damage products can themselves cause further lesions (Dedon *et al.*, 1998). Thus, oxidative damage can exert widespread effects in the cell, both directly and indirectly.

2.2. Chemical and Physical Agents

In addition to normal metabolic sources of free radicals, environmental compounds can divert single electrons to generate oxygen radicals. These redox-cycling compounds undergo enzymatic reduction and are then reoxidized by O_2, a cyclic process that generates a flux of superoxide catalytically (Kappus and Sies, 1981; Fridovich, 1995). Redox-cycling agents can thus act as efficient sources of oxidative stress. The variety of superoxide-generating compounds is large and includes PQ and many types of quinones, naphthquinones, and nitroquinolones as well as diverse other compounds (Kappus and Sies, 1981). Several physical agents can also impose oxidative stress: ionizing radiation produces free radicals by radiolysis of water, while ultraviolet light produces H_2O_2 through photochemical reactions involving various chromophores, including the amino acid tryptophan (Gilbert and Colton, 1999; Halliwell and Gutteridge, 1999).

2.3. Photosynthesis

Aerobic metabolism is not the only source of oxidative stress. If photosynthetic cells are exposed to light in excess of their synthetic capabilities, light-harvesting antennae can transfer excitation energy to ground-state oxygen and yield singlet oxygen, another highly reactive species (though not actually a free radical). In addition, light-driven electron transport systems may divert electrons to oxygen instead of $NADP^+$, resulting in the same reactive derivatives as found for aerobic respiration. This light-dependent production of oxygen derivatives is termed photooxidative stress (Gilbert and Colton, 1999).

2.4. Denitrification

During bacterial denitrification, the aerobic conversion of nitrate into N_2, nitric oxide ($NO^{\bullet}$) is formed as a product of nitrite reduction (Zumft, 1997). The accumulation of this toxic intermediate is minimized by the catalytic conversion of $NO^{\bullet}$ into nitrous oxide (N_2O) by the enzyme nitric oxide reductase. If

NO$^{•}$ production during denitrification is not controlled, enough of this radical can accumulate to kill the denitrifying cells themselves (Tosques *et al.*, 1996).

2.5. Immune and Inflammatory Responses

The bacteriostatic and bactericidal effects of free radicals have been exploited by macrophages and other phagocytic cells to attack microbial infections. Upon phagocytosis of bacteria, macrophages and neutrophils are immediately activated by the bacterial lipopolysaccharide coat to produce copious amounts of superoxide giving rise to a flux of H_2O_2 (Babior, 1992). In a longer time scale (hours), macrophages produce very high amounts of NO$^{•}$ (MacMicking *et al.*, 1997), and many other cell types produce NO$^{•}$ during inflammatory responses. Analogous responses may be found in invertebrates (Ottaviani *et al.*, 1993) and plants (Klessig *et al.* 2000; Garcia-Olmedo *et al.* 2001). In addition, NO$^{•}$ reacts very rapidly with superoxide to generate another unstable and even more reactive compound, peroxynitrite (ONOO$^-$) (Halliwell *et al.*, 1999). Thus, the ultimate effects of one reactive species may be entwined with the effects of another.

3. LOW-MOLECULAR-WEIGHT ANTIOXIDANTS: THE FIRST LINE OF DEFENCE

Various small molecules act to neutralize free radicals in both prokaryotes and eukaryotes (Gilbert and Colton, 1999; Halliwell *et al.*, 1999). Such compounds include ascorbate (vitamin C) and α-tocopherol (vitamin E), but their synthesis does not appear to be regulated by oxidative stress responses in microbes. In contrast, glutathione (γ-L-glutamyl-L-cysteinylglycine; GSH) is a key component of globally regulated systems, as we will see. GSH is present in bacteria, fungi, animals and plants as the major low-molecular-weight thiol, typically at millimolar concentrations (Carmel-Harel and Storz, 2000). GSH acts as a chemical scavenger of radicals such as $^{•}$OH, and also as an H atom donor to restore macromolecules that have been attacked by free radicals. GSH in eukaryotic cells, including yeast, is also a cofactor for the H_2O_2-destroying enzyme GSH peroxidase and can be enzymatically conjugated to oxidative products to mark them for disposal from the cell (Carmel-Harel and Storz, 2000; Grant, 2001). Finally, GSH supports the glutaredoxin pathway, which helps maintain protein thiols in the cell (see below). GSH consumed in these reactions is replenished by reduction by NADPH-dependent GSH reductase, or by resynthesis (Carmel-Harel and Storz, 2000; Grant, 2001). In organisms where GSH is not present, a related peptide thiol is usually found, together with the respective thiol reductase (Flohe *et al.*, 1999).

4. RESPONSE SYSTEMS: REGULATORS AND EFFECTORS

Many organisms exhibit adaptive responses to oxidative stress. That is, exposure of the cells to sublethal oxidative stress enhances the resistance to subsequent, higher levels of oxidative stress (Demple, 1999). These adaptive responses depend on new protein synthesis, and different sets of proteins may be synthesized upon exposure to different types of oxidative agents. The induction of some key proteins in response to oxidative stress has been known for some time. For example, SOD activity increases in *E. coli* grown in high levels of oxygen or exposed to the redox-cycling agents (Fridovich, 1995). These agents also induce glucose-6-phosphate dehydrogenase, evidently to replenish NADPH used up in antioxidant reactions (e.g., by GSH reductase) (Kao and Hassan, 1985). Exposure to H_2O_2 induces catalase activity in many organisms (Gilbert and Colton, 1999) and increases GSH reductase levels in *E. coli* and *S. typhimurium* (Christman *et al.*, 1985). We now know that these inductions reflect the activation of large, co-regulated groups of genes encoding many additional proteins.

The number, identity and degree of activation of the proteins induced by oxidative stress have been studied by two-dimensional gel electrophoresis (Morgan *et al.*, 1986; Greenberg and Demple, 1989). This type of analysis has been applied to mutant strains that either overexpress or fail to induce sets of oxidative stress proteins, which has led to the isolation of genes coding for global regulators (Christman *et al.*, 1985; Greenberg *et al.*, 1990). It should be noted, however, that some protein inductions can be overlooked by focusing solely on two-dimensional gel analysis, and the numbers established by this method should be taken only as a lower limit. Nevertheless, the complexity of some responses at this level is truly daunting. For example, nearly 100 oxidative stress proteins were identified in *E. coli* by the two-dimensional gel approach alone (Greenberg and Demple, 1989). In the yeast *Saccharomyces cerevisiae*, over 250 H_2O_2-inducible proteins were detected by this method (Godon *et al.*, 1998). When the results of genetic and genomic studies are added (e.g., Pomposiello *et al.*, 2001), the numbers are much greater in both bacteria and yeast. The inability of any one method to identify the complete set of inducible genes and proteins, and the complexity of oxidative stress responses, demand that multiple approaches be employed.

Bacterial gene function is regulated mainly at the transcriptional level. Cells have evolved transcriptional modulators that sense oxidative stress and activate genes whose products avert or repair the damage caused by ROS. These transcriptional modulators usually activate multiple and unlinked promoters, which as co-regulated groups constitute regulons. In most cases, one or more activated genes decrease the activating stimulus in a type of negative feedback, which often makes the responses self-regulating (for reviews, see Storz and Imlay, 1999; Pomposiello and Demple, 2001).

4.1. The OxyR System

OxyR is a redox-sensitive protein of the LysR family of DNA-binding transcriptional modulators (Storz and Altuvia, 1994). Two-dimensional gel analysis showed that the induction of eight to nine proteins in response to H_2O_2 depends on the *oxyR* gene (Christman *et al.*, 1985). OxyR activates the transcription of *katG* (encoding a catalase), *ahpFC* (alkyl hydroperoxide reductase), *dps* (protective DNA binding protein), *gorA* (GSH reductase), *grxA* (glutaredoxin-A) and *oxyS* (small, untranslated RNA that may regulate other genes post-transcriptionally). This list was recently expanded using genomic methods (Zheng *et al.*, 2001).

OxyR is a homotetramer in solution, and the protein exists in reduced and oxidized forms that exhibit redox-regulated DNA binding (Storz *et al.*, 1990; Toledano *et al.*, 1994). Only oxidized OxyR binds tightly to the promoters of target genes such as *katG* or *ahpFC*, where it activates transcription, evidently through contacts with the α subunit of RNA polymerase. Reduced OxyR binds tightly only to the *oxyR* promoter itself, as does oxidized OxyR, which limits the expression of this regulatory protein under all circumstances. Active, oxidized OxyR has an intramolecular disulphide bond between cysteines 199 and 208 (Zheng *et al.*, 1998a), which is evidently formed by direct reaction with H_2O_2 or another oxidant. Reduced, inactive OxyR is maintained by thiol-disulphide exchange proteins such as glutaredoxin (see below), which is dependent on GSH (Åslund and Beckwith, 1999; Carmel-Harel and Storz, 2000). Thus, the induction of *gor* and *grxA* may ultimately down-regulate the response by regenerating reduced OxyR. A recent crystallographic analysis indicates dramatic structural differences between oxidized and reduced OxyR (Choi *et al.*, 2001), which provides a lead-in for understanding differential DNA binding by the two forms and perhaps the mechanism of selective oxidation of the protein.

4.2. The SoxR/SoxS System

The SoxR protein is a redox-sensing transcriptional activator that belongs to the MerR family of DNA-binding proteins (Hidalgo *et al.*, 1997). SoxR is a homodimer of 17-kDa subunits, each one containing a redox-active [2Fe-2S] centre. The DNA-binding activity of SoxR does not depend on the presence (Hidalgo and Demple, 1994) or oxidation state (Gaudu and Weiss, 1996) of the iron-sulphur clusters, but only SoxR with oxidized [2Fe-2S] centres activates transcription (Ding *et al.*, 1996; Gaudu and Weiss, 1996). Activation of SoxR through the one-electron oxidation of its [2Fe-2S] centres therefore corresponds to an allosteric transition in the DNA–protein complex of SoxR with its target, the *soxS* promoter. This activating transition does not substantially increase binding by RNA polymerase (Hidalgo and Demple, 1994), but rather the formation of the 'open' complex essential for initiating transcription is stimulated (Hidalgo *et al.*, 1995). It is not yet known how the [2Fe-2S] centres of SoxR are maintained in the reduced state, but one or more reductase enzymes is presumably responsible (Ding and Demple, 1997; Gaudu *et al.*, 1997).

Nitric oxide activates SoxR by a different mechanism (Nunoshiba *et al.*, 1993a, 1995). NO$^{•}$ reacts with the SoxR [2Fe-2S] centres to displace sulphide atoms and generate mixed dinitrosyl-iron-cysteine complexes anchored to the protein (Ding and Demple, 2000). The nitrosylated form of SoxR, which is observed both in intact, NO$^{•}$-treated bacteria and with the purified protein reacted with pure NO$^{•}$ *in vitro*, has transcriptional activity close or identical to oxidized SoxR. Although nitrosylated SoxR is quite stable *in vitro*, the modification is rapidly removed *in vivo* without apparent turnover of the protein (Ding and Demple, 2000). This observation suggests that cellular activities exist to mediate rapid removal of dinitrosyl-iron-cysteine complexes from proteins and their replacement by undamaged iron-sulphur centres. This type of nitrosylation represents a fundamentally new, reversible signal tranduction pathway for nitric oxide.

Activation of SoxR stimulates the transcription of only one gene, *soxS*, which encodes a second transcriptional modulator (Hidalgo *et al.*, 1997). The SoxS protein, a 13-kDa monomer homologous to the C-terminal portion of AraC protein, amplifies the activation signal (Amabile-Cuevas and Demple, 1991) by inducing the transcription of ≥60 promoters (Pomposiello *et al.*, 2001). This binding activates transcription by recruiting RNA polymerase to these genes (Li and Demple, 1994). The SoxS-regulated genes in *E. coli* include *sodA* (Mn-containing SOD), *zwf* (glucose-6-phosphate dehydrogenase), *micF* (an antisense RNA that inhibits translation of the porin OmpF), *nfo* (endonuclease IV, a repair enzyme for oxidative DNA damage), *fpr* (ferredoxin reductase), *acrAB* (cellular efflux pumps), and *fumC* (redox-resistant fumarase). Some of these functions (e.g. the *zwf* gene product) could aid in switching off SoxR by contributing to reductive pathways (Liochev and Fridovich, 1992). SoxS in other organisms may induce different genes: activation of *S. typhimurium soxRS* induces Mn-SOD and endonuclease IV, but not fumarase or glucose-6-phosphate dehydrogenase (E. Martins and B. Demple, unpublished data). Like many other regulatory proteins, both SoxS and SoxR limit their own expression by binding the promoters of their own structural genes (Nunoshiba *et al.*, 1993b; Hidalgo *et al.*, 1998).

4.3. The Sigma S Regulon

The alternative sigma factor, sigma S (σ^s), encoded by the *rpoS* gene, is expressed during nutrient starvation or in stationary phase in many bacterial species (Hengge-Aronis, 1999). RNA polymerase containing σ^s activates the expression of several genes that counteract oxidative stress, and starved or stationary-phase bacteria are more resistant to oxidative stress than are cells growing exponentially. The σ^s-activated genes include *katE* (catalase-hydroperoxidase HPII) and *katG* (HPI), *dps* (protective DNA binding protein), *xthA* (exonuclease III, another DNA repair enzyme for oxidative damage), and *gorA* (GSH reductase). The signals controlling σ^s activity are poorly understood but involve distinct transcriptional, translational and proteolytic steps (Hengge-Aronis, 1999).

4.4. The Fur Regulon

Antioxidant defence genes are also controlled by proteins of the Fur family. *E. coli* Fur protein is a repressor governing a system of genes involved in iron uptake, and the system is switched on when bacteria are grown in medium with limiting Fe concentrations (Escolar *et al.*, 1999). It is unclear whether Fur senses iron levels directly or through some other signal. Fur in *E. coli* also regulates the *sodA*-encoded Mn-SOD (Tardat and Touati, 1993).

Recently, a direct regulatory connection between oxidative stress and iron metabolism was shown by Zheng *et al.*, who demonstrated independent transcriptional activation of *fur* under *soxRS* and *oxyR* control (Zheng *et al.*, 1999). The exact physiological relevance of this observation requires further exploration.

More complex regulation involving Fur-related proteins has been described in *Bacillus subtilis* (Bsat *et al.*, 1998), in which the PerP protein regulates genes encoding a catalase, an alkyl hydroperoxide reductase, and a Dps-like protein. Although this collection of gene products is reminiscent of the *E. coli oxyR* system, the regulatory mechanism is quite different. PerP is a repressor, possibly with a metal co-repressor, so that gene induction depends on lowered DNA binding by the regulator, i.e. derepression rather than positive control. Mutant strains lacking PerP have increased resistance to H_2O_2 (Bsat *et al.*, 1998).

5. BIOCHEMICAL HOMEOSTASIS

The diverse sources of oxidative stress have probably modelled the evolution of adaptations to aerobic life (Gilbert and Colton, 1999; Halliwell and Gutteridge, 1999). Consistent with this idea, aerobic organisms display both constitutive and inducible defences against free radicals. In addition to the small molecules mentioned earlier, several regulated enzymes scavenge reactive species or repair oxidative damage to important cellular components. The increased expression of these activities in oxidative stress responses is a kind of cellular or biochemical homeostasis, as they restore conditions to tolerable levels of reactive species and oxidative damage (Fig. 1). We will consider a few key examples.

5.1. Superoxide Dismutase

SOD catalyses the dismutation (redistribution of electrons between molecules) of superoxide to form H_2O_2 and O_2, a reaction that plays a central role in protection against oxidative damage (Fridovich, 1995). All aerobic organisms have at least one form of SOD or a small molecule equivalent. *E. coli* has three isozymes, encoded by the *sodA*, *sodB* and *sodC* genes (Imlay and Imlay, 1996). The products of the *sodA* and *sodB* genes are cytoplasmic, whereas the *sodC* product is periplasmic. The three SODs of *E. coli* differ in the metals at their active sites: the SodA protein contains manganese, SodB contains iron, and SodC contains copper and zinc. *S. typhimurium* contains a second copper-zinc SOD in the periplasm, which is important for virulence (Fang *et al.*, 1999).

It is estimated that the high catalytic efficiency of SOD maintains a steady-state level of superoxide in *E. coli* of $<10^{-9}$ M during aerobic growth (Fridovich, 1995); this level corresponds to <1 $O_2^{\bullet-}$ molecule per cell. Bacterial strains lacking both of the cytoplasmic SOD enzymes have an estimated 10^4-fold higher steady-state level of superoxide (Imlay and Fridovich, 1991). Such SOD-deficient strains suffer increased oxidative DNA damage during aerobic growth, which results in an elevated mutation rate. The aerobic growth of SOD-deficient *E. coli* in rich media is only slightly impaired, but in minimal media growth is abolished unless amino acids, particularly the branched-chain types, are provided. This conditional multiple auxotrophy, which is not observed during anaerobic growth, is due to oxidative inactivation of biosynthetic enzymes that contain iron-sulphur centres (Gardner and Fridovich, 1991; Flint *et al.*, 1993).

The regulation of the *sod* genes is a good example of homeostatic fine-tuning (Fig. 1). The *sodA* gene alone is controlled by at least six global regulators responding to diverse environmental conditions (Compan and Touati, 1993). Expression of *sodA* is derepressed in response to the transition from anaerobic to aerobic growth under control of the Fnr and ArcAB systems. Further activation of *sodA* transcription in aerobic conditions can be triggered by oxidative stress, a response mediated by the SoxRS system. Thus, the expression of *sodA* is responsive to a dynamic range of oxidative conditions to maintain superoxide levels within a narrow range as the production of this free radical varies greatly. Additionally, *sodA* is activated in response to various xenobiotic compounds and antibiotics under MarRAB control, and the gene is

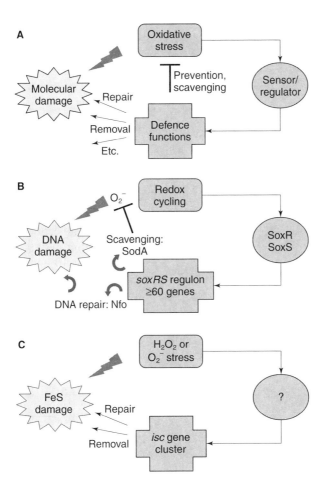

Figure 1 Models for biochemical homeostasis. A, general scheme. Oxidative stress damages many different molecules in the cell, but the stress also activates defence pathways via redox-sensing gene control systems. The resulting activation of defence gene expression turns on many activities that repair or remove damage, eliminate reactive species by scavenging, etc. In some cases, the induced functions also switch off the regulatory system (not shown; see text). B, an example of homeostasis from the *soxRS* regulon. Redox-cycling agents elevate superoxide (O_2^-) levels, which leads to DNA damage (see text). These same agents also activate the SoxR/SoxS regulatory system which turns on several dozen genes. Among these are genes encoding the DNA repair enzyme endonuclease IV (Nfo protein) to correct oxidative damage, and Mn-containing superoxide dismutase (SodA) to eliminate O_2^- and thus prevent the damage. C, damage to protein iron-sulphur (FeS) clusters and homeostatic response. Both H_2O_2 and O_2^- act through unknown pathways to trigger expression of the *isc* gene cluster. The Isc proteins include functions that regenerate functional FeS clusters by rebuilding them after oxidative damage.

repressed by Fur and integration host factor, although in a subtler manner (Compan and Touati, 1993). One might speculate that these additional controls are a generic response to other environmental conditions that threaten to increase superoxide production, but this point has not been extensively explored. Finally, modest post-transcriptional regulation of *sodA* has been reported (Schrum and Hassan, 1994), but the molecular mechanism of this regulation remains obscure.

5.2. Catalases and Peroxidases

Catalases and peroxidases are haem-containing enzymes that eliminate H_2O_2 by related mechanisms (Gilbert and Colton, 1999; Halliwell and Gutteridge, 1999). Catalases redistribute electrons among H_2O_2 molecules by alternating two-electron oxidation and reduction, which generates oxygen and water, respectively. Peroxidases oxidize an organic compound while generating H_2O. *E. coli* has two catalases (Schellhorn, 1995): the OxyR-regulated catalase-hydroperoxidase I (HPI), encoded by the *katG* gene, and the predominant form during exponential growth; and HPII, as mentioned earlier, the expression of which (along with HPI) is increased by σ^s during starvation or in stationary phase. The AhpFC proteins constitute an NADPH-dependent alkyl hydroperoxide reductase important for resistance to organic peroxides (Storz *et al.*, 1989). The AhpC protein is a member of a large family of thiol-specific antioxidant proteins with broad antioxidant roles (Chae *et al.*, 1994).

The peroxidases play homeostatic roles analogous to those ascribed to SOD enzymes. Elevated H_2O_2 levels trigger the OxyR protein, which leads to increased expression of HPI and AhpFC (Christman *et al.*, 1985). These increases eliminate peroxides more rapidly to restore a tolerable steady state to the cell. Modulated expression of HPI and AhpFC not only deals with environmentally generated oxidative stress, but also maintains the H_2O_2 produced by aerobic metabolism within very narrow limits ($0.15\ \mu M \pm 0.05\ \mu M$) (Gonzalez-Flecha and Demple, 1995). This control serves to limit lipid peroxidation and mutagenic DNA damage (Gonzalez-Flecha and Demple, 1997).

5.3. Flavohaemoglobin

The bacterial protein flavohaemoglobin, which contains a flavoprotein domain joined to a haemoglobin-like domain, was initially suspected to function in oxygen transport (Zhu and Riggs, 1992) or as a new type of oxygen sensor (Poole *et al.*, 1994). However, the *hmp* gene encoding this protein in *E. coli* was found to respond to NO• exposure independently of SoxRS (Poole *et al.*, 1996), and the enzyme was fortuitously discovered to have enzymatic activity in oxidizing NO• directly to nitrate, i.e. a nitric oxide dioxygenase (Gardner *et al.*, 1998; Hausladen *et al.*, 1998). This appears to be a fundamental detoxifying reaction: flavohaemoglobin-deficient *S. typhimurium* (*hmp* mutants) are hypersensitive to NO•-generating agents (Crawford and Goldberg, 1998). As with SOD and catalase regulation, induction of flavohaemoglobin by NO• is homeostatic in the sense that increased expression of the enzyme should reduce the steady-state level of nitric oxide.

5.4. Thiol-disulphide Balance: Thioredoxins and Glutaredoxins

A family of small proteins, the thioredoxins and glutaredoxins, act as efficient thiol donors to many cellular proteins (Åslund and Beckwith, 1999). These proteins have conserved pairs of cysteine residues that are oxidized to cystine disulphides as other proteins are reduced. Reduced thioredoxin is in turn regenerated by a reductase using NADPH; reduced glutaredoxin is regenerated using GSH as a reducing donor (Åslund and Beckwith, 1999). These proteins connect to biochemical homeostasis in two ways. First, glutaredoxin-1 in *E. coli* is the main reductant for OxyR, with a secondary role for the thioredoxins (Zheng *et al.*, 1998b; Åslund *et al.*, 1999). When oxidative stress is imposed by excess H_2O_2, OxyR is rapidly oxidized and activated, and somewhat more slowly reduced and inactivated again by glutaredoxin-1. The extent and persistence of OxyR activation therefore depends directly on both the level of H_2O_2 and the availability of reduced glutaredoxin-1, and indirectly on the level of reduced GSH (Carmel-Harel and Storz, 2000; Prieto-Alamo *et al.*, 2000). Second, the *grxA* gene encoding glutaredoxin-1 and the *gor* gene encoding GSH reductase are activated by OxyR, so that more extensive thiol oxidation leads to greater protection against disulphide formation in the cytoplasm (Carmel-Harel and Storz, 2000; Grant, 2001).

Although they are no longer free-living microbes, chloroplasts also exploit thiol-disulphide chemistry to sense redox status and mediate gene expression. In the chloroplasts of *Clamydomonas reinhardtii*, translation of the *psbA* mRNA encoding a photosystem component is controlled by the RB47 binding protein, which is active only in the reduced state (Fong *et al.*, 2000). The redox status of RB47 in turn depends on a chloroplast protein disulphide isomerase (Kim and Mayfield, 1997). The isomerase acts in thiol-disulphide exchange to transmit the oxidation state of GSH to RB47. When active photosynthesis maintains generally reducing conditions, reduced GSH maintains active RB47 and more photosystem components are produced. When photosynthesis is off, GSH is oxidized and this is transmitted to shut off RB47 and thereby translation of *psbA*. In this way, thiol-disulphide status is linked not to a stress response, but to photosynthetic activity (Demple, 1998).

5.5. Reactions and Repair of Iron-sulphur Centres

Another type of biochemical homeostasis involves iron-sulphur (FeS) centre in proteins. As noted above, reactions of [2Fe-2S] centres underlie signal transduction in the *E. coli* SoxR protein (Hidalgo *et al.*, 1997). Upon oxidative stress imposed by redox-cycling agents or SOD deficiency, the SoxR [2Fe-2S] centres are oxidized (formally, converted from the 2+/3+ state to the 3+/3+ state), which triggers SoxR as a transcriptional activator. In the case of Fnr protein, oxidation converts a [4Fe-4S] centre to [2Fe-2S], with loss of the repressor's binding activity (Beinert and Kiley, 1999). Thus, redox reactions of iron-sulphur centres have been exploited to conduct intracellular signalling (Beinert and Kiley, 1999).

In many other FeS proteins, particularly certain of the [4Fe-4S] type such as aconitase, such oxidation damages the FeS centre and inactivates protein function (Gardner and Fridovich, 1991; Flint *et al.*, 1993). In these cases, *soxRS*-inducible functions such as the *fpr* gene product, ferredoxin:NADPH oxidoreductase, may again reduce the metal centre and restore function. In some cases, oxidation may initiate the destruction of FeS centres, and new centres may have to be synthesized (Gardner, 1997). This problem may be counteracted by the products of the *isc* gene cluster, which have functions analogous to *Azotobacter vinelandii* enzymes such as NifS. NifS helps regenerate FeS clusters of various types by providing sulphide from L-cysteine (Zheng and Dean, 1994). Other *A. vinelandii nif* genes are required to maintain FeS centres in nitrogenase, although the specific functions are not all known (Zheng *et al.*, 1998a), and many of these genes have counterparts in the *E. coli isc* cluster (Schwartz *et al.*, 2000). Recent results indicate that the *E. coli isc* gene cluster is induced by oxidative stress caused by H_2O_2 (Zheng *et al.*, 2001) or PQ (P.J. Pomposiello and B. Demple, unpublished data). Interestingly, these inductions appear to be independent of the *oxyR* and *soxRS* loci, respectively.

Genomic methods recently identified a gene of unknown function, *yggX*, as induced under *soxRS* control (Pomposiello *et al.*, 2001), and we have demonstrated that *yggX* mutant strains have pronounced hypersensitivity to PQ in gradient plates (P.J. Pomposiello and B. Demple, in preparation). Independently, a recent report provides evidence that YggX protein is involved in either preventing or repairing oxidative damage to protein FeS centres in *E. coli* (Gralnick and Downs, 2001). Thus, the overall role of the *soxRS* regulon in coping with damaged FeS centres may be quite broad.

Nitric oxide also reacts avidly with FeS centres, and in the case of SoxR this is also employed as an activating signal (Nunoshiba *et al.*, 1993a; Ding and Demple, 2000). Most noteworthy is that the nitrosylated FeS centres in NO•-activated SoxR turn over very rapidly in intact bacteria (Ding and Demple, 2000). Since the nitrosylated centres of SoxR are relatively stable *in vitro*, this observation implies that some active process *in vivo* rapidly eliminates nitrosylated FeS centres. New [2Fe-2S] centres could then be rapidly resynthesized in SoxR by the mechanisms suggested above. Again, the activation of some of these functions as part of the *soxRS* system makes sense in light of the threat to FeS centres posed by NO˙.

6. INSIGHTS FROM GENOMICS STUDIES

6.1. Advantages and Limitations

The application of recent genomic techniques to study gene regulation under stress conditions has the potential of enhancing our understanding of microbial physiology. Among these whole-genome approaches, transcriptional profiling using gene arrays allows the simultaneous measurement of RNA levels from many or all genes of an organism. Transcriptional profiling has provided new insights into the heat shock response (Richmond *et al.*, 1999), bacterial growth in minimal and rich media (Tao *et al.*, 1999), acid stress (Arnold *et al.*, 2001), superoxide stress (Pomposiello *et al.*, 2001), and H_2O_2 stress (Zheng *et al.*, 2001). Transcriptional profiling has also been used to identify targets of regulatory proteins like IHF (Arfin *et al.*, 2000), NtrC (Zimmer *et al.*, 2000), MarA (Barbosa and Levy, 2000) and SoxS (Pomposiello *et al.*, 2001). The main assumption in the use of transcriptional profiling for modelling metabolism is that the transcription of a gene obeys the rules of cellular economy, i.e. gene products are made when needed.

Genomic studies have the remarkable advantage of breadth, but also a number of built-in limitations. First, commercially available gene arrays are a fixed platform that does not admit modifications. Commercial arrays typically exclude untranslated RNAs, which are clearly involved in many stress responses. Second, transcriptional profiling reveals only the relative steady-state levels of mRNAs, not absolute levels, and without any information about post-transcriptional processes or actual protein expression. Third, despite the almost complete coverage of the genome, transcriptional profiling experiments consistently fail to detect changes in genes already known to be modulated by the stimuli of interest. For example, the use of gene arrays to analyse the heat shock response revealed only 23 of the 51 known genes (Richmond *et al.*, 1999), while genomic analysis of the *soxRS* regulon identified only nine of the 16 previously known genes present in the array. Finally, the results can vary significantly between transcriptional profiling experiments in the same laboratory, and almost certainly between laboratories. Thus, although transcriptional profiling can be a very efficient screening method, the behaviour of specific genes of interest must be substantiated by genetic and biochemical methods, and additional approaches must be employed to detect regulated genes comprehensively.

Genomic studies in transcriptional regulation involve large populations of genes in which a single quantity (expression ratio) is assigned to each gene. The expression ratio in a large population of genes varies across a very wide range, which necessitates statistical analysis of the data. Therefore, the final answer to the question of which set of genes is induced or down-regulated in a particular physiological condition will depend heavily on the statistical thresholds applied in the specific analysis. A brief review of the literature reveals a wide variation in analytical methods (Tao *et al.*, 1999; Barbosa and Levy, 2000; Arnold *et al.*, 2001; Pomposiello *et al.*, 2001; Zheng *et al.*, 2001), which complicates the comparisons among independent studies. Part of the answer to this problem is the 'virtual' publication of the raw data, made public on the World Wide Web, to allow comparison among independent studies. Additionally, the scientific community should agree on common statistical standards for transcriptional profiling studies.

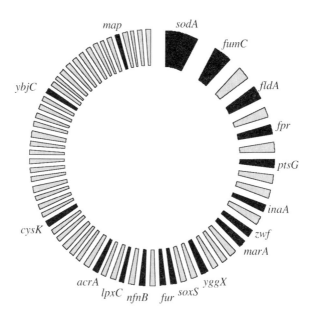

Figure 2   A schematic representation of the superoxide stimulon. Each segment represents one of the 66 genes activated by superoxide stress, as revealed by whole genome transcriptional profiling (Pomposiello *et al.*, 2001). The size of each slice is in direct proportion to the activation of that gene by treatment with paraquat (PQ), a superoxide-generating agent. The black segments represent genes with demonstrated SoxS-dependent expression.

6.2. Oxidative Stress Responses and Maintenance of NADPH Pools

Recently, we have used genomic tools to study the genetic response of *E. coli* to superoxide stress, using the redox-cycling agent PQ (Pomposiello *et al.*, 2001) (Fig. 2). PQ is reduced intracellularly at the expense of NADPH in a reaction catalysed by at least four oxido-reductases (Liochev *et al.*, 1994). PQ reduced by one electron is oxidized by O_2 to form superoxide, resulting in a redox cycle that produces a flux of superoxide (Kappus and Sies, 1981). Thus, the cell faces a double threat under exposure to PQ and other redox-cycling agents: the deleterious effects of superoxide itself, and the decreased level of NADPH that limits biosynthetic capabilities (Liochev *et al.*, 1994). Equilibration of NADPH with NADH would generalize this limitation of cellular reducing power.

Previous observations suggested that treatment of growing cells with PQ induces pathways that replenish reducing power. First, PQ activates the expression of glucose-6-phosphate dehydrogenase, the first enzyme of the pentose phosphate pathway, in a *soxRS*-dependent manner (Hidalgo *et al.*, 1997). This pathway generates NADPH and is required for resistance to redox-cycling agents (J.T. Greenberg, Ph.D. thesis, Harvard University). Second, PQ also activates the *soxRS*-dependent expression of two enzymes of the tricarboxylic acid (TCA) cycle, fumarase C and aconitase (Hidalgo and Demple, 1996). These enzymes support the reduction of $NA(P)D^+$ to NAD(P)H.

In our studies, exposure to PQ activated the expression of additional genes whose products help to replenish reducing power (Pomposiello *et al.*, 2001). These genes code for proteins involved in sugar transport (*ptsG*, *gatABD*, *malEK* and *lamB*), glycolysis (*pgi*), amino acid transport and degradation (*artIP*, *tnaA* and *dadX*), and the TCA cycle (*gltA*, *sdhB* and *sucD*). A similar redirection of carbon metabolism to pathways that reconstitute NADPH was observed in *S. cerevisiae* after treatment with hydrogen peroxide and proteomic analysis by two-dimensional gels (Godon *et al.*, 1998). Thus, regenerating NAD(P)H may be a fundamental and general aspect of cellular responses to oxidative stress.

6.3. Adaptation and Repair Pathways under Oxidative Stress

Challenge of growing *E. coli* cells with PQ induces the genes coding for nine ribosomal proteins (Pomposiello *et al.*, 2001). This increase in ribosomal building blocks was not predicted, in that the bacterial growth rate was not significantly affected during treatment with PQ. Another gene coding for a translational regulator, *fmt* is also induced by PQ. The product of this gene, methionyl-tRNA formyltranferase, is an important factor in translational initiation (Guillon *et al.*, 1992). One possibility is that increased translational capacity counterbalances a faster turnover of proteins due to oxidative damage and increased degradation. In this scenario, enhanced synthesis would be required to maintain the fast growth rate.

The response of *E. coli* to superoxide stress may also function to maintain the metabolic rate, as indicated by the activation of genes coding for products involved in crucial anabolic and catabolic pathways (Pomposiello *et al.*, 2001). These genes include *nuoI* and *nuoK* coding for subunits of NADPH dehydrogenase, the first electron acceptor of the respiratory chain. Interestingly, 11 out of the 12 genes coding for NADPH dehydrogenase subunits showed some degree of activation in duplicate PQ-exposure experiments, albeit below the statistical threshold to register as induced in the combined data.

The activation of the *dadX* and *murF* genes, coding for proteins involved in peptidoglycan synthesis, and the *lpxC* gene, coding for an enzyme involved in lipopolysaccharide synthesis, suggests that the repair mechanisms triggered by oxidative stress extend to extracytoplasmatic structures (Pomposiello *et al.*, 2001).

6.4. Modulation of Metabolism by Oxidative Stress in Yeasts

Microbial responses to oxidative stress are by no means limited to bacteria. The yeasts *Saccharomyces cerevisiae* and *Schizosaccharomyces pombe* have received particular attention recently. Physiological and genetic analysis (Jamieson, 1998; Toone and Jones, 1998) shows that distinct regulatory systems govern adaptive resistance to H_2O_2 compared with redox-cycling (superoxide-generating compounds). The response to H_2O_2 involves a large number of inducible proteins, including some of the same defence functions mentioned above: catalase, SOD, glucose-6-phosphate dehydrogenase and GSH reductase (Godon *et al.*, 1998). Modulated expression of metabolic enzymes in H_2O_2-treated yeast is proposed to remodel metabolism to increase the regeneration of NADPH at the expense of glycolysis. Expression of a portion of the H_2O_2-inducible proteins in yeast (e.g. GSH) is controlled by Yap1, a yeast homologue of the c-Jun/c-Fos transcription activators of mammalian cells, perhaps by direct oxidation of Yap1 (Delaunay *et al.*, 2000).

ACKNOWLEDGEMENTS

The authors are grateful to colleagues in our laboratory and in the microbiology community for many useful discussions. This work has been supported by grants from the National Cancer Institute of the National Institutes of Health of the USA.

REFERENCES

Amábile-Cuevas, C.F. and Demple, B. (1991) Molecular characterization of the *soxRS* genes of *Escherichia coli*: two genes control a superoxide stress regulon. *Nucl. Acids Res.* **19**, 4479–4484.

Arfin, S.M., Long, A.D., Ito, E.T., Tolleri, L., Riehle, M.M. Paegle, E.S. and Hatfield, G.W. (2000) Global gene expression profiling in *Escherichia coli* K12. The effects of integration host factor. *J. Biol. Chem.* **275**, 29672–29684.

Arnold, C.N., McElhanon, J., Lee, A., Leonhart, R. and Siegele, D.A. (2001) Global analysis of *Escherichia coli* gene expression during the acetate-induced acid tolerance response. *J. Bacteriol.* **183**, 2178–2186.

Åslund, F. and Beckwith, J. (1999) The thioredoxin superfamily: redundancy, specificity, and gray-area genomics. *J. Bacteriol.* **181**, 1375–1379.

Åslund, F., Zheng, M., Beckwith, J. and Storz, G. (1999) Regulation of the OxyR transcription factor by hydrogen peroxide and the cellular thiol-disulfide status. *Proc. Natl Acad. Sci. USA* **96**, 6161–6165.

Babior, B.M. (1992) The respiratory burst oxidase. *Enzymol. Rel. Areas Mol. Biol.* **65**, 49–65.

Barbosa, T.M. and Levy, S.B. (2000) Differential expression of over 60 chromosomal genes in *Escherichia coli* by constitutive expression of MarA. *J. Bacteriol.* **182**, 3467–3474.

Beinert, H. and Kiley, P.J. (1999) Fe-S proteins in sensing and regulatory functions. *Curr. Opin. Chem. Biol.* **3**, 152–157.

Bsat, N., Herbig, A., Casillas-Martinez, L., Setlow, P. and Helmann, J.D. (1998) *Bacillus subtilis* contains multiple Fur homologues: identification of the iron uptake (Fur) and peroxide regulon (PerR) repressors. *Mol. Microbiol.* **29**, 189–198.

Carmel-Harel, O. and Storz, G. (2000) Roles of the glutathione- and thioredoxin-dependent reduction systems in the *Escherichia coli* and *Saccharomyces cerevisiae* responses to oxidative stress. *Annu. Rev. Microbiol* **54**, 439–461.

Chae, H.Z., Robison, K., Poole, L.B., Church, G., Storz, G. and Rhee, S.G. (1994) Cloning and sequencing of thiol-specific antioxidant from mammalian brain: alkyl hydroperoxide reductase and thiol-specific antioxidant define a large family of antioxidant enzymes. *Proc. Natl Acad. Sci. USA* **91**, 7017–7021.

Chance, B., Sies, H. and Boveris, A. (1979) Hydroperoxide metabolism in mammalian organs. *Physiol. Rev.* **59**, 527–605.

Chaudhary, A.K., Nokubo, M., Reddy, G.R., Yeola, S.N., Morrow, J.D., Blair, I.A. and Marnett, L.J. (1994) Detection of endogenous malondialdehyde-deoxyguanosine adducts in human liver. *Science* **265**, 1580–1582.

Choi, H., Kim, S., Mukhopadhyay, P., Cho, S., Woo, J., Storz, G. and Ryu, S. (2001) Structural basis of the redox switch in the OxyR transcription factor. *Cell* **105**, 103–113.

Christman, M.F., Morgan, R.W., Jacobson, F.S. and Ames, B.N. (1985) Positive control of a regulon for defenses against oxidative stress and some heat-shock proteins in *Salmonella typhimurium*. *Cell* **41**, 753–762.

Compan, I. and Touati, D. (1993) Interaction of six global transcription regulators in expression of manganese superoxide dismutase in *Escherichia coli* K-12. *J. Bacteriol.* **175**, 1687–1696.

Crawford, M.J. and Goldberg, D.E. (1998) Role for the *Salmonella* flavohemoglobin in protection from nitric oxide. *J. Biol. Chem.* **273**, 12543–12547.

Dedon, P.C., Plastaras, J.P., Rouzer, C.A. and Marnett, L.J. (1998) Indirect mutagenesis by oxidative DNA damage: formation of the pyrimidopurinone adduct of deoxyguanosine by base propenal. *Proc. Natl Acad. Sci. USA* **95**, 11113–11116.

Delaunay, A., Isnard, A.D. and Toledano, M.B. (2000) H_2O_2 sensing through oxidation of the Yap1 transcription factor. *EMBO J.* **19**, 5157–5166.

Demple, B. (1998) A bridge to control. *Science* **279**, 1655–1656.

Demple, B. (1999) Radical ideas: genetic responses to oxidative stress. *Clin. Exp. Pharmacol. Physiol.* **26**, 64–68.

Demple, B. and Harrison, L. (1994) Repair of oxidative damage to DNA: enzymology and biology. *Ann. Rev. Biochem.* **63**, 915–948.

Ding, H.G. and Demple, B. (1997) *In vivo* kinetics of a redox-regulated transcriptional switch. *Proc. Natl Acad. Sci. USA* **94**, 8445–8449.

Ding, H. and Demple, B. (2000) Direct nitric oxide signal transduction via nitrosylation of iron-sulfur centers in the SoxR transcription activator. *Proc. Natl Acad. Sci. USA* **97**, 5146–5150.

Ding, H., Hidalgo, E. and Demple, B. (1996) The redox state of the [2Fe-2S] clusters in SoxR protein regulates its activity as a transcription factor. *J. Biol. Chem.* **271**, 33173–33175.

Escolar, L., Perez-Martin, J. and de Lorenzo, V. (1999) Opening the iron box: transcriptional metalloregulation by the Fur protein. *J. Bacteriol.* **181**, 6223–6229.

Fang, F.C., DeGroote, M.C., Foster, J.W., Baumler, A.J., Ochsner, U., Testerman, T., Bearson, S., Giard, J.C., Xu, Y., Campbell, G. and Laessig, T. (1999) Virulent *Salmonella typhimurium* has two periplasmic Cu, Zn-superoxide dismutases. *Proc. Natl Acad. Sci. USA* **96**, 7502–7507.

Flint, D.H., Tuminello, J.F. and Emptage, M.H. (1993) The inactivation of Fe-S cluster containing hydro-lyases by superoxide. *J. Biol. Chem.* **268**, 22369–22376.

Flohe, L., Hecht, H.J. and Steinert, P. (1999) Glutathione and trypanothione in parasitic hydroperoxide metabolism. *Free Rad. Biol. Med.* **27**, 966–984.

Fong, C. L., Lentz, A. and Mayfield, S.P. (2000) Disulfide bond formation between RNA binding domains is used to regulate mRNA binding activity of the chloroplast poly(A)-binding protein. *J. Biol. Chem.* **275**, 8275–8278.

Fridovich, I. (1995) Superoxide radical and superoxide dismutases. *Ann. Rev. Biochem.* **64**, 97–112.

Garcia-Olmedo, F., Rodriguez-Palenzuela, P., Molina, A., Alamillo, J.M., Lopez-Solanilla, E., Berrocal-Lobo, M. and Poza-Carrion, C. (2001) Antibiotic activities of peptides, hydrogen peroxide and peroxynitrite in plant defence. *FEBS Lett.* **498**, 219–222.

Gardner, P.R. (1997) Superoxide-driven aconitase FE-S center cycling. *Biosci. Rep.* **17**, 33–42.

Gardner, P.R. and Fridovich, I. (1991) Superoxide sensitivity of the *Escherichia coli* aconitase. *Journal of Biological Chemistry* **266**, 19328–19333.

Gardner, P.R., Gardner, A.M., Martin, L.A. and Salzman, A.L. (1998) Nitric oxide dioxygenase: an enzymic function for flavohemoglobin. *Proc. Natl Acad. Sci. USA* **95**, 10378–10383.

Gaudu, P. and Weiss, B. (1996) SoxR, a [2Fe-2S] transcription factor, is active only in its oxidized form. *Proc. Natl Acad. Sci. USA* **93**, 10094–10098.

Gaudu, P., Moon, N. and Weiss, B. (1997) Regulation of the soxRS oxidative stress regulon. Reversible oxidation of the Fe-S centers of SoxR in vivo. *J. Biol. Chem.* **272**, 5082–5086.

Gilbert, D.L. and Colton, C.A. (eds). (1999) *Reactive Oxygen Species in Biological Systems: An Interdisciplinary Approach.* Kluwer Academic/Plenum Publishers, New York.

Godon, C., Lagniel, G., Lee, J., Buhler, J.M., Kieffer, S., Perrot, M., Boucherie, H., Toledano, M.B. and Labarre, J. (1998) The H_2O_2 stimulon in *Saccharomyces cerevisiae*. *J. Biol. Chem.* **273**, 22480–22489.

Gonzalez-Flecha, B. and Demple, B. (1995) Metabolic sources of hydrogen peroxide in aerobically growing *Escherichia coli*. *J. Biol. Chem.* **270**, 13681–13687.

Gonzalez-Flecha, B. and Demple, B. (1997) Homeostatic regulation of intracellular hydrogen peroxide concentration in aerobically growing *Escherichia coli*. *J. Bacteriol.* **179**, 382–388.

Gralnick, J. and Downs, D. (2001) Protection from superoxide damage associated with an increased level of the YggX protein in *Salmonella enterica*. *Proc. Natl Acad. Sci. USA* **98**, 8030–8035.

Grant, C. (2001) MicroReview: Role of the glutathione/glutaredoxin and thioredoxin systems in yeast growth and response to stress conditions. *Mol. Microbiol.* **39**, 533–541.

Greenberg, J.T. and Demple, B. (1989) A global response induced in *Escherichia coli* by redox-cycling agents overlaps with that induced by peroxide stress. *J. Bacteriol.* **171**, 3933–3939.

Greenberg, J.T., Monach, P., Chou, J.H., Josephy, P.D. and Demple, B. (1990) Positive control of a global antioxidant defense regulon activated by superoxide-generating agents in *Escherichia coli*. *Proc. Natl Acad. Sci. USA* **87**, 6181–6185.

Guillon, J.M., Mechulam, Y., Schmitter, J.M., Blanquet, S. and Fayat, G. (1992) Disruption of the gene for Met-tRNA(fMet) formyltransferase severely impairs growth of *Escherichia coli*. *J. Bacteriol.* **174**, 4294–4301.

Halliwell, B. and Gutteridge, J.M.C. (1999) *Free Radicals in Biology and Medicine*. Oxford, Oxford University Press.

Halliwell, B., Zhao, K. and Whiteman, M. (1999) Nitric oxide and peroxynitrite. The ugly, the uglier and the not so good: a personal view of recent controversies. *Free Rad. Res.* **31**, 651–669.

Hausladen, A., Gow, A.J. and Stamler, J.S. (1998) Nitrosative stress: metabolic pathway involving the flavohemoglobin. *Proc. Natl Acad. Sci. USA* **95**, 14100–14105.

Hecker, M. and Volker, U. (1998) Non-specific, general and multiple stress resistance of growth-restricted *Bacillus subtilis* cells by the expression of the sigmaB regulon. *Mol. Microbiol.* **29**, 1129–1136.

Hengge-Aronis, R. (1999) Interplay of global regulators and cell physiology in the general stress response of *Escherichia coli*. *Curr. Opin. Microbiol.* **2**, 148–152.

Hidalgo, E. and Demple, B. (1994) An iron-sulfur center essential for transcriptional activation by the redox-sensing SoxR protein. *EMBO J.* **13**, 138–146.

Hidalgo, E. and Demple, B. (1996). Adaptive responses to oxidative stress: the *soxRS* and *oxyR* regulons. In: *Regulation of Gene Expression in Escherichia coli*. (E.C.C. Lin and A.S. Lynch, eds), pp. 435–452. R.G. Landes, Austin, Tx.

Hidalgo, E., Bollinger, Jr, J.M., Bradley, T.M., Walsh, C.T. and Demple, B. (1995) Binuclear [2Fe-2S] clusters in the *Escherichia coli* SoxR protein and role of the metal centers in transcription. *J. Biol. Chem.* **270**, 20908–20914.

Hidalgo, E., Ding, H. and Demple, B. (1997) Redox signal transduction via iron-sulfur clusters in the SoxR transcription activator. *Trends Biochem. Sci.* **22**, 207–210.

Hidalgo, E., Leautaud, V. and Demple, B. (1998) The redox-regulated SoxR protein acts from a single DNA site as a repressor and an allosteric activator. *EMBO J.* **17**, 2629–2636.

Imlay, J.A. and Fridovich, I. (1991) Assay of metabolic superoxide production in *Escherichia coli*. *J. Biol. Chem.* **266**, 6957–6965.

Imlay, K.R. and Imlay, J.A. (1996) Cloning and analysis of *sodC*, encoding the copper-zinc superoxide dismutase of *Escherichia coli*. *J. Bacteriol.* **178**, 2564–2571.

Jamieson, D.J. (1998) Oxidative stress responses of the yeast *Saccharomyces cerevisiae*. *Yeast* **14**, 1511–1527.

Kao, S.M. and Hassan, H.M. (1985) Biochemical characterization of a paraquat-tolerant mutant of *Escherichia coli*. *J. Biol. Chem.* **260**, 10478–10481.

Kappus, H. and Sies, H. (1981) Toxic drug effects associated with oxygen metabolism: redox cycling and lipid peroxidation. *Experientia* **37**, 1233–1241.

Kim, J. and Mayfield, S.P. (1997) Protein disulfide isomerase as a regulator of chloroplast translational activation. *Science* **278**, 1954–1957.

Klessig, D.F., Durner, J., Noad, R., Navarre, D.A., Wendehenne, D., Kumar, D., Zhou, J.M., Shah, J., Zhang, S., Kachroo, P., Trifa, Y., Pontier, D., Lam, E. and Silva, H. (2000) Nitric oxide and salicylic acid signaling in plant defense. *Proc. Natl Acad. Sci. USA* **97**, 8849–8855.

Li, Z. and Demple, B. (1994) SoxS, an activator of superoxide stress genes in *Escherichia coli*. Purification and interaction with DNA. *J. Biol. Chem.* **269**, 18371–18377.

Liochev, S.I. and Fridovich, I. (1992) Fumarase C, the stable fumarase of *Escherichia coli*, is controlled by the *soxRS* regulon. *Proc. Natl Acad. Sci. USA* **89**, 5892–5896.

Liochev, S.I., Hausladen, A., Beyer Jr, W.F. and Fridovich, I. (1994) NADPH: ferredoxin oxidoreductase acts as a paraquat diaphorase and is a member of the *soxRS* regulon. *Proc. Natl Acad. Sci. USA* **91**, 1328–1331.

MacMicking, J., Xie, Q.W. and Nathan, C. (1997) Nitric oxide and macrophage function. *Ann. Rev. Immunol.* **15**, 323–350.

Morgan, R.W., Christman, M.F., Jacobson, F.S., Storz, G. and Ames, B.N. (1986) Hydrogen peroxide-inducible proteins in *Salmonella typhimurium* overlap with heat shock and other stress proteins. *Proc. Natl Acad. Sci. USA* **83**, 8059–8063.

Nunoshiba, T., deRojas-Walker, T., Wishnok, J.S., Tannenbaum, S.R. and Demple, B. (1993a) Activation by nitric oxide of an oxidative-stress response that defends *Escherichia coli* against activated macrophages. *Proc. Natl Acad. Sci. USA* **90**, 9993–9997.

Nunoshiba, T., Hidalgo, E., Li, Z. and Demple, B. (1993b) Negative autoregulation by the *Escherichia coli* SoxS protein: a dampening mechanism for the *soxRS* redox stress response. *J. Bacteriol.* **175**, 7492–7494.

Nunoshiba, T., DeRojas-Walker, T., Tannenbaum, S.R. and Demple, B. (1995) Roles of nitric oxide in inducible resistance of *Escherichia coli* to activated murine macrophages. *Infect. Immun.* **63**, 794–798.

Ottaviani, E., Paeman, L.R., Cadet, P. and Stefano, G.B. (1993) Evidence for nitric oxide production and utilization as a bacteriocidal agent by invertebrate immunocytes. *Eur. J. Pharmacol.* **248**, 319–324.

Pomposiello, P.J. and Demple, B. (2001) Redox-operated genetic switches: the SoxR and OxyR transcription factors. *Trends Biotechnol.* **19**, 109–114.

Pomposiello, P.J., Bennik, M.H. and Demple, B. (2001) Genome-wide transcriptional profiling of the *Escherichia coli* responses to superoxide stress and sodium salicylate. *J. Bacteriol.* **183**, 3890–3902.

Poole, R.K., Ioannidis, N. and Orii, Y. (1994) Reactions of the *Escherichia coli* flavo-haemoglobin (Hmp) with oxygen and reduced nicotinamide adenine dinucleotide: evidence for oxygen switching of flavin oxidoreduction and a mechanism for oxygen sensing. *Proc. R. Soc. Lond. B Biol. Sci.* **255**, 251–258.

Poole, R.K., Anjum, M.F., Membrillo-Hernández, J., Kim, S.O., Hughes, M.N. and Stewart, V. (1996) Nitric oxide, nitrite, and Fnr regulation of hmp (flavohemoglobin) gene expression in *Escherichia coli* K-12. *J. Bacteriol.* **178**, 5487–5492.

Prieto-Alamo, M.J., Jurado, J., Gallardo-Madueno, R., Monje-Casas, F., Holmgren, A. and Pueyo, C. (2000) Transcriptional regulation of glutaredoxin and thioredoxin pathways and related enzymes in response to oxidative stress. *J. Biol. Chem.* **275**, 13398–13405.

Richmond, C.S., Glasner, J.D., Mau, R., Jin, H. and Blattner, F.R. (1999) Genome-wide expression profiling in *Escherichia coli* K-12. *Nucl. Acids Res.* **27**, 3821–3835.

Schellhorn, H.E. (1995) Regulation of hydroperoxidase (catalase) expression in *Escherichia coli*. *FEMS Microbiol. Lett.* **131**, 113–119.

Schrum, L.W. and Hassan, H.M. (1994) The effects of fur on the transcriptional and post-transcriptional regulation of MnSOD gene (*sodA*) in *Escherichia coli*. *Arch. Biochem. Biophys.* **309**, 288–292.

Schwartz, C.J., Djaman, O., Imlay, J.A. and Kiley, P.J. (2000) The cysteine desulfurase, IscS, has a major role in in vivo Fe-S cluster formation in *Escherichia coli*. *Proc. Natl Acad. Sci. USA* **97**, 9009–9014.

Sies, H. (1991) Oxidative stress: introduction. In: *Oxidative Stress: Oxidants and Antioxidants* (H. Sies, ed), pp. xv–xxii. Academic Press, London.

Storz, G. and Altuvia, S. (1994) OxyR regulon. *Meth. Enzymol.* **234**, 217–223.

Storz, G. and Imlay, J.A. (1999) Oxidative stress. *Curr. Opin. Microbiol.* **2**, 188–194.

Storz, G., Jacobson, F.S., Tartaglia, L.A., Morgan, R.W., Silveira, L.A. and Ames, B.N. (1989) An alkyl hydroperoxide reductase induced by oxidative stress in *Salmonella typhimurium* and *Escherichia coli*: genetic characterization and cloning of *ahp*. *J. Bacteriol.* **171**, 2049–2055.

Storz, G., Tartaglia, L.A. and Ames, B.N. (1990) Transcriptional regulator of oxidative stress-inducible genes: direct activation by oxidation. *Science* **248**, 189–194.

Tao, H., Bausch, C., Richmond, C., Blattner, F.R. and Conway, T. (1999) Functional genomics: expression analysis of *Escherichia coli* growing on minimal and rich media. *J. Bacteriol.* **181**, 6425–6440.

Tardat, B. and Touati, D. (1993) Iron and oxygen regulation of *Escherichia coli* MnSOD expression: competition between the global regulators Fur and ArcA for binding to DNA. *Mol. Microbiol.* **9**, 53–63.

Toledano, M.B., Kullik, I., Trinh, F., Baird, P.T. Schneider, T.D. and Storz, G. (1994) Redox-dependent shift of OxyR-DNA contacts along an extended DNA-binding site: a mechanism for differential promoter selection. *Cell* **78**, 897–909.

Toone, W.M. and Jones, N. (1998) Stress-activated signalling pathways in yeast. *Genes Cells* **3**, 485–498.

Tosques, I.E., Shi, J. and Shapleigh, J.P. (1996) Cloning and characterization of nnrR, whose product is required for the expression of proteins involved in nitric oxide metabolism in *Rhodobacter sphaeroides* 2.4.3. *J. Bacteriol.* **178**, 4958–4964.

Zheng, L. and Dean, D.R. (1994) Catalytic formation of a nitrogenase iron-sulfur cluster. *J. Biol. Chem.* **269**, 18723–18726.

Zheng, M., Åslund, F. and Storz, G. (1998a) Activation of the OxyR transcription factor by reversible disulfide bond formation. *Science* **279**, 1718–1721.

Zheng, L., Cash, V.L., Flint, D.H. and Dean, D.R. (1998b) Assembly of iron-sulfur clusters. Identification of an iscSUA-hscBA-fdx gene cluster from *Azotobacter vinelandii*. *J. Biol. Chem.* **273**, 13264–13272.

Zheng, M., Doan, B., Schneider, T.D. and Storz, G. (1999) OxyR and SoxRS regulation of *fur*. *J. Bacteriol.* **181**, 4639–4643.

Zheng, M., Wang, X., Templeton, L.J., Smulski, D.R., LaRossa, R.A. and Storz, G. (2001) DNA microarray-mediated transcriptional profiling of the *Escherichia coli* response to hydrogen peroxide. *J. Bacteriol.* **183**, 4562–4570.

Zhu, H. and Riggs, A.F. (1992) Yeast flavohemoglobin is an ancient protein related to globins and a reductase family. *Proc. Natl Acad. Sci. USA* **89**, 5015–5019.

Zimmer, D.P., Soupene, E., Lee, H.L., Wendisch, V.F., Khodursky, A.B., Peter, B.J., Bender, R.A. and Kustu, S. (2000) Nitrogen regulatory protein C-controlled genes of *Escherichia coli*: scavenging as a defense against nitrogen limitation. *Proc. Natl Acad. Sci. USA* **97**, 14674–14679.

Zumft, W.G. (1997) Cell biology and molecular basis of denitrification. *Microbiol. Mol. Biol. Rev.* **61**, 533–616.

Author Index

Subject Index